Mathematical Visualization

Springer
Berlin
Heidelberg
New York
Barcelona
Hong Kong
London
Milan
Paris
Singapore
Tokyo

Hans-Christian Hege
Konrad Polthier (Eds.)

Mathematical Visualization

Algorithms, Applications and Numerics

With 187 Figures, 46 in Color and 12 Tables

Springer

Hans-Christian Hege
Wissenschaftliche Visualisierung
Konrad-Zuse-Zentrum für Informationstechnik Berlin
Takustraße 7
D-14195 Berlin, Germany
e-mail: hege@zib.de

Konrad Polthier
Fachbereich 3, Mathematik
Technische Universität Berlin
Straße des 17. Juni 136
D-10623 Berlin, Germany
e-mail: polthier@math.tu-berlin.de

Library of Congress Cataloging-in-Publication Data

Mathematical visualization : algorithms, applications, and numerics /
Hans-Christian Hege. p. cm.
Includes bibliographical references.

1. Mathematics--Graphic methods--Data processing--Congresses.
2. Computer graphics--Data processing--Congresses.
I. Hege, Hans-Christian, 1954- II. Polthier, Konrad.
QA90.M34 1998 98-36657
510.'.285'66--dc21 CIP

Mathematics Subject Classification (1991): 65-04, 53-04, 65S05, 68U05

ISBN 978-3-642-08373-0

Cover figure by H. Löffelmann, T. Kučera, and E. Gröller, Institute of Computer Graphics, Vienna University of Technology, Wien (Austria).

Cover design: *design & production* GmbH, Heidelberg

Photo composition output by Text & Grafik, B.E.S. GmbH, Heidelberg

Preface

Mathematical Visualization is a young field in the interdisciplinary area of numerics, geometry, and computer graphics. It develops powerful visualization tools for mathematical research and utilizes mathematical techniques for computer graphics and scientific visualization.

The present book is the second in a series of publications on this subject. The articles were presented at the international workshop "Visualization and Mathematics", held from September 16-19, 1997 in Berlin-Dahlem (Germany). Well-known experts contributed latest research material to this volume. Each paper was carefully reviewed and evaluated by an international program committee. The articles cover many topics of mathematical visualization, comprising computer graphical techniques and visualization methods, handling of meshes and polygonal data representations, as well as application of visualization techniques in geometry and numerics.

We organized the material in the following five sections although many articles can not uniquely be associated with a single category:

- Meshes, Multilevel Approximation, and Visualization
- Geometry and Numerics
- Graphics Algorithms and Implementations
- Geometric Visualization Techniques
- Vector Fields and Flow Visualization.

The themes represent most active research topics. Specifically there are new methods and experimental results for surfaces with given curvature properties, the use of Morse theory in the validation of triangle nets, and Clifford algebras for approximation of vector fields. Promising trends are new developments in the numerics on discrete geometries, the study of adaptive and hierarchical techniques in space and time, and new visualization methods for displaying mathematical structures.

We hope the book unveals new insight into the evolving and fascinating area, and the reader will become acquainted with recent developments.

We thank all authors for their contributions and all members of the program committee for their efforts and thorough reviews. Special thanks to Axel Friedrich for his help in preparing the final LaTeX manuscript. The personal interest and continuous support of Dr. Martin Peters from Springer-Verlag is very much appreciated.

Berlin, 1998 *Hans-Christian Hege*[1] · *Konrad Polthier*[2]

[1] Konrad-Zuse-Zentrum für Informationstechnik Berlin (ZIB), Department of Scientific Visualization

[2] Technische Universität in Berlin (TUB), Fachbereich Mathematik and Sonderforschungsbereich "Differential Geometry and Quantum Physics"

Table of Contents

IV Geometric Visualization Techniques

V Vector Fields and Flow Visualization

List of Contributors

Ulrike Axen
University of Illinois
Dep. of Computer Science
MC-258
Urbana, IL 61801
USA
axen@cs.uiuc.edu
www.ncsa.uiuc.edu/VEG/audio/people/ua

Ken Brodlie
University of Leeds
School of Computer Studies
Leeds LS2 9JT
United Kingdom
kwb@scs.leeds.ac.uk
www.scs.leeds.ac.uk/kwb

Paolo Cignoni
Instituto di Elaborazione
dell'Informazione Consiglio
Nazionale delle Ricerche
Via S. Maria 46
56126 Pisa
Italy
cignoni@iei.pi.cnr.it
miles.cnuce.cnr.it/cg/peopleCigno.html

Jürgen Döllner
Universität Münster
Institut für Informatik
Einsteinstr. 62
48149 Münster
Germany
dollner@uni-muenster.de
wwwmath.uni-muenster.de/~dollner

Gerd Dziuk
Universität Freiburg
Inst. für Angewandte Mathematik
Hermann-Herder-Str. 10
79104 Freiburg
Germany
gerd@mathematik.uni-freiburg.de
www-mathematik.uni-freiburg.de/
homepages/gerd.html

Herbert Edelsbrunner
University of Illinois
Department of Computer Science
MC-258
Urbana, IL 61801
USA
edelsbrunner@cs.uiuc.edu
www.cs.uiuc.edu

Thomas Ertl
Universität Erlangen
IMMD 9
Am Weichselgarten 9
91058 Erlangen
Germany
ertl@informatik.uni-erlangen.de
www9.informatik.uni-erlangen.de/
Persons/Ertl

George Francis
University of Illinois
Department of Mathematics
1407 W Green St
Urbana, IL 61801
USA
gfrancis@uiuc.edu
www.math.uiuc.edu/~gfrancis

Georg Glaeser
University of Applied Arts
O. Kokoschka-Platz 2
1010 Vienna
Austria
gg@picasso.tuwien.ac.at
www.geometrie.tuwien.ac.at/glaeser

Eduard Gröller
Vienna University of Technology
Karlsplatz 13
1040 Vienna
Austria
groeller@cg.tuwien.ac.at
www.cg.tuwien.ac.at/staff/EduardGroeller.html

Karsten Große-Brauckmann
Universität Bonn
Mathematisches Institut
Beringstr. 1
53115 Bonn
Germany
kgb@math.uni-bonn.de
rhein.iam.uni-bonn.de/~kgb

Roberto Grosso
Universität Erlangen
IMMD IX
Am Weichselgarten 9
91058 Erlangen
Germany
grosso@informatik.uni-erlangen.de
www9.informatik.uni-erlangen.de/Persons/Grosso

Hans Hagen
Universität Kaiserslautern
Fachbereich Informatik
Postfach 3049
67653 Kaiserslautern
Germany
hagen@informatik.uni-kl.de
davinci.informatik.uni-kl.de:8000/people/hagen.html

John C. Hart
Washington State University
School of EECS
Pullman, WA 99164-2752
USA
hart@eecs.wsu.edu
www.eecs.wsu.edu/~hart

Chris Hartman
University of Illinois
Department of Mathematics
1409 W. Green St.
Urbana, IL 61801
USA
hartman@math.uiuc.edu
www.math.uiuc.edu/~hartman

Hans-Christian Hege
Konrad-Zuse-Zentrum für
Informationstechnik Berlin (ZIB)
Takustr. 7
D-14195 Berlin
Germany
hege@zib.de
www.zib.de/hege

Norbert Heußer
Universität Bonn
Sonderforschungsbereich 256
Wegelerstr. 6
53115 Bonn
Germany
kender@iam.uni-bonn.de
rhein.iam.uni-bonn.de/~kender

Klaus Hinrichs
Universität Münster
Institut für Informatik
Einsteinstr. 62
48149 Münster
Germany
Klaus.Hinrichs@informatik.uni-muenster.de
www.uni-muenster.de/informatik/u/khh

John E. Hutchinson
Australian National University
School of Mathematical Sciences
GPO Box 4
Canberra ACT 0200
Australia
john.hutchinson@anu.edu.au
www.anu.edu.au

Tim Hoffmann
Technische Universität Berlin
Fachbereich Mathematik, MA 8-3
Straße des 17. Juni 136
10623 Berlin
Germany
timh@sfb288.math.tu-berlin.de
www-sfb288.math.tu-berlin.de/~timh

Alfred Inselberg
Multidimensional Graphs Ltd.
36 A Yehuda Halevy Street
43556 Raanana
Israel
aiisreal@math.tau.ac.i

Markus Kohler
Universität Dortmund
Lehrstuhl Informatik VII
Otto-Hahn-Straße 16
D-44221 Dortmund
Germany
Markus.Kohler@cs.uni-dortmund.de
ls7-www.informatik.uni-dortmund.de/~kohler

Heinz Krüger
University of Kaiserslautern
Physics Department
Postfach 3049
67653 Kaiserslautern
Germany
krueger@physik.uni-kl.de

Wolfgang Kühn
Konrad-Zuse-Zentrum für
Informationstechnik Berlin (ZIB)
Takustrasse 7
14195 Berlin
Germany
kuehn@zib.de
www.zib.de/kuehn

Thomas Kučera
Vienna University of Technology
Institute of Computer Graphics
Karlsplatz 13/186/2
1040 Vienna
Austria
tkucera@cg.tuwien.ac.at
www.ac.tuwien.ac.at/home

Robert B. Kusner
University of Massachusetts
Dep. of Math. and GANG Center
1435G Lederle Grad. Res. Tower
Amherst MA 01003
USA
kusner@gang.umass.edu
www.gang.umass.edu/people/rk.html

Helwig Löffelmann
University of Technology
Institute of Computer Graphics
Karlsplatz 13/186/2
1060 Vienna
Austria
helwig@cg.tuwien.ac.at
www.cg.tuwien.ac.at/home

Adriano Lopes
University of Leeds
School of Computer Studies
Leeds LS2 9JT
United Kingdom
adriano@scs.leeds.ac.uk
www.scs.leeds.ac.uk/postgrads/adriano.html

Claudio Montani
Instituto di Elaborazione
dell'Informazione Consiglio
Nazionale delle Ricerche
Via S. Maria 46
56126 Pisa
Italy
montani@iei.pi.cnr.it
miles.cnuce.cnr.it/cg/peopleMonta.html

Heinrich Müller
Universität Dortmund
Informatik VII
Graphische Systeme
D-44221 Dortmund
Germany
mueller@ls7.informatik.uni-dortmund.de
ls7-www.informatik.uni-dortmund.de/~mueller

Cornelius W. A. M. van Overveld
Eindhoven Uni. of Technology
Dep. of Math. and Comp. Science
Den Dolech 2
5600 MB Eindhoven
The Netherlands
wsinkvo@win.tue.nl
www.win.tue.nl/inf

Konrad Polthier
Technische Universität Berlin
FB Mathematik, MA 8-3
Straße des 17. Juni 136
D-10623 Berlin
Germany
polthier@math.tu-berlin.de
www-sfb288.math.tu-berlin.de/~konrad

René T. Rau
Universität Tübingen
WSI/GRIS
Auf der Morgenstelle 10
72076 Tübingen
Germany
rrau@gris.uni-tuebingen.de
www.gris.uni-tuebingen.de/people/staff/rrau

Dennis Roseman
The University of Iowa
Department of Mathematics
Iowa City, IA 52242-1419
USA
roseman@math.uiowa.edu
www.math.uiowa.edu/~roseman

Hanns Ruder
Universität Tübingen
Theoretische Astrophysik
Auf der Morgenstelle 10
72076 Tübingen
Germany
ruder@tat.physik.uni-tuebingen.de
aorta.tat.physik.uni-tuebingen.de

Martin Rumpf
Universität Bonn
Institut für Angewandte
Mathematik
Wegelerstr. 6
D-53115 Bonn
Germany
rumpf@iam.uni-bonn.de
www.iam.uni-bonn.de/~rumpf

Detlef Ruprecht
Andersen Consulting
Otto-Volger-Straße 15
65843 Frankfurt
Germany

Gerik Scheuermann
Universität Kaiserslautern
Postfach 3049
67653 Kaiserslautern
Germany
scheuer@informatik.uni-kl.de
davinci.informatik.uni-kl.de:8000/~scheuer

Markus Schmies
Technische Universität Berlin
FB Mathematik, MA 8-3, Sfb 288
Straße des 17. Juni 136
D-10623 Berlin
Germany
tn13@sfb288.math.tu-berlin.de
www-sfb288.math.tu-berlin.de

Roberto Scopigno
CNUCE - Cons. Naz. delle Ricerche
Department Parallel Processing
Via S. Maria 36
56126 Pisa
Italy
r.scopigno@cnuce.cnr.it
miles.cnuce.cnr.it/cg/peopleScop.html

Hans-Peter Seidel
Universität Erlangen
IMMD - IX
Am Weichselgarten 9
91058 Erlangen
Germany
seidel@informatik.uni-erlangen.de
www9.informatik.uni-erlangen.de/
Persons/Seidel

Philipp Slusallek
University of Erlangen
IMMD IX
Am Weichselgarten 9
91058 Erlangen
Germany
slussalek@informatik.uni-erlangen.de
www9.informatik.uni-erlangen.de/
Persons/Slusallek

Detlev Stalling
Konrad-Zuse-Zentrum für
Informationstechnik Berlin (ZIB)
Takustr. 7
D-14195 Berlin
Germany
stalling@zib.de
www.zib.de/stalling

Marc Stamminger
Universität Erlangen
IMMD IX
Am Weichselgarten 9
91058 Erlangen
Germany
stamminger@informatik.uni-erlangen.de
www9.informatik.uni-erlangen.de/
Persons/Stamminger

Ivan Sterling
University of Toledo
Department of Mathematics
Toledo, OH 43606-3390
USA
isterlin@math.utoledo.edu
www.math.utoledo.edu/faculty_pages/
isterling.html

Thomas Sterling
1715 Chandler
Ann Arbor
Michigan 48105
USA

John M. Sullivan
University of Illinois
Mathematics Department
1409 W Green St
Urbana, IL 61801-2975
USA
jms@math.uiuc.edu
www.math.uiuc.edu/~jms

Alexandru C. Telea
Eindhoven Uni. of Technology
Dep. of Math. and Comp. Science
Den Dolech 2
5600 MB Eindhoven
The Netherlands
alext@win.tue.nl
www.win.tue.nl/inf

Laurent Testard
Institute d'Informatique et de
Mathématiques Appliques de
Grenoble (IMAG, LMC)
51 rue des Mathématiques, BP 53
38041 Grenbole Cedex 9
France
Laurent.Testard@imag.fr
www-lmc.imag.fr/~testard

Daniel Weiskopf
Universität Tübingen
Theoretische Astrophysik
Auf der Morgenstelle 10
72076 Tübingen
Germany
daniel.weiskopf@student.uni-tuebingen.de
www.gris.uni-tuebingen.de

Part I

Meshes, Multilevel Approximation, and Visualization

Tetrahedra Based Volume Visualization

Paolo Cignoni[1], Claudio Montani[1], and Roberto Scopigno[2]

[1] Istituto di Elaborazione dell'Informazione Consiglio Nazionale delle Ricerche, Pisa Italy
[2] CNUCE – Consiglio Nazionale delle Ricerche, Pisa Italy

Abstract. Volume Visualization techniques have advanced considerably since the first international symposium held on this topic eight years ago.
This paper briefly reviews the techniques proposed for the visualization of irregular (or scattered) volume datasets. In particular, methods which adopt simplicial decompositions of E^3 space are considered, and this choice is justified both in terms of modeling and visualization. Simplicial complexes are powerful and robust geometric structures, and a number of efficient visualization algorithms have been proposed. We show that simplicial cells (or simply tetrahedral cells since our target is 3D space) may be conceived as being the unifying kernel primitive for the visualization of not-regular meshes.

1 Introduction

A *volume dataset* consists of points in E^3 space, with one or more scalar or vector sample values associated with each point. The need for a visual representation of the content of such datasets led to a substantial research effort and to a new computer graphics field, *Volume Visualization*. Applications which produce volume data and require data visualization tools exist in many fields of research, including molecular modeling, medical imaging, mathematics, geosciences, computational fluid dynamics, and finite element analysis. Volume data can come in various grid types, e.g. regular, curvilinear, irregular or scattered [34]. Volume Visualization provides the user with the data representation structures and the rendering techniques to get insight into the data more easily than by merely representing such data as tables or sequences of 2D images.

Simplicial complexes are widely used in Volume Visualization due to the following properties: most cell complexes found in Volume Visualization can easily be decomposed into a simplicial complex (even implicitly, in order to avoid the growth of datasets); simplicial complexes can easily be rendered using various algorithms; isosurface extraction from simplicial complexes avoids ambiguities that occur with hexahedral complexes; most of the rendering algorithms for irregular cell complexes are simpler (and therefore faster) to describe and implement on tetrahedral complexes.

Moreover, some properties of simplicial complexes suggest their use as the kernel data representation structure in Scientific Visualization: they are suitable for modeling data in any dimension; they are a suitable basis for many interpolation techniques; data structure design is simplified; handling degenerated cases (i.e. coincident points in 3D space) is simpler, because a simplex degenerates to a lower dimensional simplex.

Reasons for adopting simplicial complexes in volume visualization are given in the following section. The approaches proposed for simplicial visualization are then introduced in Section 3. Isosurface fitting and direct volume rendering techniques are reviewed in Sections 4 and 5, respectively. Finally, some open issues are listed in the concluding section.

2 Volume Modeling Based on Simplicial Complexes

In this section we formally introduce simplicial complexes. We describe their properties and how they can be used to represent volumetric scalar datasets.

Consider a set $V = \{v_0, v_1, \dots, v_d\}$ of $d + 1$ linearly independent points in the k-dimensional Euclidean space $\mathbb{E}^k$, with $d \leq k$. The subset σ of $\mathbb{E}^k$, formed by all points which can be expressed as linear convex combinations of the points of V, is called a *d-simplex*.

A collection Σ of simplices is called a *d-simplicial complex* when the following conditions hold:

- for each simplex $\sigma \in \Sigma$, all the faces of σ belong to Σ;
- for each pair of simplices $\sigma, \tau \in \Sigma$, either $\sigma \cap \tau = \emptyset$ or $\sigma \cap \tau$ is a proper face of both σ and τ;
- d is the maximum of the orders of the simplices belonging to Σ (d is called the *order* of Σ).

In practice, d-simplices are used as building blocks to cover the domain. Boundary faces form the boundary of the domain, while internal faces separate such blocks from one another. If general polyhedra are used as building blocks, instead of simplices, the previous concepts can be generalized to define a *cell complex*.

One advantage of *simplicial complexes* over the more generical *cell complexes* is in the design of data representation schemes, because: a d-simplicial complex Σ is fully characterized by its combinatorial description plus the coordinates of its vertices; any simplex σ implicitly defines all its faces; the number of k faces of a d-simplex is a constant; the combinatorial structure of Σ is completely characterized by the list of its top simplices; if Σ is regular then it is characterized by the list of its d-simplices.

Delaunay complexes are a particular class of d-simplicial complexes. A d-simplicial complex Σ in $\mathbb{E}^d$ is called a *Delaunay simplicial complex* if and only if it covers the convex hull of its vertices and the hypersphere circumscribing each d-simplex of Σ does not contain any vertex of Σ in its interior. For a given set $V \subset \mathbb{E}^3$ of n points ($n \geq d + 1$), there exists a unique Delaunay simplicial complex having V as its vertices if and only if there are no $d + 2$ points of V that are cospherical.

A Delaunay simplicial complex can be built on any set of vertices V, and the shape of its simplices is the most regular of all possible simplicial complexes built on V. Moreover, efficient algorithms have been proposed which, given a set of vertices V, build the corresponding Delaunay simplicial complex [36].

From a visualization point of view, a number of rendering algorithms exist that handle simplicial complexes. The use of simplicial decompositions to manipulate and render non regular hexahedral datasets leads to an increase in the number of cells (at least 5 simplices for each hexahedral cell), but simplifies problems caused by occasional degenerate cells (e.g., non-hexahedral cells due to coincident sites) and cells with non-planar faces. Handling such cases is generally more complex when a non-simplicial cellular decomposition is used. In fact, many visualization algorithms have been described with high generality (i.e. they have been specified "on paper" for general convex cell complexes), but often they have only been implemented on simplicial complexes.

3 Visualization of Simplicial Complexes

The visualization approaches proposed for rendering volume datasets can be categorized as follows.

- *Slicing:* only a 2D subset of the information of the dataset, extracted by the use of cutting plane/surfaces, is visualized for each shot.
- *Isosurface fitting:* reconstruction and visualization of poligonal iso-surfaces, where each of them is an approximation of the subset of points where the value of the sampled field f is equal to the threshold value; see Figure 2 top for an example of an isosurface extracted from the well known *bluntfin* dataset.
- *Direct Volume Rendering:* visualization of the whole dataset at once; the scalar field value is mapped into visual attributes (color and opacity), and then images are computed by accumulating the color densities of the cell sections that are projected onto the same image parcel (or pixel). The function that maps scalar field into color and opacity is usually called the *Transfer Function* (TF), which is a useful instrument for the user to enhance or remove part of the data content.
  DVR techniques may be further divided into three classes: *ray tracing*, *scan line*, and *projective* methods; an image obtained with projective DVR is shown in Figure 2 bottom; the blue zones denote subvolumes where the field is higher.

In the next two sections isosurface fitting and DVR techniques are reviewed in the framework of the simplicial representation of volume datasets.

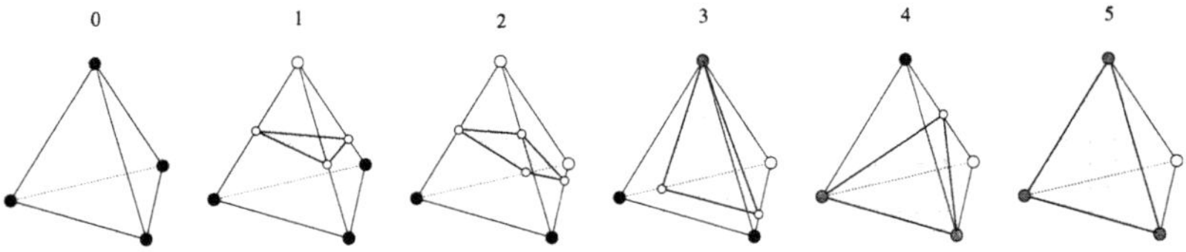

Fig. 1. Marching Tetrahedra: vertex configurations and corresponding isosurface patches.

4 Isosurface Fitting

Given a volume dataset described by a tetrahedral complex Σ with a set of scalar values w_i associated with the vertices of the complex, and a threshold value δ, the isosurface passing through the points of the volume dataset having value δ can be reconstructed by using a *per cell* approach similar to the Marching Cubes algorithm [24]. It is therefore usually called Marching Tetrahedra (MT).

The main idea of the algorithm is to traverse all the dataset cells and to compute for each cell σ crossed by the isosurface (active cell) the isosurface patch passing through σ. Each vertex v of the dataset is classified as black, gray or white if the value associated with v is, respectively, lower, equal or greater than the given threshold δ. This classification of cell vertices can generate $3^4 = 81$ different combinatorial configurations. By exploitating symmetries, the latter can be traced back to the six main classes shown in Figure 1. Note that if an isosurface passes exactly through the vertices of some tetrahedra then it must be explicitly managed in order to avoid the creation of null or duplicate triangles (see class 5 in Figure 1).

Once the class for each tetrahedron has been identified, the positions of the isosurface vertices and the associated normals are calculated by linearly interpolating on the tetrahedron edges.

Optimizations

Isosurface Normals. Gradients of the scalar field at the vertices of a tetrahedral mesh can be computed in a preprocessing step. The gradient of the scalar field within each tetrahedron t of the mesh is the gradient of the linear interpolation of the field at the vertices of t. The gradient at each vertex v of the dataset can be pre-computed as the weighted average of normalized gradients at all tetrahedra incident at v, where the weight for the contribution of a tetrahedron t is given by the solid angle of t at v[1].
The normal at a vertex of the isosurface w is computed during isosurface

[1] The solid angle of a triedral angle is $A = \alpha + \beta + \gamma - \pi$ where α, β, γ are the dihedral angles between the tetrahedron facets.

extraction by the linear interpolation of gradients at the endpoints of the cell edge where w lies.

Purging not–active cells. The identification of the set of cells crossed by the isosurface entails traversing the whole dataset, even if the isosurface being searched for only crosses a few cells. Many speedup techniques [38,19,12,23] have been proposed in order to avoid the analysis of non-active cells. An optimal solution to the search for active cells was proposed by Cignoni et al. in [6]. The method is based on a data structure called *interval tree* which encodes a set of intervals on the real line and supports optimal-time retrieval of all intervals containing a given value. If the adjacency information between cells is stored, it can be exploited to efficiently retrieve, by topological propagation, all the cells crossed by an isosurface. The cells from which the propagation starts are called *seed cells.* Techniques to choose a seed set composed of only a small subset of dataset cells are described in [1].

This approach was extended by storing the interval tree on secondary memory and accessing it with an I/O optimal algorithm without loading either the interval tree or the whole dataset in the main memory [3].

Reducing redundant computations. The redundant computation of both coordinates and normals on the vertices of the isosurface may be avoided through a common technique based on *hash indexing* [40,6]. The edges of the mesh that define the vertices of the isosurface are used as access keys to a hash table.

5 Direct Volume Rendering Algorithms

We review direct volume rendering solutions, focusing on those that work on simplicial decompositions. *Splatting* algorithms are therefore not considered. The rendering techniques are presented in order of image quality: from slow, high quality algorithms to faster and somehow less precise projective techniques exploiting graphics hardware.

5.1 Volume Shading Models

Many models for simulating the propagation, scattering and shadowing through semi-transparent media have been proposed in the literature. The first models aimed to simulate natural phenomena, such as dust [2] or clouds [20]. These models were then extended or modified [9,29,22] to fulfil scientific visualization needs. This distinction still remains, and it is therefore useful to clarify the two different objectives in modeling semi-transparent media:

- *Photorealism*: if we want to model real semi-transparent media, such as clouds, fog, dust, to reproduce their aspect with the utmost visual fidelity.

– *Comprehension*: if we want to exploit the hints that shading can add to better understand the 3D information we are visualizing.

In the first case, an accurate model of the medium should take into account phenomena such as self-shadowing and second and higher order scattering of light; computational times are so great that interactivity is often prevented. In the following, we sketch the basis of the most common model in scientific visualization, known as the *density emitter* model. A dissertation on different lighting models for direct volume rendering was proposed by Max [25].

A volume dataset lighting model can be approximated by a number of infinitely small sphere particles with a projected area A. Consider a small cylindrical slab with a base of area E, a thickness of δs and a volume of $E\delta s$. This slab is filled with particles with a given density ρ (number of particles per unit volumes). The slab therefore contains $N = \rho E \delta s$ particles. We assume that particles glow diffusely (isotropically) and that they absorb all the light that they occlude without any scattering (reflective) effects. If the thickness of the slab δs is small enough, we can assume that there is no overlap between particles along the height of the cylinder, so the total occlusion area is $NA = \rho E \delta A$ and the fraction of light blocked by the particles is $\rho A \delta s$. With $\tau(s) = \rho(s)A$ we denote the extinction coefficient that defines the rate of light that is occluded at distance s. We assume that the glowing intensity of particles at position s is $g(s)$. We can write the differential expression showing the fraction of absorbed/emitted light when the slab thickness goes to zero as:

$$\frac{dI}{ds} = g(s) - \tau(s)I(s)$$

where s is a lenght parameter, $I(s)$ is the light intensity at distance s and $\tau(s) = \rho(s)A$ is the extinction coefficient that defines the rate of light that is occluded at distance s. The negative sign in this differential equation is because we are calculating the light that is subtracted when traversing the volume. Remember that the extinction coefficient (also referred in literature with a misleading term as *opacity*) range from zero (no particles) to infinity (opaque medium). The solution of the above equation gives us the light intensity $I(s)$ that reach position s along a ray traversing a volume.

$$I(s) = I_0 T(s) + \int_0^s g(t)T'(t)dt \quad \text{where} \quad T'(k) = exp\left(-\int_k^s \tau(x)dx\right)$$

There is no simple closed form for the equation above, but efficient techniques for approximating it have been devised [25].

5.2 Ray Tracing

Ray tracing methods process the dataset in *image-order*, and accumulate color and opacities interpolated while tracing each ray. For each pixel of the image, a ray is cast and intersected with the cells of the volume data. The transfer function, which transforms the data values into opacities and colors, is then sampled and integrated along the ray. The result of this integration determines the color of the pixel. Such techniques were initially described for regular volume datasets only (which can be raytraced quite efficiently by exploiting the regular structure of the data).

Garrity presented a technique to ray trace irregular datasets which exploits the topological relationships between cells to efficiently perform color and opacity integration along the ray [15].
This approach works in two steps. In a preprocessing phase, the cell faces that are on the boundary of the dataset are detected and inserted into a bucketing structure (based on regular space subdivision). In the second phase, for each ray, the nearest intersected face is searched from the boundary faces, and cell tracing starts from the associated cell. Then, jumping from cell to cell by exploiting the connectivity of the dataset, all the cells intersected by the ray are detected. When the ray quits the dataset the list of boundary faces is browsed again to see if the ray pierces the dataset again (e.g. in the case of non–convex or multi–component datasets). Particular attention must be paid to degenerate cases (ray passing through faces, edges or vertices). The restriction to tetrahedral cells simplifies handling degenerate cases.

5.3 Scan Line

The raytracing approach fails to exploit the fact that adjacent rays probably intersect the same cells, and its computational cost is therefore excessively high. Scanline algorithms try to exploit this coherence.
The first scanline algorithm for rendering arbitrary cell complexes was proposed by Giertsen [16]. The algorithm uses a *scan plane buffer*, a data structure associated with a plane (xz) perpendicular to the viewplane and passing through a line of the viewplane. Volume cells are progressively sliced by the scan plane. Each slice is then triangulated in order to linearly interpolate the values inside each cell slice[2].
The volume cells are maintained y-ordered and the set of cells intersected by the scan plane is updated following the y-sweeping of the plane. The *scan plane buffer* (*spb*) is a structure (a bidimensional array) that discretely maintains the intersections between the cells and the scan plane. Cell slices are randomly scan-converted into z segments orthogonal to the viewplane, and the length and the opacity/color contribution of each segment are stored in the *spb*. Once all the cell slices have been scan–converted, the colors of the pixels on the current scan line are calculated by traversing the *spb* in the z direction and accumulating the opacity/color contributions contained.

[2] Note that this kind of interpolation is not rotational invariant, it thus creates aliasing effects when rotating the volume. A more correct solution, again suggested by Giertsen [16], is to decompose each cell into tetrahedral elements.

Another algorithm based on the scanline approach is *Lazy Sweep Ray Casting* (LSRC) [33,32]. Its main contribution is to avoid the use of a discrete scanplane, to prevent the possible aliasing effects that can result from datasets with high variations in cell size (differences of the order of 1:100,000 can occur). The LSRC algorithm works in two phases: a space sweep, with a sweep plane orthogonal to the viewing (XY) plane, and a second sweep on that plane with a sweep line parallel to the Z axis.

Another scanline algorithm that exploits a spatial hierarchical organization of the dataset was described by Wilhelms et al. [39]. The main difference of this approach (which like the previous algorithms uses both a scan plane orthogonal to the view plane and a scan line lying on that plane) is that it renders semi-transparent regions of space between cell faces as well as opaque polygonal surfaces immersed in the dataset. Cells are implicitly sliced when their faces are being processed. The method builds a k-d tree over the polygons (either cell faces or immersed object faces) to improve the efficiency. The hierarchy is also used to render approximate images of the dataset. The method has been parallelized on a shared memory MIMD machine. The problem of a rotationally invariant field interpolation inside each cell still holds, unless a tetrahedral decomposition is adopted.

5.4 Projective Algorithms

Projective algorithms render a tetrahedral mesh through direct cell projection and rgbα–compositing [26,31,41]. They are generally based on a two phase process: first, cells are sorted into depth; second, depth ordered cells are projected on the view plane and rgbα–composed on the frame buffer. Different approaches to manage these two phases are presented below.

Depth Sorting. To compose rgbα contributions correctly, cells have to be depth–ordered with respect to the given viewpoint. A visibility order of a set of objects, with respect to a viewpoint p, is a sequence of such objects such that, if object A obstructs object B when seen from p, then A precedes B in the sequence. The obstruction relation (usually called *infront/behind* relation) for a pair of not self-intersecting cells γ_1, γ_2 and with respect to a viewpoint p can be formally expressed as follows:

> $\gamma_1 \prec_p \gamma_2$ iff $\exists$ a ray r emanating from p and intersecting both s_1 and s_2, such that all points in $r \cap s_1$ are closer to p than any point in $r \cap s_2$.

Acyclic Simplicial Complexes. A cell complex Γ is called *acyclic* with respect to viewpoint p if and only if relation $\prec_p$ defines a partial order on the cells of Γ. In this case, it is possible to order the cells of Γ either front-to-back or back-to-front with respect to the viewpoint. It has been proved that all cell complexes in $\mathbb{E}^k$ that can be obtained by projecting the boundary of a convex polytope in $\mathbb{E}^{k+1}$ (called *projective complexes*) are acyclic with respect to any viewpoint [10]. Delaunay simplicial complexes (see definition in Section 2) are projective, and thus they are acyclic with respect to any viewpoint [10].

If, conversely, the simplicial complex is not a Delaunay one, then acyclicity is not guaranteed. Testing a simplicial complex for acyclicity is not practical. A brute-force algorithm checks acyclicity by depth-sorting the complex from all significant possible viewpoints[3]. These viewpoints are placed in all the cells generated by the partition of the space with planes passing through all the cell's faces. But the complexity of this approach, $O(f^4)$, where f is the number of faces of the complex, limits its usability.

Topological Sort. The cells of an acyclic convex complex can be sorted by exploiting face-adjacency between tetrahedra and face orientation. An algorithm, called Meshed Polyhedra Visibility Ordering (MPVO), based on this approach was proposed by Williams [41,43]. In a preprocessing phase, the MPVO algorithm constructs the adjacency graph for the given mesh and calculates the plane equation coefficients for each face. At rendering time, the algorithm works in two steps. First, it computes the occlusion relation for all pairs of cells, given the current viewpoint, and converts the adjacency graph into a direct acyclic graph (DAG). Second, a total ordering of the cells is obtained in linear time by a topological sort of the DAG. If the topological sort is obtained by a depth–first visit of the DAG, the presence of cycles can be detected and a partially correct mesh ordering can be calculated.

This algorithm can be extended to non convex meshes by sorting all the cells with boundary faces according to their centroid, and selecting them in the DFS search algorithm according to that ordering. Note that this extension may produce a wrong sort, because the occlusion relation does not always agree with the centroid distance.

Numerical Sort. Delaunay tetrahedral complexes can be depth sorted by exploiting the following property[4]: *the length of the tangent from the viewpoint to the sphere that circumscribes a tetrahedron reflects the depth ordering of the complex.*
The centers c_i and radii r_i of all the tetrahedra σ_i of the complexes must be precomputed once and stored together with the dataset. Then, to sort the complex it is sufficient to calculate the square of the distances from the current viewpoint to centers c_i, subtract from them the squared radius r_i, and sort the resulting values numerically.

Numerical sorting is more robust than topologic sorting, and running times are comparable [7]. The main drawback of this technique is that it only works if the complex is a Delaunay one and there are no degeneracies. It can therefore fail in many common datasets, such as those obtained by regularly

---

[3] Personal communication by H. Edelsbrunner to P. Williams [43]

[4] Personal communication by H. Edelsbrunner and B. Joe to N. Max et al. [26].

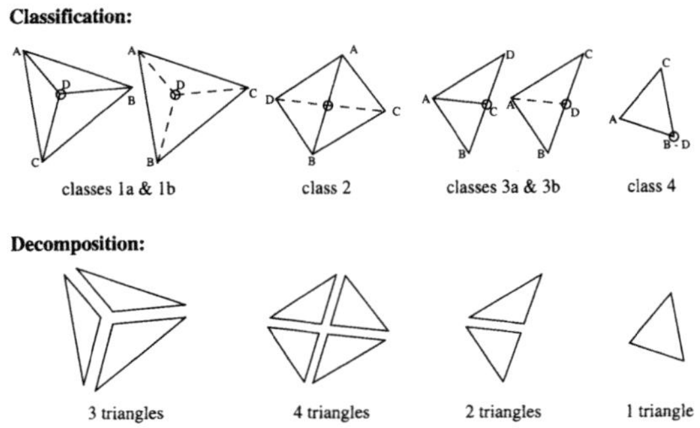

Fig. 2. Classification of the silhouette of a projected tetrahedron.

decomposing hexahedral cells into 5/6 tetrahedra (all these 5/6 tetrahedra share the same circumsphere).

Newell, Newell and Sancha (NNS). Tetrahedral meshes can also be sorted using an extension of the NNS algorithm [35]. Vertices are view–transformed, and the sorting process is again organized into two phases. The first is a preliminary sort of polyhedra according to their rearmost z component. The second step is a *fine tuning* of the sort, organized into checks of increasing computational complexity similarly to the original NSS algorithm. The goal of *fine tuning* is to find a separating plane between two cells from which the correct cell drawing order could be easily derived. The existence of cycles can be detected by tagging every overlapping cell and testing whether a cell is involved in an overlapping more than once. No solution for breaking the cycles is presented.

Classification and Projection. Once the complex has been sorted, various techniques are available to calculate the contribution of each tetrahedron to the intermediate image.
A first approach is to scan-convert each tetrahedral cell, adopting a pure software three–dimensional scan–conversion process [26]. For each cell, the color and opacity contribution is directly calculated during the 3D sampling, according to the shading model chosen.

A faster approach, Projected Tetrahedra [31], renders tetrahedral cells by approximating each cell with semi-transparent triangles, and composing such triangles using standard graphics libraries/hardware. The main idea is to classify the shape resulting from the projection of a tetrahedron (see Figure 2) in a limited number of classes which can be easily decomposed into triangles. The correct opacity/color of the thickest point of the projected tetrahedron (indicated with a small circle in Figure 2) is then computed, and the cells' contribution to the current frame is obtained by Gouraud–interpolating the projected shape.

The classification of the projected silhouette and the identification of the thickest point can be done in at least two different ways, according to the sorting algorithm used:

- *Cell classification after topological sort.* In phase 1 of the MPVO algorithm we can compute, for each tetrahedron, the number of faces which are oriented toward the observer. The class of projection can be automatically deduced from this number.
- *Cell classification after numerical or NNS sort.* In this case, each tetrahedron must be classified independently. The easiest techniques is to test the clockwise ordering of vertices of each tetrahedron face according to the viewpoint. This can be done by a simple cross product between two edges of each face.

Aliasing. Errors and visual artifacts are introduced because of the limited numerical precision used by rendering subsystems to compute rgbα compositing, and because of the linear interpolation of the opacities, while an exponential interpolation should be used to correctly approximate opacity contribution [37,35]. To reduce the latter problem, the *multipass blending* approach [37] renders and composes each triangle three times to give a quadratic interpolation. With the first two passes the quadratic interpolation of opacity is obtained by a double linear interpolation and composition of opacity; the third pass applies the color contribution. In a different solution [35], color and opacity are correctly interpolated between vertices using hardware assisted texture mapping (the texture map is in this case a two dimensional table with the values of the correct exponential opacity).

5.5 Approximate Projection Technique

An example of an approximate rendering technique is the *Incremental Slicing* method proposed by Yagel et [44]. Given the current view direction, the 2D polygonal subdivisions resulting from the slicing of that volume with a set of planes parallel to the view plane are calculated and stored. Such polygonal meshes are then rendered and composed in visibility order using graphics hardware. The number of slices is adaptive, to reduce the number of cells not intersected by any slicing plane.
Different approximate techniques for the Projected Tetrahedra algorithm were proposed by Williams [42].

6 Decreasing Complexity

Volume datasets used in current applications have a common problem, the size of the datasets, which affects both storage requirements and visualization times. Therefore, interactive image generation from very large datasets could be not feasible, even with the use of fast graphic hardware and parallelism. Approximate rendering algorithms can give only a partial solution to this problem. It is also possible to manage data complexity by adopting an *approximate representation* of the dataset. The latter approach is more general because it remains totally independent of the rendering approach. The methodology in this case is therefore to work on *data simplification* rather than on *graphics output simplification.* A comparison between these two approaches is given in [7].

6.1 Simplification Algorithms

The main approach to build an approximate representation of a tetrahedral dataset is to choose a subset of the original vertices and to build a new triangulation of (almost) the same domain. A naive *random subsampling* as proposed in [43] has serious drawbacks: there is no control on the accuracy of the simplified mesh; the technique is not adaptive, i.e. data density cannot vary over different regions of the domain. Adaptive methods that try to select the smallest set of point that approximate a dataset within a given error have been developed for the simplification of 2D surfaces [27]; efficient solutions have been obtained through incremental techniques, based on either *refinement* (refine a coarse representation by adding points [8]), or *simplification* (simplify the dataset by removing points [30]). Most of these techniques can be extended to simplify volume data, but only a few experiments have been made [4,18,28]. Another technique for reducing of the size of the dataset is the one proposed in [17]; starting from a regular dataset it permits to choose a interval of the dataset range and build a tetrahedral representation of that portion of space, called *interval set*, that is covered by the chosen range.

6.2 Multiresolution Management

The iterative application of a simplification technique with different approximation threshold produces a collection of representations, and data structures which hold different representations of the dataset in a compact manner may be devised. Multiresolution or LoD can greatly improve rendering efficiency, e.g., through suitable progressive visualization algorithms. Many approaches have recently been proposed for the multiresolution management of surfaces (see, e.g., [8] for a survey), while multiresolution volume data management is still under developed. Some steps in this direction were made in [5]. Multiresolution scheme together with spatial hierarchical organization were also used in [39] to achieve parallelizable scan-line DVR together with very fast low-quality preview images.

7 Concluding Remarks

In spite of the intensive research done on volume visualization and, in particular, on simplicial volume visualization, some issues are still open or need further investigation. We divide them into two main classes: modeling–related and rendering–related topics.

Modeling Issues. Although simplicial complexes are a powerful basic representation structure from many points of view, assumptions on the nature of these complexes are often made to make visualization easier. Below are some of the problems that require further research.

- *Conversion:* not all irregular grids that arise in visualization are simplicial complexes; efficient algorithms (or heuristics) for the decomposition of general meshes in simplicial complexes have to be devised. Some of the main problems regard how to build simplicial complexes safely: cell decomposition may not be straightforward (worst-case cells exist that cannot be decomposed into tetrahedra without adding points); it is not clear how to combine multiple (possibly interpenetrating) grids into a single complex.
- *Acyclicity:* not all simplicial complexes can be correctly depth sorted. Methods or heuritics should be developed to tesh mesh acyclicity and for making a complex acyclic. A negative result in this direction is that the problem of finding the minimal set of vertices involved in a cycle is np-complete (*feedback vertex set* problem [14]).
- *Simplification and multiresolution:* only a few preliminary attempts have taken into account the simplification of volume datasets and the definition of multiresolution data structure for their management.
- *Multi-variate meshes*: efficient methods to manage datasets which represent the evolution of a phenomena on a discrete time span and/or have many values for each vertex.
- *Volume modeling*: how to provide the user with tools to define, model and interact with simplicial complexes, for example extending solid modeling techniques to volume data. The first examples, proposed for voxel–based datasets, are: volume carving [13], volume deformation [11], and volume morphing [21].

Rendering issues. Many rendering algorithms exist, but their robustness and speed still need improvement:

- *Sorting*: projective algorithms need a robust and efficient sort algorithm. Current solutions are not general enough (e.g. a topological sort cannot manage all types of meshes) or not efficient enough (e.g. NNS is more robust but considerably slower than others); moreover, from a correctness point of view, none of the presented solutions is able to *break* a cycle.
- *Correct transfer function management*: the linear interpolation of the visual attributes inside a tetrahedron can create some aliasing effects when the transfer function is not linear in the field range spanned by the tetrahedron. This fact is rarely considered in current rendering solutions.

– *Multiresolution rendering*: new rendering approaches can be developed that exploit the availability of simplified or multiresolution representations; the design of methods for the semi-automatic *quality* vs *time* trade-off evaluation in rendering would also be an important advance.
– *Hybrid rendering (DVR + surfaces)*: the integration of surface–based rendering with direct volume rendering is not easy for most of the projective rendering methods, but is a feature required by many applications.

A general issue is how these techniques could be extended to manage the representation and visualization of datasets described in spaces larger than the usual E^3.

Finally, we wonder if the state of Volume Visualization is stable enough to include volume primitives in graphics systems, either software libraries, toolkits or hardware subsystems. In the case of regular volume datasets, this has been recently made possible by slight modifications to the graphics subsystems. The adoption of new rendering approaches based on hardware texture mapping and trilinear interpolation has produced an impressive speedup to voxel-based applications (e.g. medical imaging). Will it be possible in the near future to have a support for irregular data, possibly based on the tetrahedral primitive?

8 Acknowledgements

This work was partially financed by the Progetto Coordinato "*Modelli multirisoluzione per la visualizzazione di campi scalari multidimensionali*" of the Italian National Research Council (CNR).

References

1. L. B. Chandrajit, V. Pascucci, and D. R. Schikore, *Fast isocontouring for improved interactivity,* 1996 IEEE Volume Visualization Symposium (1996), 39–46.
2. J. F. Blinn, *Light reflection functions for simulation of clouds and dusty surfaces,* Computer Graphics (SIGGRAPH '82) **16**:3 (1982), 21–29.
3. Y. Chiang and C. T. Silva, *I/O optimal isosurface extraction,* IEEE Visualization '97 Proceedings, (R. Yagel and H. Hagen, eds.), 1997.
4. P. Cignoni, L. De Floriani, C. Montani, E. Puppo, and R. Scopigno, *Multiresolution modeling and rendering of volume data based on simplicial complexes,* Proceedings of 1994 Symposium on Volume Visualization, ACM Press, 1994, 19–26.
5. P. Cignoni, C. Montani, E. Puppo, and R. Scopigno, *Multiresolution Representation and Visualization of Volume Data,* Technical Report C97-05, Istituto CNUCE – C.N.R., Pisa, Italy, January 1997.
6. P. Cignoni, C. Montani, E. Puppo, and R. Scopigno, *Speeding up isosurface extraction using interval trees,* IEEE Trans. on Visualization and Comp. Graph. **3**:2 (1997).

7. P. Cignoni, C. Montani, D. Sarti, and R. Scopigno, *On the optimization of projective volume rendering,* Visualization in Scientific Computing 1995, Springer, Wien, 1995, 58–71.
8. L. De Floriani, P. Marzano, and E. Puppo, *Multiresolution models for topographic surface description,* The Visual Computer, **12**:7 (1996), 317–345.
9. R. A. Drebin, L. Carpenter, and P. Hanrahan, *Volume rendering,* Computer Graphics (SIGGRAPH '88 Proceedings) **22** (1988), 65–74.
10. H. Edelsbrunner, *An acyclicity theorem for cell complexes in d dimensions,* Combinatorica **10** (1990), 251–260.
11. S. F. Gibson, *Beyond volume rendering: visualization, haptic exploration, and physical modeling of voxel-based objects,* Visualization in Scientific Computing '95, Springer, Wien, 1995, 10–24.
12. R. S. Gallagher, *Span filter: an optimization scheme for volume visualization of large finite element models,* IEEE Visualization '91 Proc. (1991), 68–75.
13. T. A. Galyean and J. F. Hughes, *Sculpting: an interactive volumetric modeling technique,* Computer Graphics **25**:4 (1991), 264–274.
14. M. R. Garey and D. S. Johnson, *Computers and intractability: A guide to the theory of NP-completeness.* W. H. Freeman and Company, New York, 1979.
15. M. P. Garrity, *Raytracing irregular volume data,* Computer Graphics (San Diego Workshop on Volume Visualization), **24**:5 (1990), 35–40.
16. C. Giertsen, *Volume visualization on sparse irregular meshes,* IEEE Computer Graphics & Applications (1992), 40–48.
17. B. Guo, *A multiscale model for structured-based volume rendering,* IEEE Trans. on Visualization and Computer Graphics **1**:4 (1995), 291–301.
18. B. Hamann and J. L. Chen, *Data point selection for piecewise trilinear approximation,* Computer Aided Geometric Design **11** (1994), 477–489.
19. T. Itoh and K. Koyamada, *Automatic isosurface propagation using an Extrema Graph and sorted boundary cell lists,* IEEE Trans. on Vis. and Comp. Graph. **1**:4 (1995), 319–327.
20. J. T. Kajiya and Brian P. Von Herzen, *Ray tracing volume densities,* Computer Graphics (SIGGRAPH '84 Proceedings) **18** (1984), 165–174.
21. A. Lerios, C. D. Garfinkle, and M. Levoy, *Feature-based volume metamorphosis,* Comp. Graph. (Siggraph '95), ACM Press, 1995, 449–464.
22. M. Levoy, *Display of surfaces from volume data,* IEEE Computer Graphics and Applications **8**:3 (1988), 29–37.
23. Y. Livnat, H. V. Shen, and C. R. Johnson, *A near optimal isosurface extraction algorithm for structured and unstructured grids,* IEEE Trans. on Vis. and Comp. Graph. **2**:1 (1996), 73–84.
24. W. E. Lorensen and H. E. Cline, *Marching cubes: A high resolution 3D surface construction algorithm,* Computer Graphics (SIGGRAPH '87 Proceedings) **21** (1987), 163–170.
25. N. Max, *Optical models for direct volume rendering,* IEEE Trans. on Vis. and Comp. Graph. **1**:2 (1995), 99–108.
26. N. Max, P. Hanrahan, and R. Crawfis, *Area and volume coherence for efficient visualization of 3D scalar functions,* Computer Graphics (San Diego Workshop on Volume Visualization), **24**:5 (1990), 27–33.

27. E. PUPPO AND R. SCOPIGNO, *Simplification, LOD, and Multiresolution – Principles and Applications,* Technical Report C97-12, CNUCE, C.N.R., Pisa (Italy), June 1997. (also in: EUROGRAPHICS'97 Tutorial Notes, Eurographics Association, Aire-la-Ville (CH)).
28. K. J. RENZE AND J. H. OLIVER, *Generalized unstructured decimation,* IEEE C.G.&A. **16**:6 (1996), 24–32.
29. P. SABELLA, *A rendering algorithm for visualizing 3D scalar fields,* Computer Graphics (SIGGRAPH '88 Proceedings) **22**:4 (1988), 51–58.
30. W. J. SCHROEDER, J. A. ZARGE, AND W. E. LORENSEN, *Decimation of triangle meshes,* ACM Computer Graphics (SIGGRAPH '92 Proceedings) **26** (1992), 65–70.
31. P. SHIRLEY AND A. TUCHMAN, *A polygonal approximation to direct scalar volume rendering,* Computer Graphics (San Diego Workshop on Volume Visualization), **24**:5 (1990), 63–70.
32. C. T. SILVA AND J. S. B. MITCHELL, *The lazy sweep ray casting algorithm for rendering irregular grids,* Technical Report 11794/3600, State University of New York, Stony Brook, 1997.
33. C.T. SILVA, J. MITCHELL, AND A. KAUFMAN, *Fast rendering of irregular grids,* Proceedings 1996 Symp. on Volume Visualization (1996), 15–22.
34. D. SPERAY, S. KENNON, *Volume probes: Interactive data exploration on arbitrary grids,* Computer Graphics (San Diego Workshop on Volume Visualization), **24**:5 (1990), 5–12.
35. C. STEIN, B. BECKER, AND N. MAX, *Sorting and Hardware Assisted Rendering for Volume Visualization,* Proceedings of 1994 Symposium on Volume Visualization, ACM Press, 1994, 83–90.
36. P. SU AND R. L. S. DRYSDALE, *A comparison of sequential delaunay triangulation algorithms,* 11th ACM Computational Geometry Conf. Proc. (Vancouver, Canada), ACM Press, 1995, 61–70.
37. A. VAN GELDER AND J. WILHELMS, *Rapid exploration of curvilinear grids using direct volume rendering,* IEEE Visualization '93 Proceedings, 1993, 70-77.
38. J. WILHELMS AND A. VAN GELDER, *Octrees for faster isosurface generation,* ACM Computer Graphics **24**:5 (1990), 57–62.
39. J. WILHELMS, A. VAN GELDER, P. TARANTINO, AND J. GIBBS, *Hierarchical and parallelizable direct volume rendering for irregular and multiple grids,* Visualization '96 Proceedings, IEEE Press, 1996, 57–64.
40. J. WILHELMS AND A. VAN GELDER, *Octrees for faster isosurface generation,* ACM Transaction on Graphics **11**:3 (1992), 201–227.
41. P. L. WILLIAMS, *Visibility ordering of meshed polyhedra,* ACM Transaction on Graphics **11**:2 (1992), 103–126.
42. P. L. WILLIAMS, *Interactive splatting of nonrectilinear volumes,* A.E. Kaufman and G.M. Nielson, editors, Visualization '92 Proceedings, IEEE Computer Society Press, 1992, 37–45.
43. P. L. WILLIAMS, *Interactive Direct Volume Rendering of Curvilinear and Unstructured Data.* PhD thesis, University of Illinois at Urbana–Champaign, 1993.
44. R. YAGEL, D. M. REED, A. LAW, P. W. SHI, AND N. SHAREEF, *Hardware assisted volume rendering of unstructured grids by incremental slicing,* Proceedings 1996 Symp. on Volume Visualization, 1996, 55–62.

Mesh Optimization and Multilevel Finite Element Approximations

Roberto Grosso and Thomas Ertl

Lehrstuhl für Graphische Datenverarbeitung (IMMD9),
Universität Erlangen-Nürnberg, Erlangen, Germany

Abstract. Mesh reduction techniques are used for accelerating the visualization process for large datasets. Typical examples are scalar or vector valued functions defined on complex 2 or 3 dimensional meshes. Grosso et al. presented a method for mesh optimization based on finite elements approximations with the L_2 norm and adaptive local mesh refinement. Starting with a very coarse triangulation of the functional domain a hierarchy of highly non-uniform tetrahedral (or triangular in 2D) meshes is generated adaptively by local refinement. This process is driven by controlling the local error of the piecewise linear finite element approximation of the function on each mesh element. In this paper we extend the algorithm to the Sobolev space H^1, where the error norm allows for gradient information to be included. This improves the convergence of the algorithm in regions, where the function has high frequency oscillations. In order to analyze the properties of the optimized meshes we consider iso-surfaces of volume data.

1 Introduction

The visualization of scalar functions, which are discretized by high resolution meshes with complex topology, is an important topic of research in scientific visualization. Due to the increasing computing capacities, the improvement of the numerical algorithms and measurement techniques many applications in visualization deal with large data sets. Typical examples range from two dimensional geometric data such as satellite images or laser scans to three dimensional scalar fields on regular grids like medical datasets or even time dependent flow fields on unstructured grids of a complex CFD topology. This data usually contains redundant information which makes the extraction and rendering of geometric primitives like height fields, iso-surfaces or streamlines very compute intensive. Two strategies are pursued for handling such problems: data reduction and mesh optimization, or multiresolution representation of the data for flexible level-of-detail control.

In this paper we present a technique based on linear approximations in Hilbert spaces and the finite element method to generate sequences of nested approximating spaces for a given function. This method is an extension of the method presented by Grosso et al. [10] for the L_2 norm to more general approximation spaces. In contrast to other techniques based on wavelets and multiresolution analysis [13,6,9], the algorithm starts at the coarsest level and proceeds to the finer ones. We interpret this process as a *data driven* mesh generation, since we iteratively generate a hierarchy of meshes by adaptive refinement depending on the error distribution across the mesh elements. After each iteration step of the algorithm, a complete hierarchy of nested spaces is available, with the *best approximation* of the underlying function known at highest level of the hierarchy. Besides to the standard linear space L_2, in this work we are mainly concerned with the Sobolev space H^1 of piecewise linear functions. The error norm defined in the space allows the integration of a priori knowledge of the smoothness of the function into the approximation problem. Once the best approximation is known for a nested sequence of spaces, a multiresolution decomposition of the function is obtained using hierarchical bases [21].

In contrast to wavelet based approximations, which are based *on global mesh refinement*, the nested sequences of approximating spaces generated by the algorithm are optimal due to the *adaptive refinement* of the underlying meshes. Besides the enormous compression potential of locally refined meshes, visualization algorithms applied to *full grids* at the different levels of the hierarchy are very simple and efficient. In contrast, *hierarchical grid* algorithms, as in the case of wavelets, are more inefficient: to obtain a function value from a wavelet decomposition for a given resolution a complex hierarchy have to be traversed. The global accuracy of the approximation is well defined in the underlying norm of the approximation Hilbert space, and can be controlled by the user. Additionally, control over the local contributions to the error is also given, which allows for local mesh refinement. Our approach also has the many advantages of a multiresolution representation of the data, such as level-of-detail control, compression, progressive transmission and visualization, etc.

In order to investigate, how the mesh hierarchy can be exploited for higher dimensional datasets, we decided to use it as a basis for iso-surface extraction from 3D scalar fields. The drawbacks of this widely used technique for indirect volume visualization are the time spent on the extraction of the polygonal description eventually visiting each cell of a large dataset, and the enormous amount of generated triangles, which makes interactive handling difficult even on high-end graphics workstations. In this paper we show that a mesh which is adaptively refined only in regions where the underlying function presents high frequency variations allows for improved iso-surface generation.

The following sections describe the related work, the theoretical background of the finite element approximation, some of the implementation details of the algorithm and elaborate on the iso-surface application. We conclude with the analysis of the experimental results and with some ideas for future work.

2 Mesh Reduction Techniques

A widely used multiresolution analysis technique is the wavelet decomposition [7,16]. The use of multiresolution analysis has successfully been extended to meshes of arbitrary topology by Lounsbery et al. [13] and Eck et al. [6]. Besides many advantages such as compression, level-of-detail control and the generation of multiresolution representations some aspects remain open. One problem is, that the wavelet refinement cannot be performed adaptively, resulting in a re-meshing as a post-processing step, after the wavelet decomposition has been performed. Re-meshing is necessary to avoid hanging nodes, when adding wavelet coefficients in the reconstruction step. Another problem is, that the multiresolution analysis starts at the finest resolution and proceeds to the coarser levels. For practical explorative applications the reverse direction would be preferred. The orthogonal basis of the wavelet analysis guarantees a best least-squares approximation with respect to the L_2 norm. Because of the successive adding of orthogonal subspaces during wavelet analysis, an error estimation between levels can be given [9].

The wavelet based methods approach the approximation problem from a functional point of view. These methods allow for a control of the overall accuracy of the approximation and give information about the local triangulation error. In contrast, the approaches of Schroeder et al. [17], Klein et al. [12], Turk [19], Hoppe [11] and Cohen et al. [5] for 2D-meshes and Cignoni et al. [4] and Lürig et al. [14] for 3D-meshes can be considered geometric. These approaches analyze every single node with some local criteria in order to decide whether it can be eliminated or not followed by a re-triangulation. A widely used geometric error criterion is the Hausdorff or Frechet distance, which measures the distance between manifolds. This approach also allows the construction of a hierarchy, but there is no simple recursion relation between two levels, which is one of the strong property of nested multiresolution spaces. A very interesting approach which combines the so far separated tasks of iso-surface extraction and polygon decimation in one processing step was presented by Ohlberger et al. [15]. This approach is based on the extraction of iso-surface from nested grids. The function values at different levels of the hierarchy are obtained by interpolation.

3 Linear Approximation in Hilbert Spaces

Mesh optimization techniques with wavelets and multiresolution analysis are based on the approximation of the underlying function in the L_2 sense in the corresponding wavelet spaces. The major drawback of these techniques is that they are limited to the L_2 error norm. This difficulty can be solved, if the linear approximation problem is formulated in abstract vector spaces equipped with more general inner products and corresponding norms.

3.1 Linear Approximation in Hilbert Spaces

In order to measure the quality of an approximation defined over a domain Ω we assume that the function to be approximated is an element of a linear function space equipped with a norm. The norm is induced by an inner product. More specifically, we work with *Hilbert* spaces, where a typical example is the space $L_2(\Omega)$ consisting of all square integrable functions over Ω. The inner product $(f, g)_{L_2} = \int_\Omega fg dx$ induces the norm given by $||u||_{L_2} = (u, u)_{L_2}^{1/2}$. The approximation problem can be formulated as follows. Suppose S is a linear subspace of a Hilbert space $\mathcal{E}$ with the inner product $(\cdot, \cdot)_\mathcal{E}$. An element $u \in S$ is the *best approximation* to an element $f \in \mathcal{E}$, if the orthogonality condition

$$(f - u, v)_\mathcal{E} = 0 \qquad \text{for all } v \in S \tag{1}$$

holds. The problem to find the best approximation can of course be reformulated into the well known form of a *least squares approximation*, where the best approximation $u \in S$ of the function $f \in \mathcal{E}$ is given by

$$||f - u||_\mathcal{E} = \inf_{v \in S} ||f - v||_\mathcal{E} .$$

3.2 The Hilbert Space H^1

The function space L_2 space is not always well suited for approximating functions which are in some regions rapidly oscillating with small amplitudes. Smooth functions may be in the L_2 norm a good approximation, which means that small details may be lost, even in the case of a *good* global accuracy of the approximation. Reciprocally, smooth functions may be well approximation in L_2 by rapidly oscillating functions. In order to avoid these effects other norms which include differentiability information can be used. In particular we will work with the Sobolev space $H^1(\Omega)$ equipped with the inner product

$$(u, v)_{H^1} = (u, v)_{L_2} + (\nabla u, \nabla v)_{L_2} .$$

This space consists of all functions which are square integrable and whose gradients are also square integrable. The meaning of this norm will probably become more clear if we measure the distance between two functions $f, g \in H^1$

$$||f - g||_{H^1}^2 = ||f - g||_{L_2}^2 + ||\nabla f - \nabla g||_{L_2}^2$$

which means that we are comparing not only the functions but also their gradients.

3.3 The Finite Element Approach

Finite element analysis is a numerical method for solving partial differential equations which is widely used in science and engineering. This section extends the basic underlying ideas with respect to hierarchical approximation, local mesh refinement and error analysis.

We discretize (1) by computing the Ritz projection according to the inner products defined in $H^1(\Omega)$. The domain of definition Ω is assumed to be a polygon for 2D meshes or a polyhedron for 3D meshes. We now consider $S \subset H_1$ to be an N-dimensional finite element subspace consisting of all piecewise *linear* functions with respect to a triangulation $\mathcal{T}$ of Ω. A basis $\{\varphi_i\}_{i=1}^N$ of S consists of the *hat* functions which are associated with each vertex. Their support is restricted to the elements of the triangulation containing the corresponding vertex. The *discretized form* of the equation (1) is given by the system of linear equations

$$AU = F \tag{2}$$

where

$$A_{ij} = (\varphi_j, \varphi_i)_{H^1}, \quad F_i = (f, \varphi_i)_{H^1} \text{ and } u = \sum U_i \varphi_i .$$

Up to now we have just reformulated the problem of the best approximation for a finite element space equipped with a Sobolev inner product into the solution of a linear system. The method we proposed actually generates a hierarchy of nested approximation spaces by adaptive mesh refinement. It consists of the following three steps

- *Mesh Refinement:* We need a mesh refinement algorithm which generates a hierarchy of triangulations of the domain by local refinement operations. The algorithms used will be described in section 3.4.
- *Error Analysis:* The key ingredient of an adaptive approximation method is an efficient and robust measurement of the approximation error. This method should provide a bound for the overall accuracy of the approximation and give information about the distribution of the error among the individual mesh elements, thus forming the basis of the adaptive local mesh refinement. The error measurement step will be described in Section 3.5.
- *Multilevel Preconditioning:* An efficient iterative method for solving the large linear system (2) will be addressed at the end of section 4.

3.4 Adaptive Mesh Refinement

A partition $\mathcal{T}$ of a polygonal or polyhedral domain Ω into triangles or tetrahedra is called a triangulation. The mesh refinement process we use is considered as standard method in the adaptive finite element literature [3]. We start with a coarse triangulation $\mathcal{T}_0$, and generate a sequence $\mathcal{T}_0, \cdots, \mathcal{T}_j$ of increasingly fine triangulations by successive local mesh refinement. Each triangulation in the sequence is required to be *conforming*, i.e. the intersection of two elements consists of a common face or a common edge or a common vertex or it is empty. This condition prevents *hanging nodes*, which are difficult to treat in finite element computations and problematic for rendering purposes. For the same reasons the triangulation sequence has to be *stable* with respect to some measure of degeneracy. For example, one requires that all interior angles are bounded away from zero, which is essential for the stability of the numerical computations. Finally, in order to build a hierarchy of nested spaces the triangulation sequence has to satisfy the *nestedness* condition, which means that an element in a triangulation is obtained by subdividing an element in a coarser triangulation of the sequence.

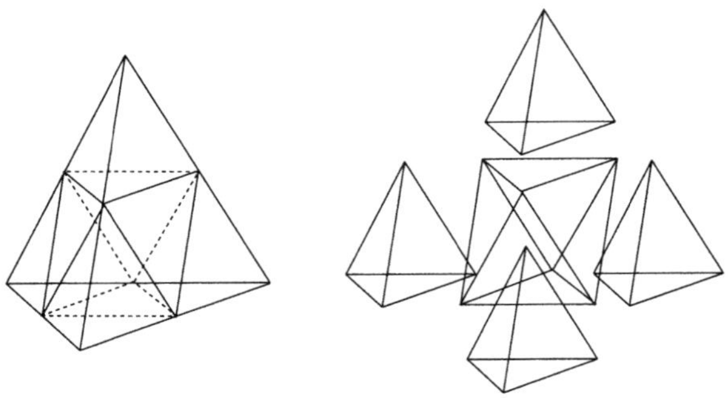

Fig. 1. Regular or red refinement of a tetrahedron

We use an algorithm which combines regular (red) and irregular (green) mesh refinement as introduced for 2D meshes by Bank et al. [1] and its extension to three dimensional tetrahedral meshes of Bey [2] and Go Ong [8]. These refinement algorithms are carried out in three steps. First, a refinement rule for a single element has to be defined, such that successive refinements produce stable and consistent triangulations. Such a refinement rule is called *red* or *regular*. Second, a set of *green* or *irregular* refinement rules is defined for the elements which share a common edge with regular refined elements. These refinement rules are *local*. Green refined tetrahedra are just inserted to satisfy border conditions and to avoid hanging nodes. In order to avoid stability problems green refined tetrahedra must not be refined again. If a subdivision is required, the originally green refined tetrahedron must be re-refined with the red rule. Finally, these local rules are combined and rearranged into a *global refinement algorithm* which guarantees for stability and conformity.

For 2D meshes the red (regular) refinement rule divides a triangle into four congruent ones by connecting the midpoints of its edges. The green (irregular) refinement consists of simple bisections connecting one edge midpoint with the opposite vertex. The regular refinement rule for tetrahedra first cuts off four sub-tetrahedra at the vertices as shown in Fig. 1. The subdivision of the remaining octahedron is not unique and depends on the choice of one of the three possible diagonals. The strategy proposed by Bey [2] is based on affine transformations to a reference tetrahedron and produces stable regular refinements. In order to compute the green closure or irregular refinement we use a *full set of refinement rules*, i.e. there always exists a refinement rule for a given edge and face refinement pattern.

3.5 Error Analysis

In order to obtain information about the accuracy of the approximation, we are interested in an efficient method to measure the error $||f - u||$. Because we have *a priori* knowledge of the function to be approximated, we evaluate the error by a direct integration. In order to obtain information on the local contributions to the global error we proceed as follows. Let $\mathcal{T}$ be a triangulation of the domain of definition Ω and Δ an element in $\mathcal{T}$. The error can be written in the form

$$\begin{aligned} ||f - u||_{H^1} &= \int_\Omega (f-u)^2 + (\nabla f - \nabla u)^2 d\Omega \\ &= \sum_{\Delta \in \mathcal{T}} E_\Delta \end{aligned} \tag{3}$$

where the E_Δ are given by

$$E_\Delta = \int_\Delta (f-u)^2 + (\nabla f - \nabla u)^2 d\Omega \,. \tag{4}$$

For the computation of the E_Δ coefficients we use a four points third order integration formula for triangles and a five points third order integration formula for tetrahedra [18]. The function values at the integration points are obtained by interpolation.

4 Algorithm

In this section we present an iterative algorithm to generate sequences of nested approximating spaces for a given data set. The general structure of the algorithm is the same as in [10]. After each iteration a new sequence of nested spaces which are obtained by local mesh refinement is derived from the previous hierarchy of spaces. The first step is to construct the initial, intentionally coarse triangulation of the domain. Depending on the geometry of the domain this may be a very complex task. Based on the initial triangulation the first approximation is computed by using a direct solver for the linear system (2). Then, we enter a loop where the mesh is iteratively refined depending on the global accuracy of the approximation. At each step of this process, which we call *data driven* mesh generation, mesh elements are marked for refinement, if their local error coefficients E_Δ in (4) exceed a certain threshold. According to these marks the mesh will be adaptively refined. Finally, the new approximation is computed by an iterative multilevel solver.

The result of the iteration step k is a sequence of triangulations $\mathcal{T}_0, \cdots, \mathcal{T}_{j_k}$ satisfying the nestedness condition. These triangulations are used to construct the hierarchy of nested piecewise linear finite element spaces:

$$S_0 \subset S_1 \subset \cdots \subset S_{j_k} \tag{5}$$

The best approximation u_S for the level S_{j_k} is efficiently computed by the multilevel linear solver. At each iteration step a few of the upper level meshes might be modified due to the substitution of irregular refinements by regular ones. Once the solution is known at the finest level of resolution, an approximation to f at the other levels of the hierarchy is obtained by a *hierarchical bases* decomposition [21]. In this case, there exists a simple recursion relation between the approximations at different levels of the hierarchy, which is a nice property of the nested spaces (5).

During the successive refinement steps the number of vertices in the meshes will increase. For a large number of vertices, especially for 3D applications, the linear system (2) will become very large, with the matrix A being sparse but not diagonal. Standard iterative methods such as Gauss-Seidel or even conjugate gradient iteration, are not appropriate. We solve this linear system (2) with a conjugate gradient method combined with a multilevel BPX preconditioner. The very efficient method makes use of the hierarchy (5) for the solution of the linear system. We remark that the condition numbers obtained for the BPX preconditioner are independent of the space dimension in contrast to the two dimensional hierarchical basis method proposed by Yserentant [21,22,20]. Thus, the method works well for triangular and tetrahedral meshes.

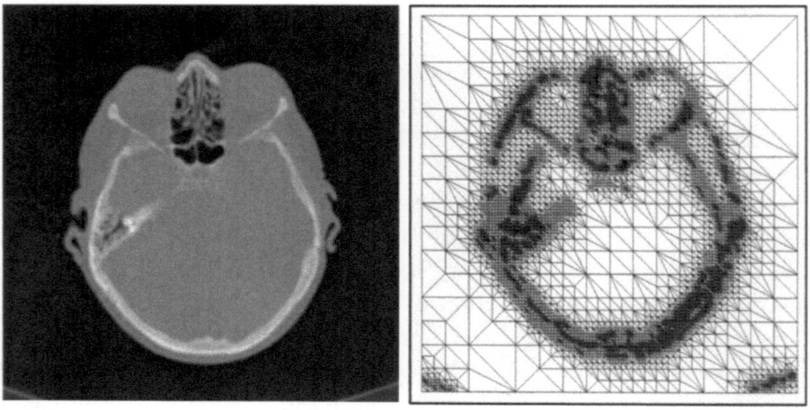

Fig. 2. Visualization of the optimized mesh of a 512^2 CT data set

5 Results

Since it is difficult to depict details from large 3D meshes we demonstrate the quality of the mesh optimization by using a 512^2 slice of a CT data set. The quality of the generated meshes is demonstrated in Fig. 2 where the effect of the local mesh refinement operations can be clearly seen. The quality of the error estimator can be appreciated from the fact that in the outer parts triangles are not refined. The input triangulation in the example consists of two triangles. The triangulation shown in the image has 6350 vertices. Furthermore, triangles are very *nice* in the sense that no thin triangles or vertices with large number of edges are present in the meshes.

The extension of this example to 3D leads to the extraction of iso-surfaces from a stack of MRI slices. In order to show the strengths of the presented iso-surface algorithm several experiments have been carried out with a MRI head-scan with 128^3 voxels of 16 bit precision. The generated surfaces show a sufficiently complex structure (see color plate 11, p. 372) to judge their quality with respect to surfaces resulting from a standard marching cubes algorithm.

Data Set	Norm	vertices	triangles	time (s)
Original		2,097,152	182,280	6.18
Head 1	H^1	38,719	71,841	1.01
Head 2	H^1	75,566	117,417	2.08
Head 3	H^1	126,810	158,508	2.57
Head 4	L_2	71,608	115,332	2.05
Orig. Abd.		47,448,064	1,471,254	126.5
Abd. 1	H^1	127,852	274,240	3.35

Table 1. Results for the iso-surface extraction

As we can see in Table 1 the time needed to generate the triangles is decreasing dramatically compared to the standard marching cubes algorithm. Looking at a special case we see, that in the first H^1 image the number of triangles is about one third and the processing time is about 6 times faster. A standard procedure would need the time of the marching cubes and the time of the polygon reduction to achieve a similar result. The resulting image (Fig. 11(a)) has lost not much quality compared to the iso-surface generated with the marching cubes algorithm at full resolution (Fig. 11(e)). In order to compare the Sobolev norm with the L_2 norm we consider the data sets corresponding to Head 2 and Head 4 in Tab. 1 (Figs. 3(a) and 3(b) and Figs. 11(b) and 11(d)). In this case the number of vertices and the number of generated triangles are comparable. The Sobolev norm produces smoother results and follows the contours better than the L_2 norm as one can see at the eyebrow and at the cheek near the ear.

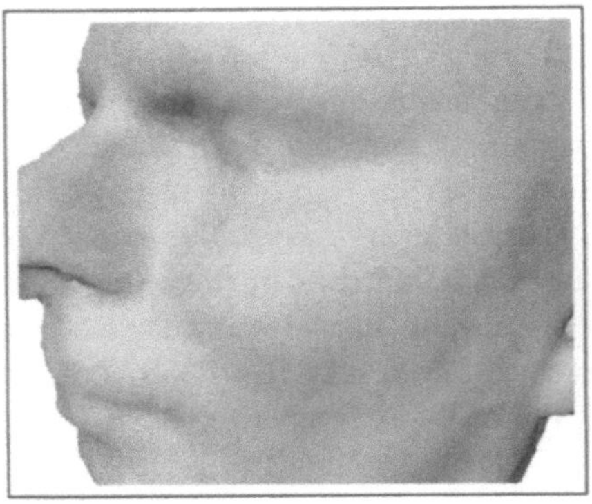

(a) Head 2, H^1 norm

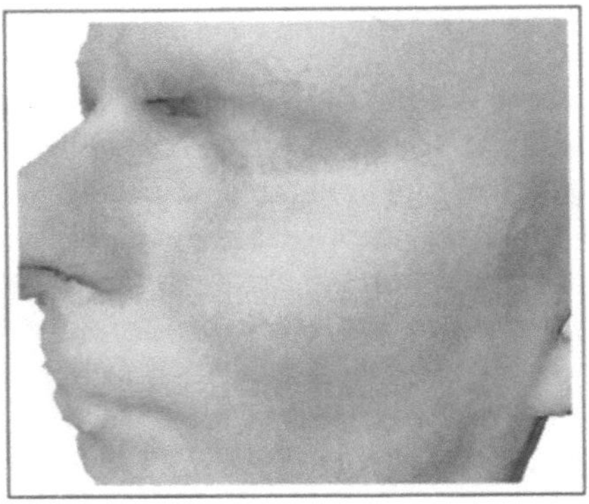

(b) Head 4, L_2 norm

Fig. 3. Iso-surfaces from a MRI scan of a human head

The power of the method becomes more evident in the case of large data sets. We have carried out experiments with a $(512)^2 * 181$ CAT abdomen scan at 8 bit precision. The CAT scan of the male abdomen is a courtesy of the Visualization Laboratory of the State University of New York at Stony Brook. In Fig. 4 an iso-surface extracted from a reduced data set is compared to the same surface extracted from the original data. The generation times are given in Tab. 1. The performance improvement is about factor 37 by a high quality image.

6 Conclusions

We have presented a new method to compute linear approximations of a given function in the Hilbert space H^1 combined with adaptive mesh refinement and error analysis techniques. This method solves the problem of mesh optimization for arbitrary unstructured two and three dimensional grids. We have shown, that the algorithm produces adaptively refined grids which are well behaved and suitable for accelerating a class of visualization and rendering methods. For the example of iso-surface extraction we have given timing results and error measurements which demonstrate that the underlying grids allow for a considerable speedup of existing algorithms while maintaining the desired accuracy. A more detailed comparison and analysis (theoretical and experimental) of the results produced using the Sobolev and the L_2 norm still have to be done.

A major advantage of this approach is that the mesh hierarchy is constructed from coarse to fine with the computational complexity of the approximation of the function on the coarse grids being nearly independent

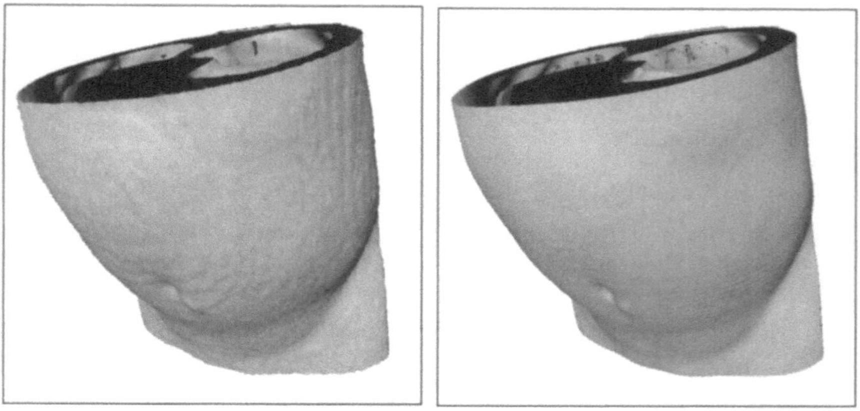

(a) Abd. 1, H^1 norm (b) Orig. Abd.

Fig. 4. Iso-surfaces from a CT scan of a male abdomen

of the actual data size. Intermediate levels can be easily reconstructed from the initial mesh based on the well defined refinement rules and a few integers indicating the simplex-ids to be regularly subdivided. This will make the proposed method very attractive for progressive transmission and display across a network without any preprocessing.

References

1. R. E. Bank, A. H. Sherman, and A. Weiser, *Refinement Algorithms and Data Structures for Regular Local Mesh Refinement*, Scientific Computing, (R. Stepleman, ed.), IMACS/North Holland, Amsterdam, 1983, pp. 3–17,
2. J. Bey *Tetrahedral Mesh Refinement*, Computing, **55**, 1995, 355–378.
3. F. Bornemann, B. Erdmann, and R. Kornhuber, *Adaptive Multilevel Methods in Three Space Dimensions*, Int. J. Numer. Meth. Eng., **36**, 1993, 3187–3203.
4. P. Cignoni, L. De Floriani, C. Montani, E. Puppo, and R. Scopigno, *Multiresolution Modeling and Visualization of Volume Data based on Simplicial Complexes*, Proceedings 1994 Symposium on Volume Visualization, 1994, pp. 19–26.
5. J. Cohen, A. Varshney, G. Turk, H. Weber, P. Agrawal, F. Brooks, and W. Wright, *Simplification Envelopes*, Proceedings SIGGRAPH '96, 1996, pp. 119–128.
6. M. Eck, T. DeRose, T. Duchamp, H. Hoppe, M. Lounsbery, and W. Stuetzle, *Multiresolution Analysis of Arbitrary Meshes*, Proceedings SIGGRAPH '95, 1995, pp. 173–182.

7. A. Finkelstein and D. Salesin, *Multiresolution Curves*, Proceedings SIGGRAPH '94, 1994, pp. 261–268.
8. M. E. Go Ong, *Hierarchical Basis Preconditioning for Second Order Elliptic Problems in Three Dimensions*, PhD thesis, University of California, Los Angeles, 1989.
9. M. H. Gross, R. Gatti, and O. Stadt, *Fast Multiresolution for Surface Meshing*, Proceedings IEEE Visualization '95, 1995, pp. 135–142,
10. R. Grosso, Ch. Lürig, and Th. Ertl, *The Multilevel Finite Element Method for Adaptive Mesh Optimization and Visualization of Volume Data*, Proceedings of the Visualization '97 Conference, Phoenix AZ, U.S.A., October 1997.
11. H. Hoppe, *Progressive Meshes*, Proceedings SIGGRAPH '96, 1996, pp. 99–108.
12. R. Klein, G. Liebich, and W. Strasser, *Mesh Reduction with Error Control*, Proceedings IEEE Visualization '96, 1996, 311–318.
13. M. Lounsbery, T. DeRose, and J. Warren, *Multiresolution Analysis for Surfaces of Arbitrary Topological Type*, Technical Report TR 93-10-05b, Department of Computer Science and Engineering, University of Washington, January 1994.
14. Ch. Lürig and Th. Ertl, *Adaptive Iso-Surface Generation*, in 3D Image Analysis and Synthesis '96, (B. Girod, H. Niemann, and H.-P. Seidel, eds.), 1996, pp. 183–190.
15. M. Ohlberger and M. Rumpf, *Hierarchical and adaptive visualization on nested grids*, In Proceedings of the Eighth Eurographics Workshop on Visualization in Scientific Computing, Boulogne sur Mer, France, April 1997.
16. P. Schröder and W. Sweldens, *Spherical Wavelets: Efficiently Representing Functions on the Sphere*, Proceedings SIGGRAPH '95, 1995, pp. 161–172.
17. W. Schroeder, J. A. Zarge, and W. E. Lorensen, *Decimation of Triangle Meshes*, Proceedings SIGGRAPH '92, 1992, pp. 65–70.
18. A. H. Stroud, *Approximate Calculation of Multiple Integrals*, Prentice Hall, 1971.
19. G. Turk, *Re-tiling Polygonal Surfaces*, Computer Graphics, **26**, 1992, 55–64.
20. J. Xu, *Iterative methods by space decomposition and subspace correction*, SIAM Review, **34**, 1992, 581–613.
21. H. Yserentant, *Hierarchical Bases*, In R. E. O'Malley, editor, ICIAM 91, SIAM, Philadelphia, 1992.
22. L. Yserentant, *On the multi-level splitting of finite element spaces,* Numer. Math, **49**, 1986, 379–412.

Efficient Visualization of Data on Sparse Grids

Norbert Heußer, Martin Rumpf

Institut für Angewandte Mathematik, Universität Bonn, Germany

Abstract. Sparse grids are nowadays frequently used in numerical simulation. With their help the number of unknown in discrete PDE or approximation problems can drastically be decreased. On sparse grids in d space dimensions $O(N \log(N)^{d-1})$ nodes are required to achieve nearly the same approximation quality for sufficiently smooth data as on a standard finite difference grid with N^d nodes. This allows numerical methods with small error tolerances on a corresponding very fine mesh width. Mapping the complete sparse grid data on an N^d standard grid in order to analyse them visually would burst a currently available work station storage for large $O(N \log(N)^{d-1})$ numbers of nodal values provided by numerical code on a sparse grid. We present an efficient, procedural approach to sparse grids for the visual post-processing. A procedural interface addresses data on grid cells only temporarily, only if actually requested by the visualization methods. The cell access procedures extract local numerical data directly from the users sparse grid data base. No additional memory is needed. The procedural approach is combined with a multiresolution strategy based on a recursive traversal of the grid hierarchy. It includes hierarchical searching for features such as isosurfaces and a locally adaptive stopping on coarser grids in areas of sufficient smoothness. This data smoothness corresponds to a user prescribed error tolerance. Examples underline the benefits of the presented approach.

1 Introduction

A variety of efficient visualization methods substantially supports the understanding of physical processes which have been simulated using advanced numerical methods on current computing hardware. Most of the currently developed graphic tools are based on structured or simple unstructured numerical grids [5,15,11,20]. Recent efficient numerical methods come along with more sophisticated grid structures. This gap between the numerical data formats and the graphically supported structures is a well known [12,19] problem in scientific visualization.

An important and highly efficient class of numerical methods are based on so called sparse grids [2,7,22]. The set of nodes of a sparse grid is a subset of the full grid nodes (cf. Fig. 1). A regular sparse grid in three dimensions consists of $O(N \log(N)^2)$ nodes in contrast to the N^3 nodes which are necessary for the corresponding full grid. The interpolation error on a sparse grid in the three dimensional case is estimated by

$$|u - u_I^s| \leq C \log(h)^2 h^2$$

where u_I^s is the sparse grid interpolation and C depends on higher derivatives of u. We compare this with the error estimate

$$|u - u_I| \leq Ch^2$$

for the standard interpolation u_I on a full grid.

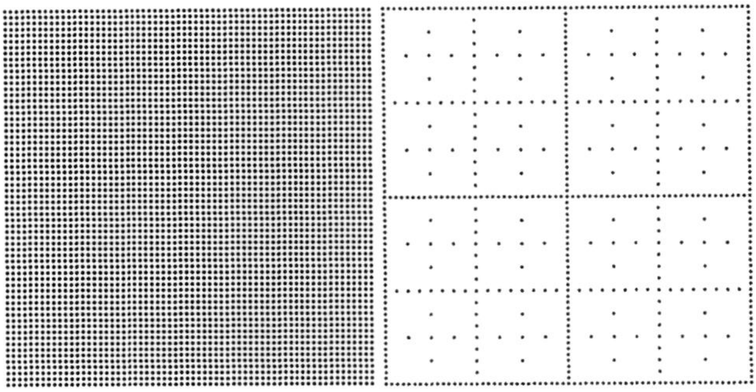

Fig. 1. The nodes of a regular 64^2 grid in 2D on the left compared to the subset of nodes contained in the corresponding 6 level sparse grid on the right.

This drastical reduction of unknowns has made sparse grid approaches prominent in numerical simulation with very good approximation quality [1,8,9]. Once the user has calculated numerical data on a sparse grid he wants to explore the results visually in a post process. The very first attempt to do so, consists of an in advance mapping of sparse grid data onto the corresponding standard N^3 grid. Therefore one stores actually available data values on the $O(N \log(N)^2)$ sparse grid nodes and applies a sparse grid interpolation scheme to obtain values on the additional $O(N^3)$ grid nodes. Numerical methods, especially multi grid methods on sparse grids, with a computing cost proportional to the number of sparse grid unknowns [22], enable the usage of underlying sparse grids with $O(N \log(N)^2)$ data which exceeds memory of a typical workstation.

But in this case the storing of interpolated values on the whole set of N^3 nodes of the corresponding standard grid is definitely ruled out. Therefore we ask for an approach which is able to represent the full sparse grid data resolution visually without any in advance data conversion. Data should preferably be addressed directly in the users data structures. Here we pick up the idea of a procedural interface to numerical data presented in [18].

The efficiency of typical visualization methods, especially in 3D, such as color shading on slices and isosurface extraction, can be significantly increased taking into account a multiresolution approach. Recently multiresolution techniques have become popular in visualization [3,6,10,13]. This basically contains a hierarchical grid traversal in search of features (e. g. as isosurfaces), to be visualized [21] and some local error measurement combined with the stopping on coarser grid levels in areas of small error contribution [4]. In [17] a hierarchical and adaptive approach is presented. It works on arbitrary nested grids consisting of elements being tensor products of simplices with the corresponding tensor product function spaces defined on them. Implementational aspects are discussed in [16].

Here these ideas are applied to sparse grid visualization. The aim is to support the user with a visualization interface for true data on large adaptive or non adaptive sparse grids. The scope primarily does not include sparse grid data compression of general data for the purpose of visualization. The procedural approach frees us from standard expensive data structures and makes large simulation applications feasible for the first time. Efficiency is achieved by appropriately transfering the above adaptive and hierarchical concept. We carefully have to respect the tensor structure of the considered sparse grids.

We overlay the sparse grid with a standard hierarchy of regular hexahedral grids. Grid cells are successively bisected, cyclically cutting them in the directions of the coordinate axes. Typical visualization methods, such as isosurface rendering or color shading on 2D domains and slices in 3D, are based on a hierarchical mesh traversal in depth first order. In each bisection step of this recursion the values on the new nodes are recovered from the sparse grid data base in a sophisticated way. The recursive processing of child cells is restricted to areas of interest, i. e. the neighborhood of the isosurface or the considered slice. A simple error control indicates if data are sufficiently resolved. On the finest grid level, depending on the underlying sparse grid and the user controlled error criteria, standard local rendering procedures generate graphic primitives [14] which are then transferred to the graphics hardware.

The presented concept is optimal concerning storage requirements. The methods work directly on the sparse grid data base of size $O(N\log(N)^2)$. The implementation of visualization methods is efficient under the hypothesis of using the interpolation onto a standard grid. To achieve in general an optimal cost of $O(N\log(N)^2)$ for typical rendering methods one would require graphic primitives which correspond closely to the tensor product structure of the sparse grids. This is not subject of the present discussion.

The paper is organized as follows. Section 2 and 3 discuss interpolation schemes on sparse grids which are essential for the procedural access to the overlayed standard grids. Then in Section 4 the hierarchical grid traversal in the visualization is explained and in Section 5 estimates for the higher order offset terms are given. Furthermore Section 6 discusses significant improvements of the basic idea to achieve efficiency. Finally in the appendix different examples underline the applicability of the presented concept. We gratefully acknowledge the support of T. Schiekofer from Bonn who provided the numerical sparse grid code. We linked our procedural interface to this software package.

Let us remark that although we mainly focus on the 3D case, analogous results hold in 2D. To simplify the presentation most of the schematic figures deal with the 2D case.

2 Brief Review of Functions on Sparse Grids

For a detailed description of sparse grids we refer to [22]. Here we restrict ourselves to a brief description of the interpolation which corresponds to the sparse grid node reduction. Let us suppose that the underlying equidistant rectangular grid for a domain $\Omega = [0,1]^3$ consists of N^3 nodes for $N = 2^n + 1$ and a corresponding meshwidth $h_n = 2^{-n}$. This grid can be interpreted as an n level hierarchy of cells successively divided into eight child cells. If we consider only bisection of cells we would end up with $3n$ grid levels, where we cyclically bisect in the three principle directions. In addition we want to restrict ourself to functions with zero boundary values. A generalization to non homogeneous boundary values is straightforward.

At first we examine the case of hierarchical grids in one space dimension. Later on they are the "factors" in the higher dimensional tensor product spaces. Defining nodes $\hat{x}_i^j = (2j-1) * h_i$ for $j > 0$ and $\hat{x}_i^0 = 0$ in 1D and corresponding 1D basis functions

$$\hat{\varphi}_i^j(x) = \begin{cases} 1 + \frac{1}{h_i}(x - \hat{x}_i^j) & \text{if } x \in [\hat{x}_i^j - h_i, \hat{x}_i^j] \\ 1 - \frac{1}{h_i}(x - \hat{x}_i^j) & \text{if } x \in [\hat{x}_i^j, \hat{x}_i^j + h_i] \\ 0 & \text{else} \end{cases}$$

for $i = 1, \dots, n$ and $j = 1, \dots, 2^{i-1}$ we get the following hierarchical representation for an interpolation of a function u in the space of the piecewise linear functions on the finest full grid of meshwidth h_n over the interval $[0,1]$:

$$\hat{u}_I = \sum_{i=1}^{n} \sum_{j=1}^{2^{i-1}} \hat{u}_i^j \hat{\varphi}_i^j$$

Here the $\hat{u}_i^j$ are the hierarchical coefficients for the basis functions $\hat{\varphi}_i^j$, which can recursively be interpreted as offset values at positions $\hat{x}_i^j$ on the level $i-1$ interpolation. Now we extend this notation to the three dimensional case and obtain nodes and basis functions

$$x_{i_1,i_2,i_3}^{j_1,j_2,j_3} = \sum_{m=1}^{3} \hat{x}_{i_m}^{j_m} e_m \qquad \varphi_{i_1,i_2,i_3}^{j_1,j_2,j_3}(x) = \prod_{m=1}^{3} \hat{\varphi}_{i_m}^{j_m}(x_m)$$

where e_m is the m'th coordinate vector. With these definitions the full grid interpolant u_I of a function u over Ω is given as

$$
\begin{aligned}
u_I(x) &= \sum_{i_1,i_2,i_3=1}^{n} \underbrace{\sum_{1\le j_k \le 2^{i_k}} u_{i_1,i_2,i_3}^{j_1,j_2,j_3} \cdot \varphi_{i_1,i_2,i_3}^{j_1,j_2,j_3}(x)}_{=: u_{i_1,i_2,i_3}(x)} \\
&= \sum_{i_1,i_2,i_3=1}^{n} u_{i_1,i_2,i_3}(x)
\end{aligned}
$$

Each u_{i_1,i_2,i_3} represents the hierarchical offset function on a grid $\mathcal{M}_{i_1,i_2,i_3}$ of fineness h_{i_1}, h_{i_2}, h_{i_3} in the x,y, respectively z direction. These offsets depend on the nodal values on new nodes which appear first on each grid level. Hierarchical interpolation is then defined as the sum over all these contributions. If we consider only contributions on grids $\mathcal{M}_{i_1,i_2,i_3}$ with $i_1 + i_2 + i_3 \le n+2$ we end up with the sparse grid interpolation where the nodes of the corresponding grids define the sparse grid nodes (cf. Fig.2). The approximation results on sparse grids (cf. Section 1) state that for smooth functions u this restriction does not significantly influence the approximation error. But it dramatically reduces the number of nodal values. In the above

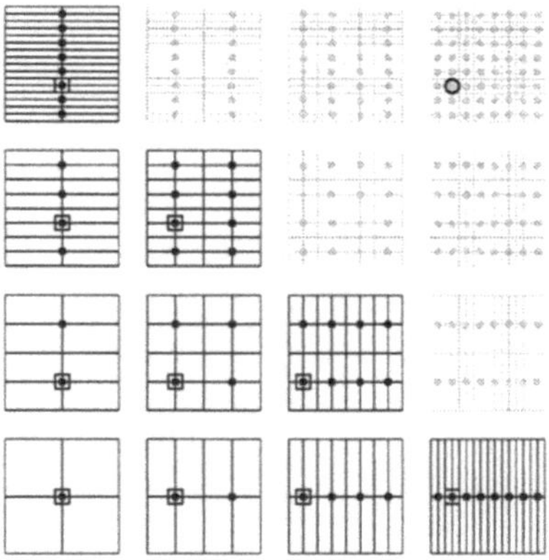

Fig. 2. The hierarchical grids $\mathcal{M}_{i_1,i_2}$ with marked sparse grid nodes in the 2D case are ordered in a scheme. Dots indicate the new nodes added on each of these grids, the grey meshes are grids $\mathcal{M}_{i_1,i_2}$ for $i_1 + i_2 > n+2$ where values on new nodes are completely interpolated. The little boxes mark sparse grid nodes, whose values contribute to the interpolation of the full grid node marked by a circle on the top right.

notation the sparse grid interpolation is

$$u_I^s(x) = \sum_{i_1+i_2+i_3 \le n+2}^{n} u_{i_1,i_2,i_3}(x)\,.$$

The interpolation value at any point can be computed recursively adding all the local contributions on successively finer and finer grids $\mathcal{M}_{i_1,i_2,i_3}$. If $i_1 + i_2 + i_3 > n + 2$ an increasing part of nodal values on the new nodes of $\mathcal{M}_{i_1,i_2,i_3}$ is already interpolated and not retrieved from the data base.

3 Recursive Sparse Grid Interpolation

One major drawback with the above sparse grid interpolation is that it operates non local. If we ask for a function value at some point $X \in \Omega$ on an n level sparse grid it depends on all nodal values from all sparse grid levels for which X is contained in the support of the corresponding basis function (cf. Fig. 2).

Let us assume that we overlay an n level sparse grid with its n level full grid and recursively traverse the full grid hierarchy. Than each time we step from a coarser i'th level cell to one of its finer $(i + 1)$'th level children we have to evaluate all these at least $O(i)$ sparse grid basis function contributions to the required function values on the new child nodes. This effort is unavoidable if we ask for the function values on a single n'th level cell. But the key point of the approach presented here is that we can reuse earlier computed contributions on other cells choosing an appropriate caching strategy.

At first we examine the 2D case. Let us suppose that for $m_k \le i_k$ a cell $E = \prod_{k=1,2}[\hat{x}_{m_k}^{j_k}, \hat{x}_{m_k}^{j_k} + h_{i_k}]$ of the full grid $\mathcal{M}_{i_1,i_2}$ is given with $i_1 \le i_2 \le i_1 + 1$. Now we bisect this cell in one refinement step without any restriction in direction x (bisection in y direction is handled analogously). This corresponds to one step in the recursive traversal of the corresponding full grid hierarchy where we process cells on grids $\mathcal{M}_{i_1,i_2}$ along the diagonal in the grid scheme of Fig. 2 in depth first order. That corresponds to *cylcic* bisection in the different directions. The hierarchical contribution visible on the new grid level, i.e. on $\mathcal{M}_{i_1+1,i_2}$, has to be added as a hierarchical offset. We obtain contributions from one 1D–offset functions (cf. Fig. 3)

$$t_{i_1+1,i_2}^{q,\cdot}$$

with $q = 2^{i_1-m_1}(2j_1 - 1) + 1$ and $t_{i_1,i_2}^{j_1,\cdot}$ corresponding to the additional grid line due to the bisection and given by

$$t_{i_1,i_2}^{j_1,\cdot} = \sum_{\substack{k \le i_2 \\ k \le n+1-i_1}} \sum_{l=1}^{2^k} u_{i_1,k}^{j_1,l} \cdot \varphi_{i_1,k}^{j_1,l}\,.$$

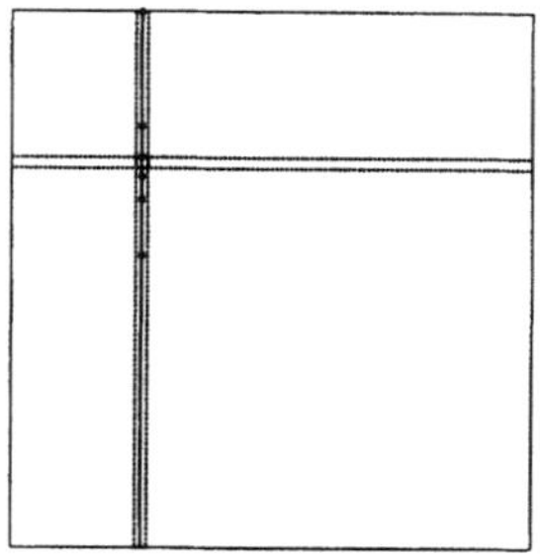

Fig. 3. Sketch of the 1D–offset functios (indicated by the thick line, where dots mark the sparse grid nodes of interest) which contribute to the hierarchical function values on the child cells of a specific cell (the intersection of the two dotted rectangles).

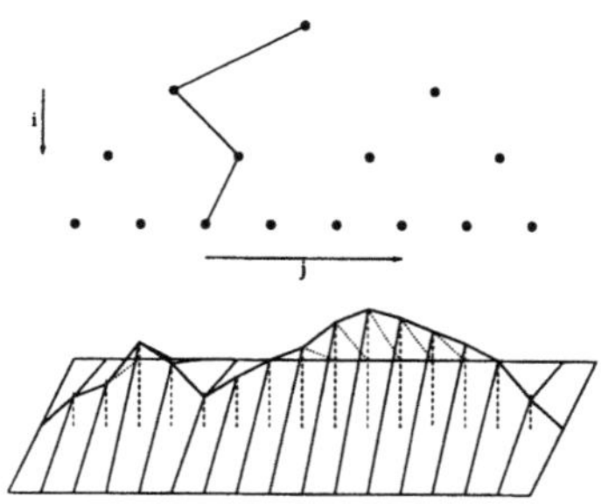

Fig. 4. 1D–offset function on a 2D sparse grid (bottom) and the graph of involved sparse grid nodes (top).

and $t^{\cdot,j_1}_{i_1,i_2}$ is defined analogously. Fig. 4 depicts a characteristic example.

The involved basis functions $\varphi^{j_1,l}_{i_1,k}$ have local support in one directions and non local support only in the remaining direction. Restricting ourself to a specific cell only a small part of the hierarchical values, i. e. $\max\{1, \min\{i_1, n+1-i_2\}\}$, is of interest for the hierarchical offset on the two new nodes on the child cells. In the 1D graph in Fig. 4, dots represent all sparse grid nodes contributing to such an offset function. Those influencing the value on one specific cell are connected by lines.

Using cell bisection in one direction in 3D contributions from basis functions which have local support in one direction and non local support in the remaining two directions have to be considered. They can be gathered in 2D offset functions $t^{j_1,\cdot,\cdot}_{i_1,i_2,i_3}$, $t^{\cdot,j_2,\cdot}_{i_1,i_2,i_3}$, and $t^{\cdot,\cdot,j_3}_{i_1,i_2,i_3}$. Let us mention that these offset function can be evaluated as sums over 1D offset functions, which have the same local support in one direction, different support intervals in the second direction and non local support in the third direction.

The advantage of this type of hierarchical decomposition is that the hierarchical contribution has been split up into factors which adequately respect the tensor product structure of the sparse grids. Therefore it will turn out to be helpful for our application (cf. Section 4) and allows an efficient implementation (cf. Section 6).

4 Procedural Data Access

In what follows we discuss a hierarchical visualization concept to sparse grids. The presented approach is embedded in a frame of general graphical interfaces. In [18] a visualization interface for arbitrary meshes with general data functions on them has been proposed. A mesh is defined as a procedurally linked list of non intersecting cells. Data is accessed by user supplied procedures addressing the user data structures and returning the required data temporarily in a prescribed cell structure, mainly consisting of a reference to some cell type, the coordinate vectors for the nodes, and function data on them. In general at the same time only one cell structure is present in storage. There is no direct access to a single cell. But this is in fact not necessary for most common and frequently used visualization methods. This concepts has been extended to hierarchical grids in [16] adding hierarchical access procedures which support a procedural traversal of a nested grid hierarchy including the local estimation of higher order data contributions and visual errors. Based on that enlargement a hierarchical and adaptive post processing can be implemented with respect to the guidelines presented in [17]. Summarizing the main features, this procedural concept strictly separates the interface to the user data, which temporarily delivers information on local cell data, and the visualization itself. To run some visualization method from a general library we only have to provide the data access procedures. Strictly speaking, this is what we are going to discuss now for sparse grids.

As already discussed in Section 1 we overlay an n level sparse grid with the corresponding n level full grid. Procedurally traversing the grid hierarchy now means a depth first processing of grids $\mathcal{M}_{i_1,i_2,i_3}$ along the diagonal in the scheme of grids (cf. Fig. 2). Here we apply a step by step bisection cycling over the different directions as introduced in Section 3. Thereby, starting at some i'th level cell E we have to generate complete information on one of its children $\mathcal{C}(E)$ to fill a cell structure of the interface. The required vertex coordinates $x^{j_1,j_2,j_3}_{i_1,i_2,i_3}$ can easily be evaluated using integer arithmetics. Furthermore the function values $u^s_I(x^{j_1,j_2,j_3}_{i_1,i_2,i_3})$ on the new nodes are calculated using the above interpolation scheme for the local contribution of the 2D–offset function on the child cell (cf. Fig. 3).

To clarify the presentation and make it more self–contained we consider isosurfaces as a typical and significant visualization tool. Other algorithms such as isoline drawing or color shading on slices can be implemented similarly (cf. color plate 16, p. 375). The extraction of an isosurface from function data on Ω can benefit from the hierarchical grid structure. Instead of traversing all cells, like a standard marching cube strategy [14] does, we can recursively test for intersections on coarser level cells E to decide whether the children $\mathcal{C}(E)$ have to be visited or not [21]. Generally this leads to a cost reduction of one order of magnitude up to a logarithmic factor. In what follows we formulate an algorithm for the hierarchical isosurface extraction for an isosurface value α:

```
Inspect(α,E) {
   v = SparseGridInterpolation(E);
   if IntersectionTest(α,v,E) {
      if (C(E) ≠ ∅)
         for all Ẽ ∈ C(E)
            Inspect(α,Ẽ);
      else Extract(α,E);
   }
}
```

where *SparseGridInterpol()* is the implementation of the above sparse grid interpolation, *IntersectionTest()* checks whether α is possibly contained in the image interval of the currently considered local function on a cell, and *Extract()* finally renders the local intersection of the isosurface with a cell. The recursive procedure *Inspect()* is first called on the only cell on level 1. Color plate 18, p. 376, shows a sample isosurface extracted from a sparse grid data set at different sparse grid levels. Color plate 17, p. 376, depicts the same test data set on a 10 level sparse grid, which corresponds to a standard grid with 1025^3 grid points. Conversion to this grid would require some *eight* GBytes of memory.

5 Estimating Higher Order Function Offsets

The intersection test on a cell requires the calculation of robust data bounds. Simply taking the function values on the cell vertices into account, will not be sufficient. We might miss information apparent on finer grid levels only, e. g. strongly curved segments of an isosurface. Let us remark that the storing of $O(N^3)$ min–max values as considered in [21] is ruled out in our case.

We ask for an effectively computable estimate $\varepsilon(E)$ for each cell E of grid $\mathcal{M}_{i_1,i_2,i_3}$ such that

$$|u_I^s(X) - P_{i_1,i_2,i_3} u_I^s(X)| \leq \varepsilon(E) \quad \forall X \in E$$

where P_{i_1,i_2,i_3} is the standard trilinear projection onto the full grid $\mathcal{M}_{i_1,i_2,i_3}$. This estimate then becomes part of our hierarchical grid interface. Having evaluated $\varepsilon(E)$ the intersection test is straightforward

$$\begin{aligned} IntersectionTest(\alpha, v, E) = (&\min_{x\in\mathcal{N}(E)} v(x) - \varepsilon(E) \leq \alpha) \\ \wedge(&\max_{x\in\mathcal{N}(E)} v(x) + \varepsilon(E) \geq \alpha) \end{aligned}$$

where $\mathcal{N}(E)$ denotes the set of vertices of E. On sparse grids with the hierarchical offsets stored on the sparse grid nodes it turns our that $\varepsilon(E)$ can be expressed in terms of these offset values. Let us first consider the 1D case. Cells are now intervals divided into two child intervals in each step. We relate the 1D estimate $\hat{\varepsilon}(\hat{E})$ for an interval $\hat{E}$ with $\hat{\varepsilon}(\hat{x}_i^j)$ where $\hat{x}_i^j$ is supposed to be the center point in $\hat{E}$. Then $\hat{\varepsilon}(\hat{x}_n^j) = \left|\hat{u}_n^j\right|$ is a proper estimate on the second finest grid level and on the third finest we find

$$\hat{\varepsilon}\left(\hat{x}_{n-1}^j\right) = \max\{\left|\hat{u}_{n-1}^j\right|, \frac{1}{2}\left|\hat{u}_{n-1}^j\right| + \max_{k=2j-1,2j+1} \left|\hat{u}_n^k\right|\}$$

as a robust evaluation of the higher order data offset. This idea can be extended to coarser levels using the slightly weaker estimate given by the recursion formula

$$\hat{\varepsilon}\left(\hat{x}_i^j\right) = \left|\hat{u}_i^j\right| + \max_{k=2j-1,2j+1} \hat{\varepsilon}\left(\hat{x}_{i+1}^k\right)$$

for levels $i = 1, \dots, n-2$. Now we transfer this type of estimate to sparse grids using the tensor product structure of the sparse grids. We calculate ε values on sparse grid nodes such that

$$|u_I^s(X) - P_{i_1,i_2,i_3} u_I^s(X)| \le \varepsilon(x_{i_1,i_2,i_3}^{j_1,j_2,j_3}) \quad \forall X \in \operatorname{supp}(\varphi_{i_1,i_2,i_3}^{j_1,j_2,j_3})$$

using the analogous recursion in 3D

$$\varepsilon\left(x_{i_1,i_2,i_3}^{j_1,j_2,j_3}\right) := \left|u_{i_1,i_2,i_3}^{j_1,j_2,j_3}\right| + \sum_{\substack{k=1,2,3 \\ i_k<n}} \max_{r\in\{0,1\}} \varepsilon(x_{i_1(k),i_2(k),i_3(k)}^{j_1(k,r),j_2(k,r),j_3(k,r)})$$

for $\sum_k i_k \le n+1$ where

$$\begin{aligned} i_m(k) &= i_m + \delta_{mk} \\ j_m(k,r) &= j_m + \delta_{mk}(j_m - r) \end{aligned}$$

and $\varepsilon\left(x_{i_1,i_2,i_3}^{j_1,j_2,j_3}\right) := \left|u_{i_1,i_2,i_3}^{j_1,j_2,j_3}\right|$ for $\sum_k i_k = n+2$.

With this definition at hand we are now able to calculate an appropriate estimator for each sparse grid node in a simple hierarchical preroll over the sparse grid hierarchy.

Fig. 5 shows the recursive dependencies of ε values on sparse grid nodes. But what we need is a bound $\varepsilon(E)$ of the possible higher order function offset on a cell E in $\mathcal{M}_{i_1,i_2,i_3}$ from the full grid hierarchy. We immediately obtain a first reliable bound defining

$$\varepsilon(E) := \max_{\substack{\operatorname{supp}(\varphi_{m_1,m_2,m_3}^{j_1,j_2,j_3})\cap E\neq\emptyset \\ m_k\le i_k}} \varepsilon(x_{m_1,m_2,m_3}^{j_1,j_2,j_3}) .$$

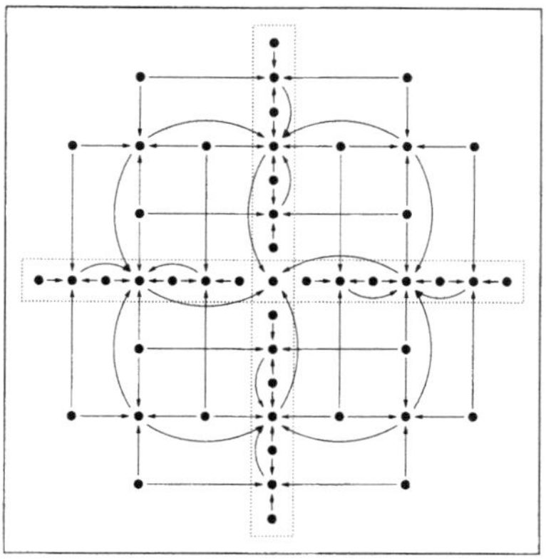

Fig. 5. The recursive dependencies of $\varepsilon(x^{j_1,j_2,j_3}_{i_1,i_2,i_3})$ values on sparse grid nodes $x^{j_1,j_2,j_3}_{i_1,i_2,i_3}$ on a 2D sparse grid

In general on typical sparse grid data bases the search for this maximum over the above set of sparse grid nodes turns out to be quite expensive. Therefore it is useful to have a second alternative definition for $\varepsilon(E)$. Let us therefore condense estimate information on classes of sparse grid nodes with the same support in x, y, or z direction and store them in 1D arrays, one for each direction. In detail we define

$$\varepsilon^{j_1,\cdot,\cdot}_{i_1,\cdot,\cdot} := \max_{\substack{i_2+i_3\leq n+2-i_1 \\ 1\leq j_2\leq 2^{i_2},1\leq j_3\leq 2^{i_3}}} \varepsilon(x^{j_1,j_2,j_3}_{i_1,i_2,i_3})$$

and proceed analogously for $\varepsilon^{\cdot,j_2,\cdot}_{\cdot,i_2,\cdot}$ and $\varepsilon^{\cdot,\cdot,j_3}_{\cdot,\cdot,i_3}$. Let us emphasize that these higher order offset bounds are no longer local. Estimate information is spread over typically thin planar layers. But it allows the following simple definition of a cell estimate

$$\varepsilon(E) := \max\{\varepsilon^{j_1,\cdot,\cdot}_{i_1,\cdot,\cdot}, \varepsilon^{\cdot,j_2,\cdot}_{\cdot,i_2,\cdot}, \varepsilon^{\cdot,\cdot,j_3}_{\cdot,\cdot,i_3}\}$$

where E is a cell in the full grid $\mathcal{M}_{i_1,i_2,i_3}$ with mid point $x^{j_1,j_2,j_3}_{i_1,i_2,i_3}$ which itself is not necessarily a sparse grid point. In the examples shown below (cf. color plate 18, p. 376) we have used this simplified cell estimate. The additional storage needed in the implementation is $3N$. Condensing estimate information along lines instead of planes would require N^2 storage which is ruled out for large sparse grids which themselves require only $O(N\log(N)^2)$ storage.

The estimate values $\varepsilon(E)$ introduced above to calculate robust data bounds for functions on coarse level cells E is in addition a useful local error indicator. During the recursive traversal of the full grid hierarchy a visualization

sparse grid level	3	4	5	6	7	8	9	10
drawing time without cache [sec]	0.3	2.8	24	102	269	-/-	-/-	-/-
drawing time with cache [sec]	0.1	0.7	5	19	46	123	406	1,097
visited elements [x1,000]	1.0	7.3	42	137	334	920	3,018	12,080
drawn elements [x1,000]	0.2	0.8	3	12	49	197	789	3,154

Table 1. Runtime measurement while drawing an isosurface

method has to check if $\varepsilon(E)$ on the current cell E is below some given threshold ε. In that case the local rendering routine, e. g. color shading or isosurface drawing, is called on E. Otherwise one proceeds with the traversal to cells of interest on finer levels. The implementation follows along the guidelines given in [17].

6 Improving Efficiency

During the procedural and recursive grid traversal in a visualization method the main computational task is the interpolation of nodal values. Here hierarchical offsets to the trilinear functions on the parent cells are evaluated based on the above 2D–offset functions. The hierarchical offset values on the involved sparse grid nodes can be stored in a compact data structures which we will call a 1D-offset line in 2D and a a set of 1D–offset lines in 3D (their sum represents the 2D–offset function). The line in the nodal graph sketched in Fig. 4 represents such an 1D–offset line. While entering cells with similar projections onto one of the coordinate lines or planes we request the same, a slightly smaller, or slightly larger set of 1D–offset lines. Therefore the efficiency of the presented approach can be significantly increased caching these 1D structures. In our implementation we use one cache for each subdivision direction in which the 1D–offset lines can be stored. Each cache entry is of a maximal length $O(n)$. If a cache entry is reused we first ask for the segment which can be recycled, cut off the remaining part and add again the new hierarchical values actually needed. Let us underline that this cache concept is part of the presented procedural interface and independent from the visualization methods which do not recognize cache hits or misses.

Furthermore a second improvement is possible which ensures high performance as long as one mainly visualizes data on coarse grid levels. For some moderate value of i we allocate memory to store values already computed once on nodes of the grid $\mathcal{M}_{i,i,i}$. Whenever these values are requested during the grid traversal, they have not to be recomputed but simply called from this array of values. Interactively inspecting the data more and more coarse grid values will already be stored such that finally the visualization combines ultimate efficiency on standard, but coarse grids with the access to very fine

background full grid level	0	5	6	7
drawing time [sec]	269	45	28	3.4

Table 2. Background full grids of different levels are activated while extracting an isosurface on a 7 level sparse grid.

sparse grid granularity. Let us finally discuss some computational results for the extraction of isosurfaces from sparse grid data sets on a SGI Indigo with R10000. Corresponding images are shown in color plate 17, p. 376. Table 1 documents the improvement obtained by the cache concept. For different levels of the considered sparse grid drawing times with and without cache, the number of elements visited by the hierarchical algorithm, and the number of elements on which primitives are actually drawn (an increase by a factor of ~ 4 from level to level instead of ~ 8 for the non hierarchical approach) are listed. On the 10'th level we count $242,389,705$ cache hits in contrast to only $2,244,975$ misses. In case of a cache hit about 11% of a 1D–offset line in the cache need to be cut off and only 17% of the new 1D–offset line need to be added in average.

Table 2 shows the improvement using storage for a coarse background full grid. We assume the values on full grids of different level to be already available.

References

1. H.-J. Bungartz, *Dünne Gitter und deren Anwendung bei der adaptiven Lösung der dreidimensionalen Poisson–Gleichung*, Ph.D. thesis, Technische Universität München, 1992.
2. H.-J. Bungartz, M. Griebel, D. Röschke, and C. Zenger, *Pointwise convergence of the combination technique for the laplace equation*, East-West Journal of Numerical Mathematics **2**:1 (1994), 21–45.
3. P. Cignoni, L. De Floriani, C. Montani, E. Puppo, and R. Scopigno, *Multiresolution modeling and visualization of volume data based on simplicial complexes*, Proceedings of the Visualization'95, 1995, pp. 19–26.
4. J. Cornhill, E. Fayyad, R. Shekhar, and R. Yagel, *Octree-based decimation of marching cubes surfaces*, Proceedings of the Visualization'96, 1996.
5. D. S. Dyer, *A dataflow toolkit for visualization*, IEEE CG&A **10**:4 (1990), 60–69.
6. A. v. Gelder, J. Gibbs, and J. Wilhelms, *Hierarchical and parallelizable direct volume rendering for irregular and multiple grids*, Proceedings of the Visualization'96, 1996.
7. M. Griebel, *Eine Kombinationsmethode für die Lösung von Dünngitter-Problemen auf Multiprozessor-Maschinen*, Numerische Algorithmen auf

Transputer-Systemen (W. Bader, Rannacher, ed.), Teubner, 1992, pp. 66–78.

8. M. Griebel and W. Huber, *Turbulence simulation on sparse grids using the combination methods*, Parallel Computational Fluid Dynamics, New Algorithms and Applications (N. Satofuka, J. Periaux, and A. Ecer, eds.), North-Holland, 1995, pp. 75–84.

9. M. Griebel and V. Thurner, *The efficient solution of fluid dynamics problems by the combination technique*, Int. J. Num. Meth. for Heat and Fluid Flow **5**:3 (1993), 251–269.

10. M. H. Gross and R. G. Staadt, *Fast multiresolution surface meshing*, Proceedings of the Visualization'95, 1995, pp. 135–142.

11. R. B. Haber, B. Lucas, and N. Collins, *A data model for scientific visualization with provisions for regular and irregular grids*, Proc. IEEE Visualization '91, 1991.

12. U. Lang, R. Lang, and R. Rühle, *Integration of visualization and scientific calculation in a software system*, Proc. IEEE Visualization '91, 1991.

13. D. Laur and P. Hanrahan, *Hierarchical splatting: A progressive refinement algorithm for volume rendering*, IEEE CG&A **25**:4 (1991), 285–288.

14. W. Lorensen and H. Cline, *Marching cubes: A high resolution 3d surface construction algorithm*, ACM Computer Graphics **21**:4 (1987), 163–169.

15. B. Lucas and et. al., *An architecture for a scientific visualization system*, Proc. IEEE Visualization '92, 1992.

16. R. Neubauer, M. Ohlberger, M. Rumpf, and R. Schwörer, *Efficient visualization of large scale data on hierarchical meshes*, Visualization in Scientific Computing '97 (W. Lefer and M. Grave, eds.), Springer, 1997.

17. M. Ohlberger and M. Rumpf, *Hierarchical and adaptive visualization on nested grids*, to appear in Computing (1997).

18. M. Rumpf, A. Schmidt, and K. G. Siebert, *Functions defining arbitrary meshes a flexible interface between numerical data and visualization routines*, Computer Graphics Forum **15**:2 (1996), 129–141.

19. L. A. Treinish, *Data structures and access software for scientific visualization*, Computer Graphics **25** (1991), 104–118.

20. C. Upson and et. al., *The application visualization system: A computational environment for scientific visualization*, IEEE CG&A **9**:4 (1989), 30–42.

21. J. Wilhelms and A. van Gelder, *Octrees for faster isosurface generation*, ACM Trans. Graph. **11**:3 (1992), 201–227.

22. C. Zenger, *Sparse grids*, Parallel Algorithms for Partial Differential Equations: Proceedings of the Sixth GAMM-Seminar, Kiel, Jan. 1990 (Hackbusch, ed.), Notes on Numerical Fluid Mechanics, vol. 31, Vieweg, 1991.

A Meta Scheme for Iterative Refinement of Meshes

Markus Kohler

Lehrstuhl für Graphische Systeme, Universität Dortmund, Germany

Abstract. Subdivision is a powerful method for smooth visualization of coarse meshes. An initial mesh of arbitrary topology is subsequently subdivided until the surface appears smooth and visually pleasing. Because of the simplicity of the algorithms and their data structure they have achieved much attention. Most schemes exhibit undesirable artifacts in the case of irregular topology. A more general description of subdivision algorithms is required to be able to avoid artifacts by changing parameters and also the topology of subdivision. We develop a partition of the algorithms to sub-algorithms and combinatoric mappings. An algebraic notation enables the description of calculations on the mesh and steering the algorithm by a configuration file. Opposite to index based notation of tensor-product surfaces the algebraic notation is even applicable for irregular and non-tensor-product surfaces. Variations of these sub-algorithms and mappings lead to known and new subdivision schemes. Further, the unified refinement scheme makes the efficient depth-first subdivision available to all these subdivision schemes.

1 Introduction

Presuming a visual 3D-object is available or easy to calculate at certain points, the task of reconstructing the remaining proportion consists in approximating or interpolating the points through a surface. An often used paradigm in CAD and visualization is to transfer a given shape of the real world into a simpler data structure by sampling the shape and reconstructing it for visualization. Conventional reconstruction is based on a functional representation (e. g. splines), which is sub-sampled at the desired level of detail resulting in a visually smooth mesh of small patches.

Contrarily, subdivision straightforwardly generates an approximating or interpolating surface that appears smooth and visually pleasing. Typical examples are the "cutting corner" algorithm [6] smoothing the corners in a certain number of steps, further, Catmull-Clark or Doo-Sabin subdivision [4,7]. Beginning with an input mesh, a sequence of meshes is generated, whereby new vertices are inserted as, preferably, simple local affine combinations of neighbouring vertices. There is no need for an internal functional representation and its complicated data structure. The simplicity of the algorithms and the associated data structure makes them attractive for interactive visualization where speed is of essence. For this reason some spline techniques supply a subdivision scheme such as *de Casteljau's Algorithm* for Bezier surfaces and *de Boor's Algorithm* for B-splines [1,10], which shows the importance of subdivision.

A major drawback of subdivision compared to spline technique is the continuously growing data structure and memory needs with increasing refinement, whereas spline techniques need almost constant memory only for the functional representation. However, if it is sufficient to display the final considerably small patches for visualization and immediately remove them from memory, the ***depth-first subdivision*** [13] allows to handle extremely large data sets even for many subdivision steps.

The discussion above showed the importance of subdivision for visualization. The investigation of this paper addresses the construction of a meta scheme for iterative refinement of meshes through the analysis of common concepts. This leads to a partition of the algorithm to sub-algorithms and mappings whose variation offers most of the spectrum of existing and new schemes in a unified manner. Meshes will be described by means of algebra. This enables to quickly set up algorithms through a configuration file and visually examine the influence of parameters and subdivision topology to extraordinary points. An environment of this kind would allow the user to choose the best suited subdivision scheme for his purposes, for instance in shape reconstruction.

The next Section encourages the proposal of a meta subdivision scheme (Section 2). Section 3 analyses the relation between objects of subsequent refinements. An algebraic description of the topology and the Doubling Operator are introduced in Section 4. The Averaging Operator is explained in Section 5. Section 6 examines the linking of the resulting objects.

2 Meta Scheme for Subdivision

The known subdivision schemes were developed for certain applications and work on specific meshes. The approximating Catmull-Clark scheme constructs a vertex for every vertex, edge and face of an arbitrary mesh according Fig. 6 (b). Each mask shows the topology of the initial mesh, where the new vertex is inserted (small circle). The new vertex is the weighted average

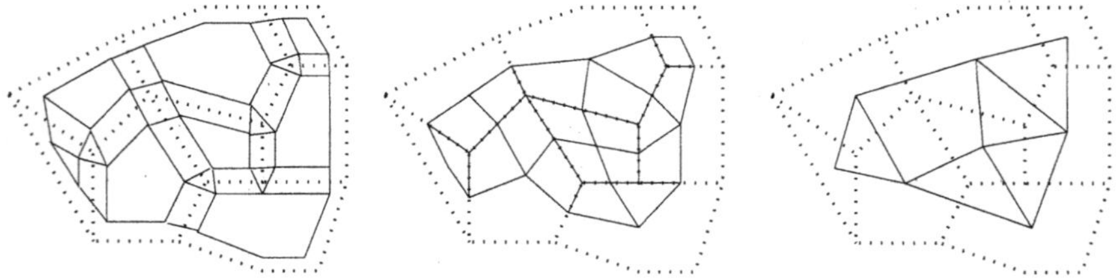

(a) (b) (c)

Fig. 1. (a) Generating a new mesh from an old with the Doo-Sabin subdivision algorithm. **(b)** Subdivision by Catmull-Clark. **(c)** Dual Surface. (the dotted lines belong to the initial mesh).

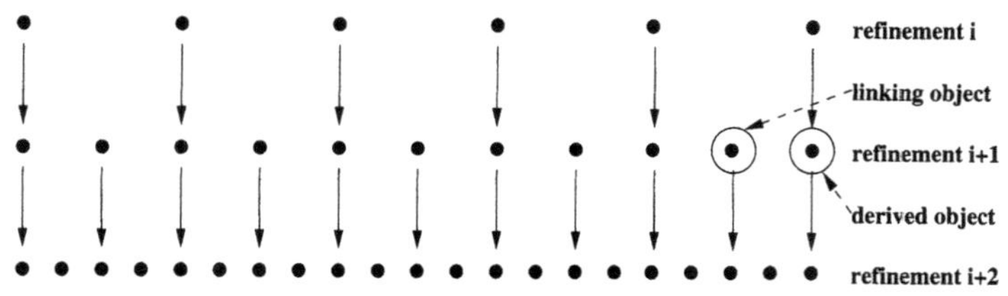

Fig. 2. There is a one-to-one relation between objects of subsequent refinements of subdivision meshes indicated by the arrows. Additional objects link the created objects to a mesh.

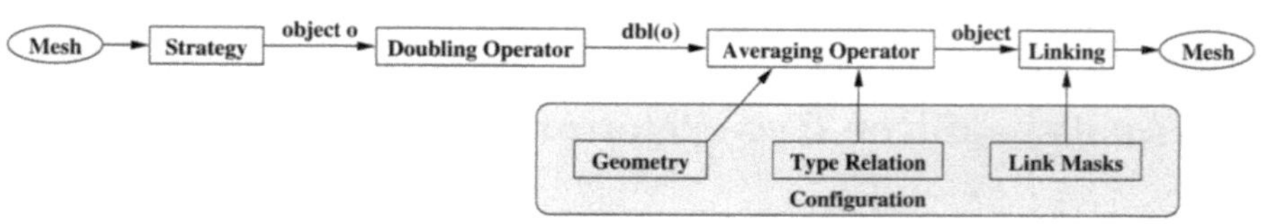

Fig. 3. Partition of the meta scheme for subdivision and its data flow.

of the mask's vertices with the weights exposed by the mask. The new vertices are connected by edges and result in the new mesh. Fig. 1 (b) shows the first subdivision of an irregular mesh. A further approximating algorithm from Doo-Sabin creates a face for every object (Fig. 1 (a), mask Fig. 6 (a)). The interpolating *Butterfly Scheme* constructs a vertex for every edge of a regular trilateral mesh according Fig. 6 (c).

Beginning with an input mesh, consisting of vertices, edges and faces, subsequently denoted as *objects* of a certain *type*, a sequence of meshes is generated. As a face and edge is based on vertices, a vertex is designated as *supporting object* or simply *support*. Also, an edge is a supporting object of a face. During the process the algorithm takes one *central object* and, preferably, local neighbouring and supporting objects of refinement i and generates one new (central) object and its support of refinement $i+1$. The process enforces a natural relation between the central objects of refinement i and $i+1$ as depicted in Fig. 2. Generally, objects are added lying between the central objects of refinement $i+1$, linking them to a mesh.

Fig. 3 proposes the data flow of a meta subdivision to be a partition of sub-algorithms. These may be configured by a description file yielding a *meta subdivision scheme*, that maintains to represent most of the known subdivision algorithms[1]. Firstly, the *Strategy* selects the mesh objects in a particular order. Besides variations explained in Kohler and Müller [13] there are two major strategies, the *Depth-first* and *Breadth-first Strategy*. For new applica-

[1] Because the averaging operator of Equation 9 is constructed as a linear sum, all subdivision schemes are represented, whose vertices may be calculated by linear combinations.

tions in image processing the *Strategy* must guarantee a unique enumeration of the objects in refinements of same topology.

The *Strategy* selects a central object that serves as base for the *Doubling Operator*, which is explained in Section 4.2. The *Doubling Operator* is a formal operator that produces tuples of objects and thus increases the overall number of objects. The tuples are equivalent to the necessary internal data structure to be able to navigate on the mesh.

The *Averaging Operator* (Section 5) creates the new objects and their supporting objects according to the *Type Relation*. The *Geometry* describes the parameters and the recursion to be applied. The *Linking* finally connects the resulting objects to a mesh (Section 6).

3 Analysis of the *Type Relation*

The *Type Relation* reflects the previously mentioned relation between the central objects (Section 2). It is a mapping from the set of types $\{\mathcal{V} = vertex, \mathcal{E} = edge, \mathcal{F} = face\}$ onto itself. There are 27 combinations of Type Relations some of them shown in Fig. 4. The *Type Relation type 1* $\rightarrow$ *type 2* indicates: If the algorithm uses a central object of *type 1* for navigation and recursion, it will create an object of *type 2* with the relevant supporting objects. We shortly enumerate the *Type Relations* of the most important subdivision schemes.

Chaikin Cutting Corners for curves. It is best to start with Chaikin scheme, as curves have only two objects (edge, vertex). Fig. 5 exhibits how new points on an edge, i. e. a new edge, is derived. The connection of the vertices further implies that a vertex is mapped onto an edge. We observe that in the Chaikin algorithm exists a correspondence between the old and newly generated objects: $\mathcal{V} \rightarrow \mathcal{E}$ and $\mathcal{E} \rightarrow \mathcal{E}$.

Doo-Sabin Scheme. It is shown in Fig. 1 (a). For every old face a new face *(f–face)* is calculated by certain rules described in [15]. For every old edge a new face *(e–face)* and for every old vertex a new face *(v–face)* is constructed. Note, the inherited correspondence between the objects and the rather natural relation described by the Type Relation of Fig. 4 (12). Fig. 6 (a) shows the masks of the Doo-Sabin subdivision for a quadrilateral mesh. The new vertex underlying the circle of the mask is calculated by a weighted average of the old vertices. All new vertices derived by the four vertex masks are linked together by edges providing the f–face. Additional edges yield the e–face and v–face (Fig. 1 (a)).

Catmull-Clark Scheme. Fig. 1 (b) shows the topology of the Catmull-Clark surface. Each object is mapped to a vertex in the new mesh. These vertices are connected by edges. Fig. 4 (1) shows the correspondence between the objects and Fig. 6 (b) the subdivision masks for a quadrilateral mesh.

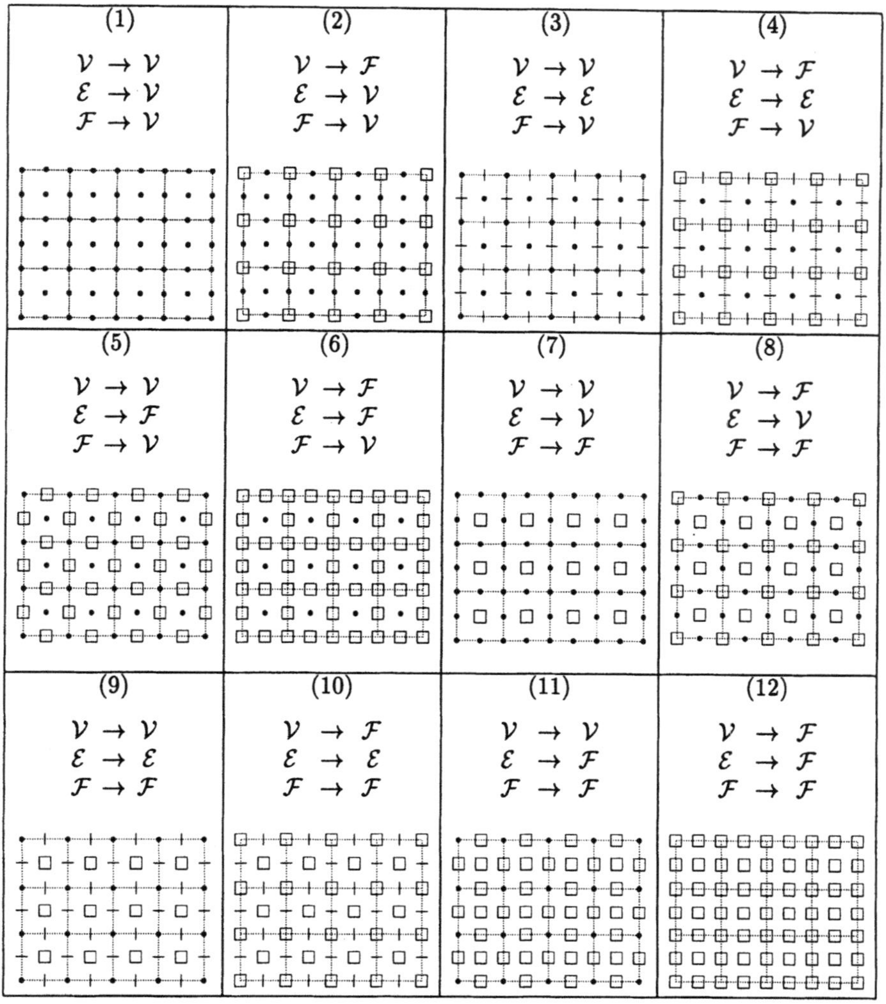

Fig. 4. Possible combinations and valid *Type Relations* illustrated for quadrilateral meshes ($\mathcal{V}$ is the vertex type, $\mathcal{E}$ is the edge type and $\mathcal{F}$ is the face type).

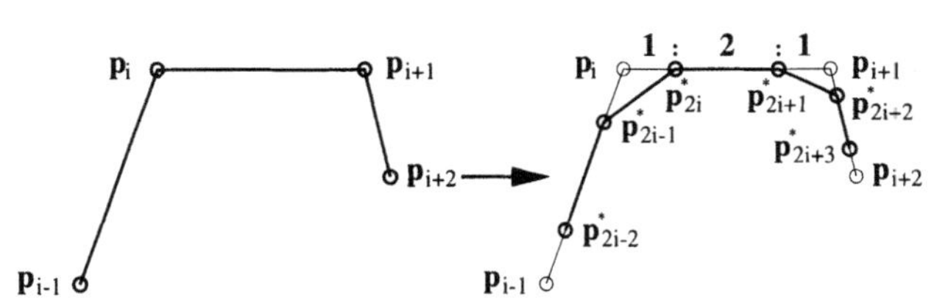

Fig. 5. Smoothing through cutting corners according Chaikin.

Butterfly Scheme. Another often reported subdivision is the *Butterfly Scheme* [1, p. 241]. Fig. 6 (c) exhibits its mask and its ten-point extension (d) [8]. The *Butterfly Scheme* and the ten-point extension have the Type Relation Fig. 4 (7).

Loop Subdivision Scheme. Loop Surfaces are introduced by the masks of Fig. 6 (e). Loop Surfaces [14] are achieved by the mapping Fig. 4 (7). The f–face results from connecting the e–vertices by edges.

Dual Surface (Fig. 1 (c)). One noticeable Type Relation should be mentioned: Given a mesh we may consider it as a graph. Its dual graph is equivalent to applying the Type Relation Fig. 4 (4) to the mesh. Constructing this *Dual Surface* after applying Catmull-Clark on a mesh yields the topology of Doo-Sabin and vice versa. For a quadrilateral, regular meshes as described by the masks of Fig. 6 (a) and (b) it may be shown, that applying the dual graph after Catmull-Clark is exactly the Doo-Sabin algorithm, if the vertices of the dual graph are mapped to the centres of gravity of the Catmull-Clark faces. This *Dual Surface* is used in the definition of the Doubling Operator of Section 4.2.

It is obvious that varying the *Type Relation's* objects may lead to 27 new subdivision schemes. However, not all of them are applicable [2]. The edge object has a direction, whereas neither the face nor the vertex has a direction. Hence, it is not obvious, how to map a face or vertex onto an edge. Mapping an edge onto a vertex or face is valid, because the vertex or face gets a directional information. Omitting the mentioned edge-face and edge-vertex mappings results in the possible valid *Type Relations* of Fig. 4.

4 Description of Topology and the Doubling Operator

Before explaining the *Doubling Operator* a description of the mesh and its topology is introduced. It is based on permutations that operate on sets of indices.

4.1 Algebraic description of the topology

A mesh consists of the objects "vertex", "edge" and "face". Pretending that the initial mesh has n vertices $X = \{\mathbf{p}_1, \ldots, \mathbf{p}_n\} \subset \mathbb{R}^d$, the vertices *contain* the geometry (d = dimension of the space). If a k-sided face consists of the vertices $\mathbf{p}_{j_1}, \ldots, \mathbf{p}_{j_k}, j_i \in \{1, \ldots, n\}$, we may identify the face with the cyclic permutation $\varphi = (j_1 \quad \ldots \quad j_k)$, that operates on the indices of the vertices: $\varphi(j_i) = j_{i+1}$ for $i < k$ and $\varphi(j_k) = j_1$. Hence, the face consists of the vertices $\mathbf{p}_{\varphi^0(j_1)}, \mathbf{p}_{\varphi^1(j_1)}, \ldots, \mathbf{p}_{\varphi^{k-1}(j_1)}$. For simplification we define the operation of φ on the geometric object (vertex) $\mathbf{p}_j$ as $\varphi(\mathbf{p}_j) := \mathbf{p}_{\varphi(j)}$. Faces and vertices will

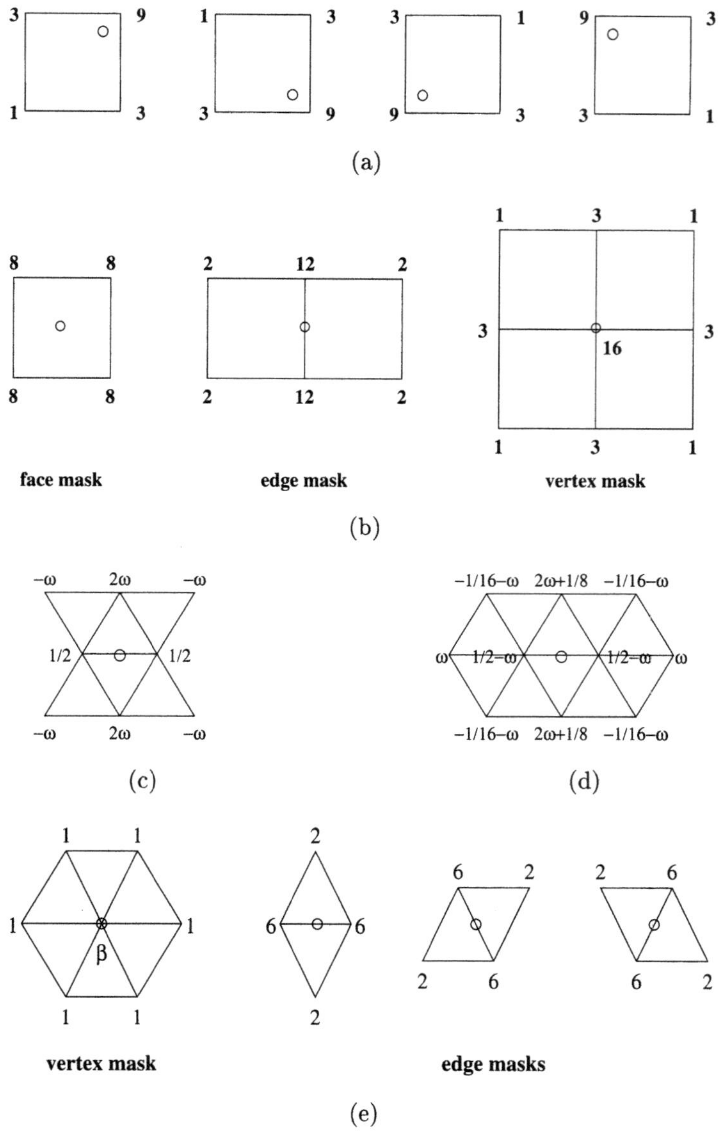

Fig. 6. **(a)** Vertex masks of the Doo-Sabin algorithm. The circle in each mask indicates the new vertex. **(b)** Vertex, edge and face mask of the Catmull-Clark algorithm. **(c)** Mask of the *Butterfly Scheme* and **(d)** the ten point stencil extension of the *Butterfly Scheme* ($\omega = 1/16$ to obtain maximum smoothness). **(e)** Vertex and edge masks for Loop Surfaces (generally $\beta = 10$).

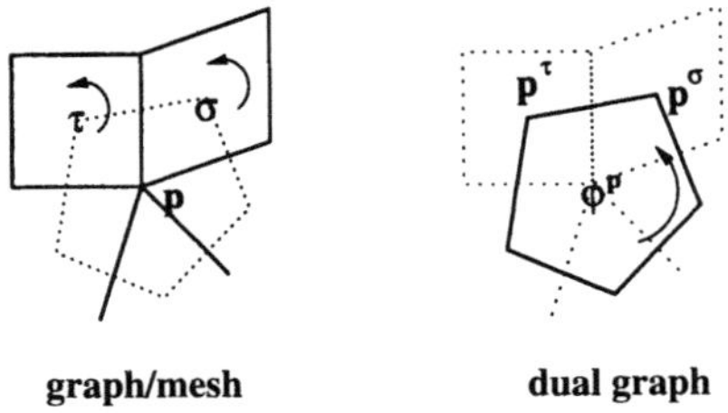

Fig. 7. Illustration of the dual operation.

be treated in the same manner (semantically and syntactically). This results in a dual operation, where the vertices operate as permutations on faces

$$\mathbf{p}(\sigma) = \tau :- \ \varphi^{\mathbf{P}}(\mathbf{p}^{\sigma}) = \mathbf{p}^{\tau} \text{ and } \sigma(\tau(\mathbf{p})) = \mathbf{p}\ . \tag{1}$$

Therefore consider Fig. 7 ($\mathbf{p}^\tau$ and $\mathbf{p}^\sigma$ are the dual objects of σ, τ and $\varphi^{\mathbf{P}}$ is dual to $\mathbf{p}$ according Fig. 4 (4)). Based on the dual graph Equation (1) formally expresses, that a vertex operating on a face with permutation σ yields in the next face with permutation τ, which is "running" around the vertex. The *order of a vertex* $\mathbf{p}$ is $ord(\mathbf{p}) := ord(\varphi^{\mathbf{P}})$, which is the total amount of adjacent edges of $\mathbf{p}$ *(valence of* $\mathbf{p}$*)*. With this algebraic model of the mesh, we can describe operations on the mesh by means of group algebras (Section 5).

4.2 Doubling Operator

The *Type Relation* of Section 3 expresses a one-to-one correspondence between certain (central) objects of the old and new mesh. In contrast to this "constant" relation the number of elements in the new mesh keeps growing compared to the old. This means that certain combinations of objects must lead to new objects. Combining elements is adequate to create the product set and mainly serves to increase the number of objects.

Concretely, during a single production step of one object in subdivision the geometric information of one element and its surrounding elements are combined to produce several new elements for the new mesh. To get all possible combinations of neighbouring objects we introduce the *Doubling Operator* which merely generates tuples of permutations and vertices.

Let B be the set of all cyclic permutations that relate to faces in the mesh. Given a n-sided face $\mathbf{f}$ represented by the permutation φ, we can combine the face and vertices to the set

$$dbl(\mathbf{f}) := \{(\varphi, \mathbf{p}) \in B \times X | \varphi(\mathbf{p}) \neq \mathbf{p}\}\ .$$

This is a more general description of the *Doubling Operator* reported by Prautzsch (course notes and [3]). Any vertex of $\mathbf{f}$ is combined with the permutation φ. This construction may be advanced to vertices operating on

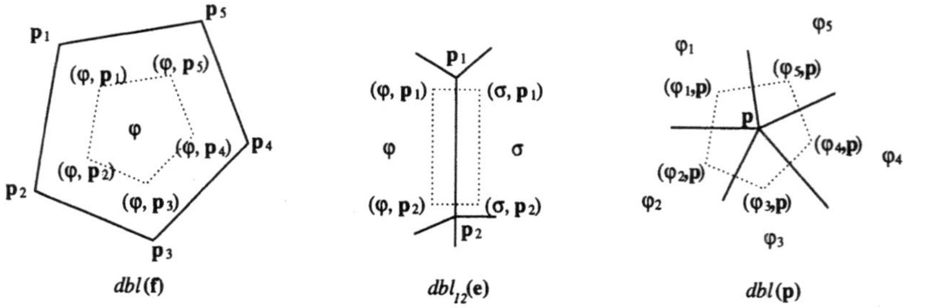

Fig. 8. Informal illustration of the doubling sets.

permutations

$$dbl(\mathbf{p}) := \{(\varphi, \mathbf{p}) \in B \times X | \varphi(\mathbf{p}) \neq \mathbf{p}\} .$$

Given an edge $\mathbf{e}$ with end points $\mathbf{p}_1, \mathbf{p}_2$ an extended set is derived by

$$\begin{aligned} dbl(\mathbf{e}) &:= dbl(\mathbf{p}_1)) \cup dbl(\mathbf{p}_2) \\ &= \{\varphi \in B | \varphi(\mathbf{p}_1) \neq \mathbf{p}_1 \text{ and } \varphi(\mathbf{p}_2) \neq \mathbf{p}_2\} \times \{\mathbf{p}_1, \mathbf{p}_2\} \dot{\cup} \qquad (2) \\ &\quad (\{(\varphi, \mathbf{p}_1) \in B \times X | \varphi(\mathbf{p}_1) \neq \mathbf{p}_1 \text{ and } \varphi(\mathbf{p}_2) = \mathbf{p}_2\} \qquad (3) \\ &\quad \cup \{(\varphi, \mathbf{p}_2) \in B \times X | \varphi(\mathbf{p}_2) \neq \mathbf{p}_2 \text{ and } \varphi(\mathbf{p}_1) = \mathbf{p}_1\}) . \qquad (4) \end{aligned}$$

We will refer to the set (2) as $dbl_{12}(\mathbf{e})$, to the set (3) as $dbl_1(\mathbf{e})$ and to the set (4) as $dbl_2(\mathbf{e})$. The Averaging operator, which is defined next, calculates new vertices of the "doubled objects". Fig. 8 is an informal illustration of the doubling sets.

5 The Averaging Operator

With the permutations operating on vertices and vertices operating on permutations, an *Averaging Operator* may be declared on any subset of $B \times X$. However, in subdivision mainly local Averaging Operators are used. They depend on the *Type Relation.* First note, that a corresponding permutation φ of a n-sided face $\mathbf{f}$ has order $ord(\varphi) = n$.

Given the *Type Relation* $\mathcal{F} \to \mathcal{F}$ we define n vertices

$$\forall (\varphi, \mathbf{p}) \in dbl(\mathbf{f}) : \mathbf{p}^*(\varphi, \mathbf{p}) := \sum_{k=0}^{ord(\varphi)-1} a_{\varphi,\mathbf{p},k} \cdot g(\varphi, \varphi^k(\mathbf{p})) . \qquad (5)$$

If the neighbouring objects (faces and their supporting objects) of $\mathbf{f}$ shall recursively influence $\mathbf{p}^*$, then $g(\varphi, \varphi^k(\mathbf{p}))$ is a summation similar to (5). Otherwise it is $g(\varphi, \varphi^k(\mathbf{p})) = \varphi^k(\mathbf{p})$. Note, even if recursion is used, (5) is a polynomial of the group algebra $\mathbb{R}[G], G =< B >$. Linking all vertex pairs $\mathbf{p}^*(\varphi, \mathbf{p})$ and $\mathbf{p}^*(\varphi, \varphi(\mathbf{p}))$ by edges yields the new face. The *Type Relation* implies, how many vertices must be generated and how to link them. As the description (5) is applicable for any face, it is sufficient to specify the *Type Relation*, the coefficients $a_{\varphi,\mathbf{p},k}$ and the recursive function g. This may be done in a description file (see Example 1).

For the *Type Relation* $\mathcal{F} \to \mathcal{V}$ we get one vertex defined by

$$\mathbf{p}^*(\varphi,\mathbf{p}) := \sum_{k=0}^{ord(\varphi)-1} b_{\varphi,\mathbf{p},k} \cdot h(\varphi, \varphi^k(\mathbf{p})) \,, \tag{6}$$

where $(\varphi,\mathbf{p})$ may be any tuple in $dbl(\mathbf{f})$. Hence, (6) must be independent of the choice of $\mathbf{p}$. Again h is a summation similar to (5). The independence of $\mathbf{p}$ forces $\mathbf{p}^*$ to have the form

$$\mathbf{p}^*(\varphi,\mathbf{p}) = \tilde{b}_\varphi \sum_{k=0}^{ord(\varphi)-1} h(\varphi, \varphi^k(\mathbf{p})) \,.$$

Given the *Type Relation* $\mathcal{V} \to \mathcal{F}$ for $\mathbf{p}$ of valence n, the n vertices of the v–face are calculated according

$$\forall(\varphi,\mathbf{p}) \in dbl(\mathbf{p}) : \mathbf{p}^*(\varphi,\mathbf{p}) := \sum_{k=0}^{ord(\mathbf{p})-1} c_{\varphi,\mathbf{p},k} \cdot u(\mathbf{p}^k(\varphi),\mathbf{p}) \,.$$

If the neighbouring objects of $\mathbf{p}$ shall be included recursively, then $u(\mathbf{p}^k(\varphi),\mathbf{p})$ is a summation similar to (5). Otherwise we choose $u(\mathbf{p}^k(\varphi),\mathbf{p}) = \mathbf{p}$. Linking all vertex pairs $\mathbf{p}^*(\varphi,\mathbf{p})$ and $\mathbf{p}^*(\mathbf{p}(\varphi),\mathbf{p})$ by edges yields the new face.

For the *Type Relation* $\mathcal{V} \to \mathcal{V}$ there is one v–vertex

$$\mathbf{p}^*(\varphi,\mathbf{p}) := \sum_{k=0}^{ord(\mathbf{p})-1} d_{\varphi,\mathbf{p},k} \cdot v(\mathbf{p}^k(\varphi),\mathbf{p}) \,, \tag{7}$$

where $(\varphi,\mathbf{p})$ may be any tuple in $dbl(\mathbf{p})$ and $v(\mathbf{p}^k(\varphi),\mathbf{p})$ is similar to $u(\mathbf{p}^k(\varphi),\mathbf{p})$. Again it is meaningful to claim, that (7) is independent from the choice of φ. This yields in

$$\mathbf{p}^*(\varphi,\mathbf{p}) := \tilde{d}_\mathbf{p} \cdot \sum_{k=0}^{ord(\mathbf{p})-1} v(\mathbf{p}^k(\varphi),\mathbf{p}) \,.$$

Given the *Type Relation* $\mathcal{E} \to \mathcal{F}$, we define four vertices by[2]

$$\forall(\varphi,\mathbf{p}) \in dbl_{12}(\mathbf{e}) : \mathbf{p}^*(\varphi,\mathbf{p}) := \sum_{k=0}^{ord(\mathbf{p})-1} e_{\varphi,\mathbf{p},k} \cdot r(\mathbf{p}^k(\varphi),\mathbf{p}) \,.$$

[2] An edge with vertices of valence n_1 and n_2 has $n_1 + n_2 - 2$ neighbouring faces. Hence, a $(n_1 + n_2 - 2)$-sided face could be generated. However, the same result may be generated by a subsequent, different subdivision step.

If the neighbouring objects of $\mathbf{p}$ shall be included recursively, then $r(\mathbf{p}^k(\varphi), \mathbf{p})$ is a summation similar to (5). Otherwise it is $r(\mathbf{p}^k(\varphi), \mathbf{p}) = \mathbf{p}$. Two vertices $\mathbf{p}^*(\varphi_1, \mathbf{p}_1)$ and $\mathbf{p}^*(\varphi_2, \mathbf{p}_2)$ are linked if either $\mathbf{p}_1 = \mathbf{p}_2$ or $\varphi_1 = \varphi_2$. This yields the new face $((\varphi_1, \mathbf{p}_1), (\varphi_2, \mathbf{p}_2) \in dbl_{12}(\mathbf{e}))$.

For the *Type Relation* $\mathcal{E} \to \mathcal{V}$ the edge is replaced by its e–vertex

$$\begin{aligned}\mathbf{p}^*(\varphi, \mathbf{p}) := & \sum_{k=0}^{ord(\mathbf{p})-1} f_{\varphi,\mathbf{p},k} \cdot s_1(\mathbf{p}^k(\varphi), \mathbf{p}) \\ & + \sum_{k=0}^{ord(\varphi(\mathbf{p}))-1} f_{\varphi,\varphi(\mathbf{p}),k} \cdot s_2((\varphi(\mathbf{p}))^k(\varphi), \varphi(\mathbf{p})) ,\end{aligned} \tag{8}$$

where $(\varphi, \mathbf{p}) \in dbl_{12}(\mathbf{e}) : \psi(\varphi(\mathbf{p})) = \mathbf{p}$ for $(\psi, \mathbf{p}) \in dbl_{12}(\mathbf{e}), \varphi \neq \psi$ and $s_1(\cdot,\cdot), s_2(\cdot,\cdot)$ are similar to $u(\cdot,\cdot)$.

The remaining *Type Relation* $\mathcal{E} \to \mathcal{E}$ is specified by two summations of type (8) one of which is applied to $(\varphi, \mathbf{p}) \in dbl_{12}(\mathbf{e})$ resulting in $\mathbf{p}_1^*(\varphi, \mathbf{p})$ and the second to $(\psi, \varphi(\mathbf{p})) \in dbl_{12}(\mathbf{e})(\varphi \neq \psi$ and $\psi(\varphi(\mathbf{p})) = \mathbf{p})$ yielding $\mathbf{p}_2^*(\psi, \varphi(\mathbf{p}))$. The above summations were explicitely written for all *Type Relations* because those forms are used for the configuration files. Although, there is a common notation for all *Type Relations*. Given an object $\mathbf{o}$ all new supporting vertices are generated by linear summations of the kind

$$\mathbf{p}^* = \sum_{(\psi,\mathbf{q}) \in dbl(\mathbf{o})} \tilde{a}_{\psi,\mathbf{q}} \cdot \tilde{g}(\psi, \mathbf{q}) . \tag{9}$$

Again $\tilde{g}$ is a recursive function of the same summation as in (9) or $\tilde{g}(\psi, \mathbf{q}) = \mathbf{q}$.

Example 1. We shortly show the description file of the Doo-Sabin algorithm:

```
f-face    :: af(f,p) #
af(f,p)   :: n :: id2(f,p) ::
             n=0     : (ord(f)+5)/(4*ord(f)) ;
             default: (3+2*cos(2*pi*n/ord(f)))/(4*ord(f)) #

v-face    :: av(f,p) #
av(f,p)   :: n :: af(f,p) ::
             n=0     : 1 ; default  : 0 #

e-face    :: ae(f,p) #
ae(f,p)   :: n :: af(f,p) ::
             n=0     : 1 ; default  : 0 #

f-face-v-face: 0 #
f-face-e-face: 0 #
e-face-v-face: 0 #
```

The statement `f-face` clearly indicates that every face **f** is mapped onto a face (the f–face). The f–face generator invokes the summation `af` for every tuple $(\varphi, \mathbf{p}) \in dbl(\mathbf{f})$ resp. `(f,p)`. The index of the summation `af` is `n`. There are two rules (`n=0` and `default`) and the recursion is `id2`. The summation `af` automatically accumulates the product for each `n`, i. e. the $\mathtt{n}^{\mathtt{th}}$ coefficient multiply `id2(f,f`$^{\mathtt{n}}$`(p))` (here `f`$^{\mathtt{n}}$`(p)` is the descriptor for $\varphi^n(\mathbf{p})$. The function `id2` is pre-defined as the projection of the second argument. The `v-face` generator calls `av` for each $(\varphi, \mathbf{p}) \in dbl(\mathbf{p})$ resp. `(f,p)`. The recursion of `av` is `af`. Through setting the first coefficient to unity and all other coefficients to zero, the summation is equivalent to `af(p`^{0}`(f),p)`. The `e-face` generator is similar to the `v-face` generator. The last three rules specify the *Link Masks*: No linking is necessary, because the objects are already linked as they use same geometry, i. e. the vertices of the v–face and e–face are the vertices of the f–face. The user is responsible to correctly setting up the configuration file.

6 Object Linking

Exactly one primary object is created for each object of the previous mesh. We denote a face as *direct neighbour* of its edges and vertices and a vertex as *direct neighbour (dn)* of its edge. There is no object between two objects derived from direct neighbours. Thus we denote these two objects as *direct succeeding neighbours (dsn).* If the rules specify same supporting vertices of two *dsn* as in Example 1, those *dsn* are already linked. Otherwise the *dsn* need to be linked together. To completely separate the calculation of the geometry (vertices) from the linking phase and because additional edges would be required, vertices are not put in between two *dsn* in order to link them together. Additional geometry may be added in the subsequent subdivision step. Similarly, if a face was used to link two *dsn*, no new vertices are introduced. Hence, linking with a face without creating new vertices is equivalent to inserting two edges. Thinking of the Type Relations, there are only five types of *dsn:* vertex-vertex, vertex-edge, vertex-face, face-edge and face-face. If the *dsn* have no common supporting object, they are linked by zero, one or two edges. Two vertices may either be linked or not, denoted by 1 or 0 in the description rule (see Example 1). A vertex and an edge, a vertex and a face, a face and an edge and a face and a face are simply not linked, or may be linked by one or two edges.

Given the *dn* $\mathbf{o}_1, \mathbf{o}_2$, two new *direct succeeding neighbours* are derived with all their supporting vertices according the Averaging Operator. For all $(\varphi, \mathbf{p}) \in dbl(\mathbf{o}_1) \cup dbl(\mathbf{o}_2)$ the two supporting objects $\mathbf{p}_1^*(\varphi, \mathbf{p}), \mathbf{p}_2^*(\varphi, \mathbf{p})$ of the *direct succeeding neighbours* are candidates to be linked. According to the *Link Mask* of the configuration file, either none, one or two links are established. If the number of links between the *dsn* is set to zero, no link is inserted. If it is set to one, the link is inserted between the nearest supporting candidates.

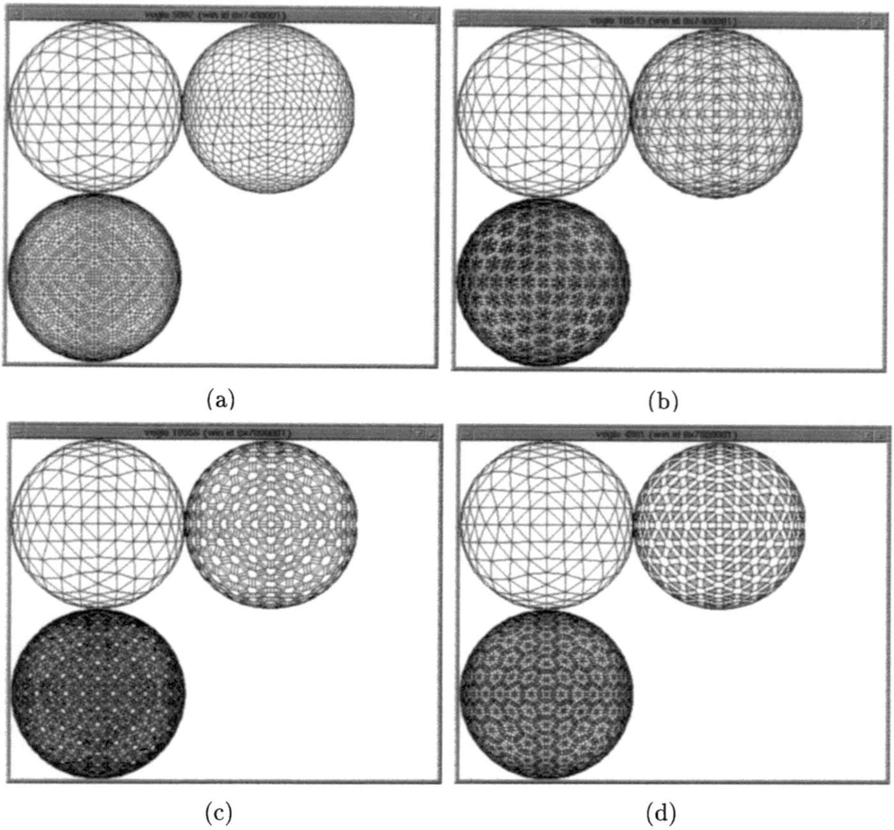

Fig. 9. Examples for Subdivision: **(a)** Catmull-Clark subdivision (1); **(b)** subdivision scheme (11); **(c)** subdivision scheme (10); **(d)** subdivision scheme (6) — see Fig. 4.

7 Conclusion

A partition of subdivision algorithms into sub-algorithms was proposed: *Strategy*, selecting the objects for processing, *Doubling Operator*, building up the data structure for navigation, the *Averaging Operator*, determining the geometry, and *Linking*, reconfiguring the objects to a mesh. The *Strategy* was discussed in [13]. With this meta refinement scheme the Strategies of [13] become available to all subdivision schemes that are producable by these means. The *Averaging Operator* and *Linking* are steered by an external configuration containing *Geometric and Topologic Description.* The algebraic modelling may be implemented straightforwardly. The operation of a permutation on vertices is equivalent to walking along the edges of a face, whereas the operation of a vertex on a permutation means walking around a vertex. The polynomials over the group algebras specify completely, how to derive the new geometry, and the description is independent from mesh topology.

There is no need for masks that strongly depend on the algorithm and the topology. The proposed refinement scheme covers most refinement schemes and unifies their implementation. Badura [2] enumerates some new subdivision schemes (Fig. 9). Arising problems with these new subdivision schemes like smoothness, convergence and shrinking borders need to be discussed. Possibly, some important results from [4,5,7,9,11,12,15–17] may be adopt for the new schemes. Further we hope that this approach will inspire for new investigations about subdivision.

Acknowledgement

I thank Gregor Badura for his help to examine subdivision algorithms and for implementing a first prototype of the meta subdivision scheme. Further, I appreciate the helpful discussions with Heinrich Müller.

References

1. S. Abramowski and H. Müller, *Geometrisches Modellieren*, Reihe Informatik, B.I. Wissenschaftsverlag, Mannheim, 1991.
2. G. R. Badura, *Generierung und Implementierung neuer Zuordnungsverfahren mittels eines kombinatorischen Ansatzes durch die objektorientierte Analyse der Unterteilungsverfahren von Catmull-Clark und Doo-Sabin*, Master's thesis, Lehrstuhl Informatik VII, Universität Dortmund, Germany, May 1996.
3. W. Boehm, H. Prautzsch, and Arner, *On triangular splines*, Constructive Approximation **3** (1987), 157–167.
4. E. Catmull and J. Clark, *Recursively generated B-spline surfaces on arbitrary topological meshes*, Computer Aided Design **10**:6 (1978), 350 – 355.
5. A. S. Cavaretta, W. Dahmen, and C. A. Micchelli, *Stationary subdivision*, Memoirs of the American Mathematical Society **93**:453 (1991), 1–186.
6. G. M. Chaikin, *An algorithm for high speed curve generation*, Computer Graphics and Image Processing **3** (1974), 346–349.
7. D. Doo and M. Sabin, *Behaviour of recursive division surfaces near extraordinary points*, Computer Aided Design **10** (1978), 356–360.
8. N. Dyn and D. Levin, *Subdivision Schemes for Surface Interpolation*, Workshop in Computational Geometry (1993), 97–118.
9. N. Dyn, D. Levin, and C. A. Micchelli, *Using parameters to increase smoothness of curves and surfaces generated by subdivision*, Computer Aided Geometric Design **7** (1990), 129–140.
10. G. Farin, *Curves and Surfaces for CAGD*, 3rd edition ed., Academic Press, San Diego, 1993.
11. M. Halstead, M. Kass, and T. DeRose, *Efficient, Fair Interpolation using Catmull-Clark Surfaces*, ACM Computer Graphics Proceedings, Annual Conference Series, August 1993, pp. 35–44.

12. L. Kobbelt, *Iterative Erzeugung glatter Interpolanten*, Dissertation, Fakultät für Informatik, Universität Karlsruhe, December 1994.

13. M. Kohler and H. Müller, *Efficient calculation of subdivision surfaces for visualization*, Visualization and Mathematics (H.-C. Hege and K. Polthier, eds.), Springer, Heidelberg, 1997, pp. 165–179.

14. C. Loop, *Smooth Subdivision Surfaces Based on Triangles*, Master's thesis, Department of Mathematics, University of Utah, 1987.

15. A. H. Nasri, *Polyhedral subdivision methods for free-form surfaces*, ACM Transactions on Graphics **6**:1 (1987), 29–73.

16. H. Prautzsch and L. Kobbelt, *Convergence of subdivision and degree elevation*, Advances in Computational Mathematics **2**:2 (1994), 143–154.

17. J. A. Roulier, *A convexity preserving grid refinement algorithm for interpolation of bivariate functions*, IEEE CG & A **7**:1 (1987), 57–62.

A Scheme for Edge-based Adaptive Tetrahedron Subdivision

Detlef Ruprecht[1] and Heinrich Müller[2]

[1] Andersen Consulting, Frankfurt, Germany
[2] Informatik VII, University of Dortmund, Dortmund, Germany

Abstract. A new scheme for adaptive refinement of tetrahedra based on an edge criterion is presented. This scheme guarantees consistent subdivision without cracks. Checks between neighboring tetrahedra of a mesh are not necessary, allowing a recursive algorithm with low space requirements and parallel implementation.

1 Introduction

Tetrahedra are a basic structural element in many applications working with solid models, e.g. finite element computations [2,4] or solid modeling [14]. They are also useful for implicit surface modeling [8] and visualization, for the deformation of solid models [15], and have even been used for flow visualization [11].

In many of these applications, there is a need to adapt the structure of tetrahedral meshes to the level of detail necessary. In FEM computations, a finer mesh is often required in the vicinity of strong force gradients [4]. For the deformation of solid models, an adaptive subdivision of tetrahedra is desired where deformation varies strongly [15]. For explicit calculation and visualization of implicit surface models, tetrahedra near the surface need to be refined [8]. For visualization of mesh data by iso-surfaces or volume rendering [12,17,6,18–20], adaptive and multilevel representations as well as mesh reduction techniques have been used for accelerating processing and rendering of large data sets. An example of particular interest for the application of this work to the field of volume visualization is the FEM-based approach of [7].

A number of methods to obtain surfaces from space subdivisions have been developed in recent years. Most of them divide the space into cubes. From these cubic meshes, surfaces can be derived with the marching cubes algorithm [10], or adaptively by tree-based methods [13,16]. Unfortunately, these methods tend to produce a large number of surface polygons and, as has been pointed out [8], cubic meshes are unsuitable for a number of applications, particularly when arbitrarily given points have to be included in the division of space. For these cases, a tetrahedral subdivision appears to be more adequate.

Methods for the construction of a tetrahedral mesh dividing a given polyhedral object are mainly known from the FEM literature [1,5]. However, little is known about adaptive subdivisions of such tetrahedralizations to fit particular precision requirements.

One approach presented by Hall and Warren [8] uses criteria for the tetrahedra. If a tetrahedron is split, e.g. because the implicit surface to be described has a zero contour within it, adjacent tetrahedra must be partially split as well to avoid cracks, i.e. discontinuities in the tetrahedral approximation of the surface. This is checked in a second pass over the tetrahedra. Finally, the respective tetrahedra are split by projecting from their centroid to the surface.

Bornemann et al. [4] and Bey [3] present a modification of a method by Zhang [21], which also uses a tetrahedron based refinement criterion causing a regular subdivision of affected tetrahedra. Adjacent tetrahedra again have to be treated in a separate step. Here, only a few simple configurations are allowed for the subdivision of adjacent tetrahedra. Tetrahedra not fitting these requirements are regularly subdivided.

In the following, we present a new algorithm for adaptive tetrahedron subdivision which has two advantages compared to the methods outlined above:

- There is no need to check neighboring tetrahedra. This allows a depth-first recursive approach with small memory requirements.
- The subdivision does not require the addition of a centroid. This reduces the number of new tetrahedra created in a recursive step and possibly also leads to an improved aspect ratio of the created tetrahedra.

The algorithm described below divides tetrahedra based on the subdivision of their edges. This guarantees consistent subdivision of neighboring tetrahedra without interdependency checks, as the edge criterion can be independently evaluated for each tetrahedron containing the edge.

The requirement that the subdivision be controlled by the subdivision of edges appears to be a useful method, as it is easy to define criteria for the splitting of edges, e.g.

- for implicit surface modeling, edges whose endpoints have opposite sign should be split.
- for deformations, if the deformation of the midpoint of an edge differs from the midpoint of the deformed edge more than some threshold, this edge should be split.
- for visualization, if the interpolation along the edge differs from the correct data values by a given threshold.

The path to the solution presented in Sect. 3 lies in a subdivision of the triangles on a tetrahedron's surface which depends only on which edges are split. As neighboring tetrahedra share surface triangles, cracks and T-junctions are effectively avoided as the subdivision of the shared triangles is independent of the tetrahedra. We will present a subdivision scheme for surface triangles, and show that for each configuration of the four surface triangles of a tetrahedron, the tetrahedron can be subdivided without the use of additional auxiliary interior points, like a centroid. The subdivision scheme for triangles is outlined in the following Sect. 2.

2 Triangle Subdivision

We use the following scheme for the subdivision of triangles, which has already proven to be useful for the adaptive subdivision of triangle meshes in geometric modeling by free form deformation [15] (Fig. 1):

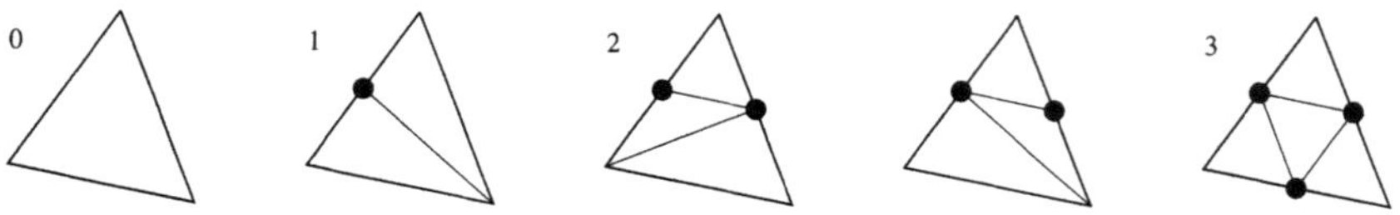

Fig. 1. Configurations for triangle subdivision.

No edges split: There is nothing to do.

One edge split: The triangle is split along the line from the splitpoint of the split edge to the opposite corner of the triangle, giving two new triangles.

Two edges split: The two splitpoints are connected. The remaining quadrilateral is divided starting from the splitpoint of the longer edge. If both diagonals have equal length, the diagonal starting from the splitpoint of the edge with the lower index is used (note that this requires that edges are ordered). This results in three new triangles.

Three edges split: All splitpoints are connected. This results in four new triangles.

For the case of two split edges, if edges are split at their midpoint, starting from the longer edge is equivalent to splitting the quadrilateral along the shorter diagonal. This helps to improve the aspect ratio of the created triangles.

As the subdivision of the triangles is solely driven by the subdivision of its edges, which are shared by neighboring triangles, this scheme ensures that the triangles fit together without cracks or T-junctions, and without having to maintain information about neighboring triangles. This allows storing the information in a rather simple data structure consisting of a list of edges which are referenced by the list of triangles. The subdivision can be performed in a *breadth-first approach* by testing each edge whether it needs to be split and then reforming all triangles.

Alternatively, a *depth-first approach* is possible where each triangle is recursively subdivided until none of the edges needs splitting. In this case, the space requirements are only in the order of the refinement depth, independent of the number of tetrahedra. However, this requires recomputing the edge criterion for each triangle an edge is part of, i.e. it doubles the number of times the edge criterion has to be evaluated.

For both methods, the processing time is linear in the number of created triangles, which is obviously optimal [15].

3 Tetrahedron Subdivision

We will now show that the triangle subdivision scheme presented above allows a subdivision of the tetrahedron for all configurations of subdivisions of the four surface triangles. This is non-obvious, and we have to take into account all possible configurations.

For a tetrahedron with its six edges, we have $2^6 = 64$ possible configurations of split edges. We will first show that after elimination of symmetrical cases, only 11 configurations remain (Fig. 2).

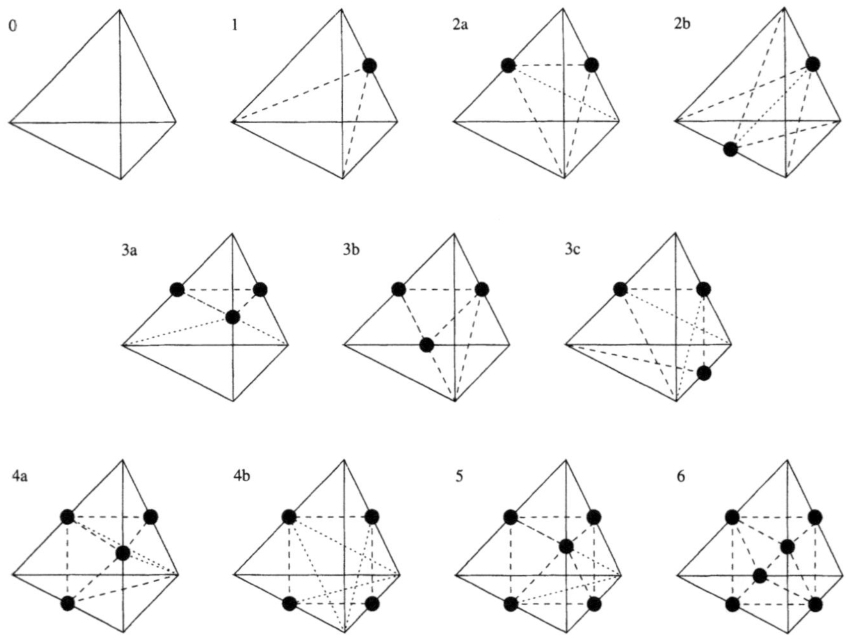

Fig. 2. Configurations for tetrahedron subdivision.

Theorem 1. *There are 11 possible configurations of split edges for a tetrahedron.*

Proof. We will determine the number of possible configurations for each number of split edges separately.

No split edges: Obviously, this is a single configuration.

One split edge: Because of the rotational symmetry of the problem, this is also a single configuration.

Two split edges: If we keep the first split edge fixed, all cases where the second split edge belongs to the same triangle as the first are symmetric. The only case left is the second split edge being diagonally opposite to the first. Thus, we get two configurations.

Three split edges: Starting with any of the two configurations for two split edges, we cannot add a third split edge without at least two of them belonging to the same triangle. So we can start considering one surface triangle with two split edges. Now, the third split edge can either belong to the same triangle, share a corner with the other two split edges, or share a corner with either one of the split edges. For the last case, it is irrelevant which one of the other edges, so we have three configurations.

Four split edges: With four split edges, we have two unsplit edges, which is symmetrical to the situation with two split edges. Thus, we get two configurations.

Five split edges: Five split edges give one unsplit edge, which leads to one configuration.

Six split edges: There is of course only one configuration with all edges split.

If we add up the seven cases above, we get $1 + 1 + 2 + 3 + 2 + 1 + 1 = 11$ configurations. □

Now, we have to show that for each of these 11 configurations, a tetrahedron subdivision is possible. This is slightly complicated by the fact that triangles with two split edges can be divided in two ways depending on the triangle geometry.

Theorem 2. *For all possible configurations, a subdivision of the tetrahedron using the triangle subdivision scheme above is possible without adding an inner point.*

Proof. We will again consider all 11 configurations in succession.

Configuration 0: There is nothing to do.

Configuration 1: The two triangles adjacent to the split edge are split in half. The two new edges, together with the tetrahedron edge diagonally opposite to the split edge, form an inner triangle of the tetrahedron which splits the tetrahedron in half.

Configuration 2a: The triangle containing the split edges is divided into three triangles. Each of these forms a new tetrahedron with the opposite corner of the original tetrahedron.

Configuration 2b: All surface triangles are split in half. By adding a new internal edge connecting the two splitpoints, we get four new tetrahedra.

Configuration 3a: This configuration is a bit more complex, as we have three triangles with two split edges each, where the geometry must be taken into account. The connecting lines of the splitpoints cut off one corner of the tetrahedron, leaving a prism. This can be split into three tetrahedra if the new edges dividing the quadrilateral form a chord, i.e. a single polyline. This is always the case, as the split point of the longest of the three split edges is common to two of the added diagonals. Regardless of the orientation of the third diagonal, it will always be connected to one of the two others.

Configuration 3b: Here, one of the triangles is divided in four by the edges connecting three splitpoints. Each of the new triangles forms a new tetrahedron with the opposite corner.

Configuration 3c: This case is again complicated by the fact that two of the triangles contain two split edges. We have to add an internal edge connecting the diagonally opposite splitpoints. Now, this edge, together with the edges created on the two triangles with one split edge each, cuts off a new tetrahedron, leaving two deformed pyramids which share one triangular surface. These two pyramids can be split into two tetrahedra each independently of each other.

Configuration 4a: This case is similar to configuration 3c in that the inner edge connecting the diagonally opposite splitpoints together with the non-ambiguous edges cut off two triangles. This again leaves two pyramids with a common triangle which can be split into tetrahedrons arbitrarily.

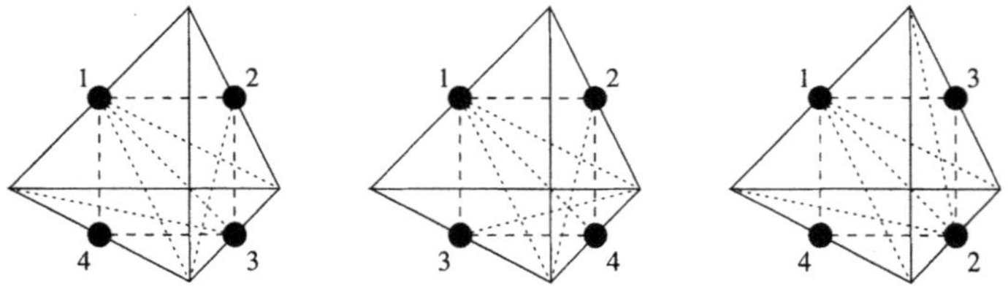

Fig. 3. Subconfigurations for configuration 4b.

Configuration 4b: This is the most complicated configuration as all four surface triangles contain two split edges. We note that the new edges connecting the four splitpoints form an inner quadrilateral which splits the tetrahedron into two prisms. These two prisms can be split into three tetrahedrons each, if their surface quadrilaterals are split by diagonals which form a chord. We have to show that regardless of the outer diagonals, we can always choose the diagonal splitting the inner quadrilateral consistently for both prisms. As the geometric criterion for the outer diagonals is based on the length of the neighboring edges, and the length is a transitive measure, we have a strict ordering of the four split edges. After removing the symmetric cases, we are left with the three subconfigurations shown in Fig. 3, where split point 1 belongs to the longest edge and split point 4 to the shortest. As we can see, for the case of Fig. 3a and Fig. 3b, at least one of the two prisms has outer diagonals meeting in one point so that the choice of the inner diagonal is arbitrary. For case 3c, both prisms constrain the orientation of the inner diagonal, but – fortunately – both produce the same constraint. Thus, we can always split the tetrahedron into six new tetrahedra.

Configuration 5: In this case, we can trivially cut off two tetrahedrons and are left with a pyramid and a prism sharing a quadrilateral. As the pyramid can be split either way, the orientation of the inner diagonal can be determined by the outer diagonals of the prism.

Configuration 6: Cutting off tetrahedra on all four corners leaves an octahedron, which can be split into four tetrahedra by adding an inner edge connecting two diagonally opposite corners of the octahedron. To minimize distortions of the created tetrahedra, the shortest of the three possible inner diagonals should be chosen.

As we have been able to give a construction for the subdivision into tetrahedra for all configurations, the proof is complete. □

With this construction of the subdivision of tetrahedra it is guaranteed that neighboring tetrahedra will fit together without cracks after the subdivision is finished, as the subdivision is driven by the surface triangles which are shared by neighboring tetrahedra. This is similar to the case of the subdivision of triangles described in Sect. 2. Again, we do not have to maintain information about neighboring triangles, so we can use a data structure and algorithms similar to the ones described in sect. 2 which give a processing time linear in the number of created tetrahedra, which is again optimal.

4 Discussion

The scheme for adaptive refinement of tetrahedra presented above appears promising for a number of applications in solid and implicit modeling, visualization, and finite element calculations. It allows the use of a simple refinement criterion defined on the edges of the tetrahedra. The required processing time is linear in the number of produced tetrahedra, which is optimal. With a depth-first algorithm similar to the one described for the adaptive subdivision of triangles elsewhere [15], the space requirement is proportional to the depth of the subdivision only. Therefore, the method does not require holding the tetrahedral model in whole in memory, making it suitable for large applications.

The properties of the created tetrahedra should be studied, e.g. if we define the aspect ratio of a tetrahedron as the ratio between the circumscribed and inscribed sphere, what distribution of aspect ratios evolves after several steps of refinement with various edges splitting criteria? It is known that if a tetrahedron is repeatedly split completely as described in configuration 6, the resulting tetrahedra belong to one of at most three congruence classes [3]. This means that the tetrahedra preserve a reasonable shape.

From our experience with the adaptive subdivision scheme for triangles, we are confident that the method will generally lead to well-behaved tetrahedra [15]. A rationale for this is the fact that for common edge splitting criteria, longer edges are more likely to be split than shorter ones, which tends to even out the edge lengths of tetrahedra, thereby improving the aspect ratio. Independently, schemes for improving the aspect ratio, like the method of Laplacian smoothing [9] can be used as a post-processing step. For some of the applications mentioned here, these methods might have to be modified so that the initial points of the tetrahedralization remain unaffected.

For applications where a surface has to be extracted from the tetrahedra, e.g. for rendering implicitly defined surfaces, an additional complication arises from the fact that even for a tetrahedralization with good aspect ratios, the derived triangulated surface might contain thin triangles. In this case, it is better to defer the smoothing step to the created triangles. A practical approach to solve this problem has been discussed by Hall and Warren [8].

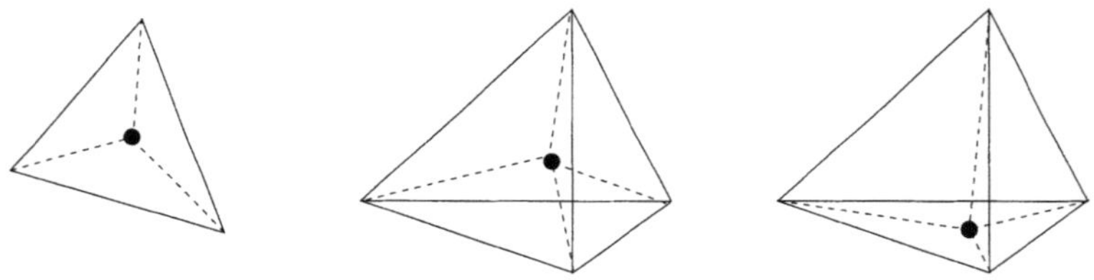

Fig. 4. Configurations for volume-driven subdivision of triangles (left) and tetrahedra. If more than one face of the tetrahedron satisfies the subdivision criterion, the configuration shown on the right is iteratively applied to the subtetrahedra incident with these faces.

Adaptive refinement by our algorithm is solely driven by the subdivision of edges which are shared by neighboring triangles. However, it might happen that there are surface or volume criteria which may enforce a subdivision although the edges do not satisfy the criterion of subdivision. For the visualization of implicit algebraic surfaces, for example, a tetrahedron may contain a zero contour of the defining function which does not intersect its edges. A solution for triangle subdivision may be a volumetric subdivision by placing a new point, for instance its centroid, inside the triangle and connect it to the vertices of the triangle (Fig. 4, left). For tetrahedra, the analogous subdivision is to insert a point inside the tetrahedron, and connect it to the four vertices (Fig. 4, middle). If in the tetrahedral case an additional criterion of subdivision for faces is available, faces which satisfy the criterion are subdivided like the triangle before. If only one face has to be subdivided, the tetrahedral subdivision is obtained by connecting the subdivision point on the face with the vertex of the tetrahedron opposite to it (Fig. 4, right). If more than one face has to be subdivided, one of the faces is selected in order to subdivide the tetrahedron according to this rule. In the resulting subdivision at most one of the subdivided faces is incident to the new tetrahedra. If one is incident, the rule is applied again to the respective tetrahedron. The volume subdivision criterion is only applied if none of the faces has to be subdivided, which in turn are only subdivided if none of the edges of the tetrahedron has to be split.

The problem with the approach just sketched is that the triangles and tetrahedra tend to become thin. If the breadth-first algorithm mentioned in sect. 2 is used rather than the depth-first approach, the unmodified edge-driven scheme may also be applied by deriving additional edge criteria from surface or volume criteria. Such criteria have also been discussed for visualization of implicit surfaces [8].

References

1. T.J. Baker, *Developments and trends in three dimensional mesh generation,* Applied Numerical Mathematics **5** (1989), 275–304.
2. R.E. Bank, A.H. Sherman, and H. Weiser, *Refinement algorithms and data structures for regular local mesh refinement,* Scientific Computing, (R. Stepleman et al., eds.), IMACS North-Holland, Amsterdam, 1983, pp. 3–17.
3. J. Bey, *Tetrahedral Grid Refinement,* Computing **55** (1995), 355–378.
4. F. Bornemann, B.Erdmann, and R. Kornhuber, *Adaptive multilevel-methods in three space dimensions,* Preprint SC 92–14, Konrad-Zuse-Zentrum für Informationstechnik Berlin (ZIB), July 1992.
5. P.L. George, *Automatic Mesh Generation,* John Wiley & Sons Ltd., 1991.
6. C. Giertsen, *Volume Visualization of sparse irregular meshes,* IEEE Computer Graphics and Applications **12**:2 (1992), 40–48.
7. R. Grosso, Ch. Lürig, and Th. Ertl, *Adaptive Multilevel Finite Elements for Mesh Optimization and Visualization of Volume Data,* to appear in: Proc. Visualization '97, IEEE Computer Science Press, 1997
8. M. Hall and J. Warren, *Adaptive polygonalization of implicitly defined surfaces,* IEEE Computer Graphics and Applications, Nov. 1990, 33–42.
9. L.R. Hermann, *Laplacian-isoparametric grid generation scheme,* Journal of the Engineering Mechanics Division of the American Society of Civil Engineers **102**/EM5 (1976), 749–756.
10. W.E. Lorensen and H.E. Cline, *Marching cubes: A high resolution 3D surface construction algorithm,* Computer Graphics **21**:4 (1987), 163–169.

11. N. Max, B. Becker, and R. Crawfis, *Flow volumes for interactive vector field visualization,* Proceedings Visualization '93, (G.M. Nielson and D. Bergeron, eds.), IEEE Computer Science Press, 1993, pp. 19–24.

12. N. Max, P. Hanrahan, and R. Crawfis, *Area and Volume Coherence for Efficient Visualization of 3D Scalar Functions,* Computer Graphics **24**:5 (1990), 27–33.

13. H. Müller and M. Stark, *Adaptive generation of surfaces in volume data,* The Visual Computer **9** (1993), 182–199.

14. A. Paoluzzi, F. Bernardini, C. Cattani, and V. Ferrucci, *Dimension-Independent Modeling with Simplicial Complexes,* ACM Transactions on Graphics **12** (1993), 56–102.

15. D. Ruprecht, R. Nagel, and H. Müller, *Spatial free form deformation with scattered data interpolation methods,* Computers & Graphics **19** (1995), 63–71.

16. M.F.W. Schmidt, *Cutting cubes – visualizing implicit surfaces by adaptive polygonalization,* The Visual Computer **10** (1993), 101–115.

17. P. Shirley and A. Tuchman, *A Polygonal Approximation to Direct Scalar Volume Rendering,* Computer Graphics **24**:5 (1990), 63–70.

18. C. Silva, J. Mitchell, and A. Kaufman, *Fast rendering of irregular grids.* Proceedings ACM Symposium on Volume Visualization, 1996, pp. 15–23.

19. J. Wilhelms, A. Gelder, P. Tarantino, and J. Gibbs, *Hierarchical and parallelizable direct volume rendering of irregular and multiple grids,* Proceedings IEEE Visualization '96, 1996, pp. 57–65.

20. R. Yagel, D. Reed, A. Law, P.-W. Shih, and N. Shareef, *Hardware assisted volume rendering of unstructured grids by incremental slicing,* Proceedings ACM Symposium on Volume Visualization 1996, pp. 55–63.

21. S. Zhang, *Multilevel Iterative Techniques,* PhD thesis, Pennsylvania State University, 1988.

Part II

Geometry and Numerics

Finite Element Approximations and the Dirichlet Problem for Surfaces of Prescribed Mean Curvature

Gerhard Dziuk[1] and John E. Hutchinson[2]

[1] Institut für Angewandte Mathematik, Universität Freiburg, Germany
[2] School of Mathematical Sciences, Australian National University, Australia

Abstract. We give a finite element procedure for the Dirichlet Problem corresponding to surfaces of prescribed mean curvature and prove an optimal convergence estimate in the H^1-norm.

1991 *Mathematics Subject Classification.* Primary: 65N30; Secondary: 49Q5, 53A10

1 *H*-Harmonic Maps

The numerical solution of the classical H-Plateau Problem consists of approximating disc-like surfaces with prescribed boundary curve and prescribed mean curvature H. For a detailed discussion of the algorithms and theory see [6] for the case of zero mean curvature, and [7] for the constant mean curvature case. In this paper we consider the associated H-Dirichlet problem.

Estimates for finite element approximations to solutions of general nonlinear elliptic systems are obtained in [4], using a continuity method involving L^∞ estimates for the discrete problem. Here we give a much shorter proof of the H^1 estimate, avoiding the need for L^∞ estimates and only assuming the discrete and smooth data are close in the H^1 sense. Our techniques apply to a wide class of nonlinear systems. We treat the case of a non-polygonal and non-convex boundary and give the explicit dependence on the non-degeneracy constant of the smooth solution being approximated. The arguments are prototypes of those used in [7] for treating the more difficult case of the (free boundary) H-Plateau Problem. The main tool for avoiding L^∞ norms in the present "borderline" case is the isoperimetric inequality due to Rado, see Remark 10.

Throughout, $\Omega\ (\subset \mathbb{R}^2)$ is a bounded domain with C^2 boundary. Function spaces will consist of functions defined over Ω with values in $\mathbb{R}^3$ unless otherwise clear from context. Constants will depend on Ω and other quantities as indicated.

By $|\cdot|_{H^1}$ is meant the H^1 seminorm, and by $\|\cdot\|_{H^1}$ the full norm. Note that by Poincaré's inequality, $|\cdot|_{H^1(\Omega)}$ is a norm on $H^1_0(\Omega)$.

* This work was partially supported by the Australian Research Council.

For vectors $a, b, c \in \mathbb{R}^3$, the *triple product* is defined by

$$[a, b, c] = a \cdot b \times c.$$

This is invariant under cyclic permutations of a, b and c, and anti-symmetric with respect to interchanging any two. It is the volume of the parallelopiped spanned by a, b and c.

Definition 1. Suppose H is a real number. A function $u \in H^2(\Omega; \mathbb{R}^3)$ is *H-harmonic* with boundary data $u^0 \in H^2(\Omega; \mathbb{R}^3)$ if

$$\Delta u = 2H u_x \times u_y \qquad \text{a.e. in } \Omega \tag{1}$$

$$u = u^0 \qquad \text{on } \partial\Omega \tag{2}$$

Example 2. Let D be the closed unit disc in $\mathbb{R}^2$. Let

$$u^0(x, y) = (x, y, 0) \colon D \to \mathbb{R}^3$$

with $0 < H < 1$. There are two solutions of (1) and (2) obtained by mapping the unit disc D conformally, i.e. stereographically projecting from a suitable point, onto the *lower* spherical caps obtained from each of the two spheres of radius $1/H$ (mean curvature H) which contain the image of $u^0|_{\partial\Omega}$. These solutions are called *small* or *large* depending on whether their images do not, or do, contain a hemisphere. (We use this example for test computations, see Tables 1 and 2.) If $-1 < H < 0$ then one similarly obtains two solutions from the upper spherical caps. If $H = 0$ then there is exactly one solution, the map $u(x, y) = (x, y, 0) \colon D \to \mathbb{R}^3$. If $H = 1$ then one obtains a solution by mapping onto the lower hemisphere of a sphere of radius 1, and onto the upper hemisphere if $H = -1$.

Equation (1) is the Euler-Lagrange system associated to the *H-Dirichlet integral*

$$D_H(u) = D_H(u; \Omega) = \frac{1}{2} \int_\Omega |\nabla u|^2 + 2HV(u), \tag{3}$$

where

$$V(u) = V(u; \Omega) := \frac{1}{3} \int_\Omega [u, u_x, u_y] \tag{4}$$

can be thought of as the signed volume of the cone over the origin obtained from the image of u. In fact, direct computation and integration by parts easily gives

$$\begin{aligned} \langle D_H{}'(u), \varphi \rangle = \langle D_H{}'(u; \Omega), \varphi \rangle &:= \frac{d}{dt}\bigg|_{t=0} D_H(u + t\varphi) \\ &= \int_\Omega \nabla u \nabla \varphi + 2H \int_\Omega [\varphi, u_x, u_y] \end{aligned} \tag{5}$$

for $u \in C^2(\bar{\Omega}; \mathbb{R}^3)$ and $\varphi \in C_0^2(\bar{\Omega}; \mathbb{R}^3)$, and hence for $u \in H^1 \cap L^\infty(\Omega; \mathbb{R}^3)$ and $\varphi \in H_0^1 \cap L^\infty(\Omega; \mathbb{R}^3)$ by a limit argument, for example see [10, Remark III.1.1]. If $u \in H^1 \cap L^\infty$ is stationary for D_H, i.e.

$$\int_\Omega \nabla u \nabla \varphi + 2H \int_\Omega [\varphi, u_x, u_y] = 0 \tag{6}$$

for all $\varphi \in H_0^1 \cap L^\infty$, then u is said to be a *weak* solution of (1).

Example 2 is fairly typical. Arguing heuristically, the energy functional $D_H(u)$ is cubic in u and thus one expects (generically) either two or no stationary points. In the former case one expects the "smaller" solution to be a local minimum and the "larger" solution to be unstable.

Indeed, one has the following result due to the combined work of Heinz, Werner, Hildebrandt, Jäger, Wente, Brezis–Coron, Struwe and Steffen. For detailed references see Struwe [10,11].

Theorem 3. *Assume $u^0 \in H^1 \cap L^\infty(\Omega; \mathbb{R}^3)$ and $H \in \mathbb{R}$ satisfy*

$$\|u^0\|_{L^\infty} |H| \le 1.$$

Then there exists $\underline{u} \in u^0 + H_0^1 \cap L^\infty$ such that

$$D_H(\underline{u}) = \min \left\{ D_H(v) : v \in u^0 + H_0^1,\ \|v\|_{L^\infty} |H| \le 1 \right\}.$$

Moreover,

$$\|\underline{u}\|_{L^\infty} \le \|u^0\|_{L^\infty} \qquad (*)$$

and $\underline{u}$ is a weak solution to (1) and (2).

If furthermore

$$\|u^0\|_{L^\infty} |H| < 1 \qquad (**)$$

then $\underline{u}$ is the unique *local minimum of D_H in $u^0 + H_0^1 \cap L^\infty$. Moreover, $\underline{u}$ is the unique weak solution of (1) and (2) which satisfies $(*)$. The function $\underline{u}$ is called the* small *solution of (1) and (2).*

Under the same assumption $(**)$ if $H \ne 0$ and u^0 is not constant, there is also a second weak solution $\bar{u}$ to (1) and (2) which satisfies

$$\|\bar{u}\|_{L^\infty} > \|u^0\|_{L^\infty}.$$

Any such solution is called a large *solution to (1) and (2).*

If $u^0 \in H^2(\Omega, \mathbb{R}^3)$ then any weak solution to (1) and (2) belongs to $H^2(\Omega, \mathbb{R}^3)$.

Remark 4.

1. The large solution need not be unique, although one would expect that this is the generic situation. An example of Wente, [11, Example IV.3.7], gives a continuum of solutions for Ω the unit disc and boundary data $u^0(x, y) = (x, 0, 0)$. See Fig. 1 for the image of the trivial small solution and of one of the large solutions on a relatively coarse grid. Rotation of $u(\Omega)$ around the u_1-axis gives a continuum of solutions.
2. The existence of a large solution is obtained by a mountain pass type argument, see [2] and [11, Theorem III.4.8]

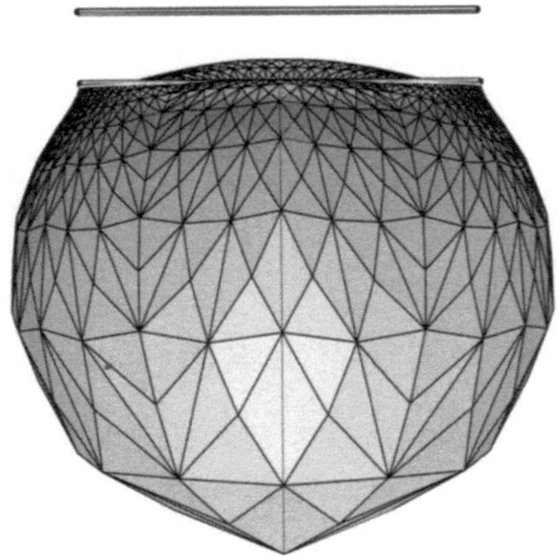

Fig. 1. Wente's example (discrete approximations); small and one of a continuum of large solutions

Remark 5 (Nondegeneracy). We will be interested in approximating functions $u \in H^2(\Omega; \mathbb{R}^3)$ which are H-harmonic and nondegenerate in the sense that the second variation $D_H''(u)$ has no zero eigenvalues. This is always true for small solutions, see [11, Lemma III.4.7].

More precisely, for $u \in H^2(\Omega; \mathbb{R}^3)$ and $\varphi, \psi \in H_0^2(\Omega; \mathbb{R}^3)$ one first easily checks by direct computation and integration by parts that

$$D_H''(u)(\varphi, \psi) = D_H''(u; \Omega)(\varphi, \psi) := \left.\frac{\partial^2}{\partial s\, \partial t}\right|_{s=t=0} D_H(u + t\varphi + s\psi)$$

$$= \int_\Omega \nabla\varphi\nabla\psi + 2H \int_\Omega [u, \varphi_x, \psi_y] + [u, \psi_x, \varphi_y] \tag{7}$$

$$= \int_\Omega \nabla\varphi\nabla\psi + 2H \int_\Omega [\psi, u_x, \varphi_y] + [\psi, \varphi_x, u_y], \tag{8}$$

see [10, Remark III.1.1] and the paragraph following (17). From (8) $D_H''(u)$ extends to a bounded symmetric bilinear functional on H_0^1, since

$$\int_\Omega |\nabla u|\,|\varphi|\,|\nabla\varphi| \le \|\nabla u\|_{L^4}\|\varphi\|_{L^4}\|\nabla\varphi\|_{L^2} \le c\|u\|_{H^2}\|\nabla\varphi\|_{L^2}^2$$

It follows that the inner product $|\cdot|_{H^1}$ induces a bounded self-adjoint linear operator $\nabla^2 D_H(u)\colon H_0^1 \to H_0^1$. The eigenvalues of $\nabla^2 D_H(u)$ are real, bounded below by $-\lambda_0$ (say) and have no accumulation point. Moreover, $\lambda_0 = \lambda_0(\Omega, H, \|u\|_{H^2})$, as follows from using (8) to estimate the Raleigh-Ritz quotient.

The *nondegeneracy constant* λ of $D_H''(u)$ is defined by

$$\lambda = \min\{\, |\gamma| : \gamma \text{ is an eigenvalue of } \nabla^2 D_H(u)\,\}$$

Then $\nabla^2 D_H(u)$ is one-one and onto iff $\lambda > 0$. Let

$$\varphi = \varphi^+ + \varphi^- \tag{9}$$

denote the $|\cdot|_{H^1}$ orthogonal decomposition of $\varphi \in H_0^1$ into members of the positive and negative spaces H^+ and H^- corresponding to the eigenvalues and eigenfunctions of $\nabla^2 D_H(u)$. Then

$$D_H''(u)(\varphi, \varphi^+ - \varphi^-) \ge \lambda|\varphi|_{H^1}^2 \tag{10}$$

for all such φ and λ is the largest such real.

From the eigenfunction equation, c.f. (8), together with the estimate for λ_0, one obtains $\varphi^- \in H^2$ and

$$\|\varphi^-\|_{H^2} \le \nu|\varphi^-|_{H^1} \tag{11}$$

where $\nu = \nu(\Omega, H, \|u\|_{H^2}, d)$ with d the dimension of H^-.

2 Discrete H-Harmonic Maps

For $h > 0$ let $\mathcal{T}_h$ be a triangulation of Ω by triangles T whose side lengths are bounded above by ch for some c independent of h and whose interior angles are bounded away from zero uniformly and independently of h. The intersection of any two different triangles is either empty, a common vertex, or a common edge.

Let

$$\Omega_h = \bigcup_{T\in\mathcal{T}_h} T.$$

Let

$$X_h = \{\, u_h \in C^0(\Omega_h;\mathbb{R}^3) : u_h|_T \in \mathbb{P}_1(T) \quad \forall T \in \mathcal{T}_h \,\},$$
$$X_{h0} = \{\, \varphi_h \in X_h : \varphi_h|_{\partial\Omega_h} = 0 \,\},$$

where $\mathbb{P}_1(T)$ is the set of polynomials over T of degree at most one.

For some $\delta > 0$ and all sufficiently small h,

$$\Omega' := \{\, \mathbf{x} \in \mathbb{R}^2 : d(\mathbf{x}, \Omega) < \delta \,\} \supset \Omega_h \cup \Omega.$$

If $u \in H^2(\Omega)$ then by the C^2 regularity of Ω there exists an extension of u to Ω', also denoted by u, such that

$$\|u\|_{H^2(\Omega')} \le c\|u\|_{H^2(\Omega)}. \tag{12}$$

Definition 6. The *discrete H-Dirichlet integral* is defined by

$$D_H(u_h; \Omega_h) = \frac{1}{2}\int_{\Omega_h} |\nabla u_h|^2 + 2HV(u_h; \Omega_h)$$

for $u_h \in X_h$.

It follows from (5) and (7) with Ω replaced by Ω_h and a limit argument, or by direct computation and noting that boundary integrals on internal edges cancel, that

$$\langle D'_H(u_h; \Omega_h), \varphi_h\rangle = \int_{\Omega_h} \nabla u_h \nabla \varphi_h + 2H\int_{\Omega_h} [\varphi_h, u_{hx}, u_{hy}], \tag{13}$$

$$D_H{}''(u_h; \Omega_h)(\varphi_h, \psi_h) =$$

$$\int_{\Omega_h} \nabla \varphi_h \nabla \psi_h + 2H\int_{\Omega_h} [u_h, \varphi_{hx}, \psi_{hy}] + [u_h, \psi_{hx}, \varphi_{hy}], \tag{14}$$

for $u_h \in X_h$ and $\varphi_h, \psi_h \in X_{h0}$.

Motivated by (6), one has

Definition 7. A function $u_h \in X_h$ is *discrete H-harmonic* if

$$\int_{\Omega_h} \nabla u_h \nabla \varphi_h + 2H\int_{\Omega_h} [u_{hx}, u_{hy}, \varphi_h] = 0, \tag{15}$$

for all $\varphi_h \in X_{h0}$.

We will prove the following.

Theorem 8. *Let $u \in H^2(\Omega; \mathbb{R}^3)$ be H-harmonic and $u = u^0$ on $\partial\Omega$ where $u^0 \in H^2(\Omega; \mathbb{R}^3)$. Assume u is nondegenerate with nondegeneracy constant λ.*

Let $u_h^0 \in X_h$ and assume $\|u^0 - u_h^0\|_{H^1(\Omega_h)} \le \alpha h$.

Then there exist constants $h_0 = h_0(\|u\|_{H^2}, \|u^0\|_{H^2}, \alpha, \Omega, d, H, \lambda)$, $\varepsilon_0 = \varepsilon_0(H, \lambda)$, and $c_0 = c_0(\|u\|_{H^2}, \|u^0\|_{H^2}, \alpha, H)$ such that if $0 < h \le h_0$ then:

1. *There exists a* unique *discrete H-harmonic function u_h such that $u_h = u_h^0$ on $\partial\Omega_h$ and*

$$\|u - u_h\|_{H^1(\Omega_h)} \le \varepsilon_0;$$

2. *Moreover,*

$$\|u - u_h\|_{H^1(\Omega_h)} \le c_0\lambda^{-1}h.$$

3 Proof of Main Theorem

With u, u^0 and u_h^0 as in the main theorem *define*

$$J_h u = u_h^0 + I_h(u - u^0) \in u_h^0 + X_{h0}, \tag{16}$$

where I_h is the standard nodal interpolation operator.

The proof of the main theorem will use the following quantitative version of the Inverse Function Theorem with $\mathcal{X} = u_h^0 + X_{h0}$, $X = X_{h0}$, $Y = X_{h0}^*$ (the dual space of X_{h0}), $x_0 = J_h u$, $f = D'_H(\cdot\,;\Omega_h)$. The proof of the lemma follows from that in [1] pp 113–114.

Lemma 9. *Let $\mathcal{X}$ be an affine Banach space with Banach space X as tangent space, and let Y be a Banach space. Suppose $x_0 \in \mathcal{X}$ and $f \in C^1(\mathcal{X}, Y)$. Assume there are positive constants α, β, δ and ε such that*

$$\begin{aligned} \|f(x_0)\|_Y &\le \delta, \\ \|f'(x_0)^{-1}\|_{L(Y,X)} &\le \alpha^{-1}, \\ \|f'(x) - f'(x_0)\|_{L(X,Y)} &\le \beta \quad \textit{for all } x \in \bar{B}_\varepsilon(x_0), \end{aligned}$$

where

$$\beta < \alpha, \quad \delta \le (\alpha - \beta)\varepsilon.$$

Then there exists a unique $x_* \in \bar{B}_\varepsilon(x_0)$ such that $f(x_*) = 0$.

Remark 10 (The Volume Functional). A fundamental result due to Wente [12] is that for any $u^0 \in H^1 \cap L^\infty(\Omega; \mathbb{R}^3)$ the functional V (and hence D_H) extends to an analytic functional on the affine space $u^0 + H_0^1(\Omega; \mathbb{R}^3)$. This is perhaps surprising, since from (5) and (7) one might expect bounds for the relevant integrals to also involve $\|\varphi\|_{L^\infty}$ and $\|u\|_{L^\infty}$ respectively.

More generally, one has the following.

For $u, v, w \in H^1 \cap L^\infty(\Omega; \mathbb{R}^3)$ *define* the trilinear functional

$$V(u, v, w) = V(u, v, w; \Omega) = \frac{1}{6} \int_\Omega [u, v_x, w_y] + [u, w_x, v_y]. \tag{17}$$

Note that $V(u) = V(u, u, u)$.

Assume now that *at least one* of u, v, w also belongs to $H_0^1(\Omega; \mathbb{R}^3)$. Then V is invariant under cyclic permutations of its arguments, as follows from integration by parts in the C^2 case and in general by a limit argument, see [10, Remark III.1.1.iii]. Since V is invariant under permutation of its first two arguments, it then follows it is invariant under *any* permutation of its arguments. Moreover under the same assumptions, from an argument similar to that in [10, proof of Theorem III.2.3] which uses an isoperimetric inequality due to Radó [8], one also has

$$V(u, v, w) \le c|u|_{H^1(\Omega)} |v|_{H^1(\Omega)} |w|_{H^1(\Omega)}. \tag{18}$$

Similar remarks and estimates apply if Ω is everywhere replaced by Ω_h.

Assume now $u, v \in H^1 \cap L^\infty(\Omega; \mathbb{R}^3)$ and $\varphi \in H_0^1 \cap L^\infty(\Omega; \mathbb{R}^3)$. It follows that

$$\langle V'(u), \varphi\rangle = 3V(u,u,\varphi) \le c|u|^2_{H^1(\Omega)}|\varphi|_{H^1(\Omega)}, \tag{19}$$

$$V''(u)(v,\varphi) = 6V(u,v,\varphi) \le c|u|_{H^1(\Omega)}|v|_{H^1(\Omega)}|\varphi|_{H^1(\Omega)}, \tag{20}$$

$$V'''(\cdot)(u,v,\varphi) = 6V(u,v,\varphi) \le c|u|_{H^1(\Omega)}|v|_{H^1(\Omega)}|\varphi|_{H^1(\Omega)}. \tag{21}$$

(In particular, if $u \in H^1 \cap L^\infty(\Omega; \mathbb{R}^3)$ these estimates allow one to define the integrals in (5) and (7) for arbitrary $\varphi, \psi \in H_0^1(\Omega; \mathbb{R}^3)$.) Similar results also hold if Ω is replaced by Ω_h.

For the remainder of this section, u is as in the Main Theorem. Extend u to Ω' as in (12) and restrict to Ω_h as necessary. Both the extension and restriction will also be denoted by u.

Lemma 11.

$$\|u - J_h u\|_{H^1(\Omega_h)} \le c_1 h$$

where $c_1 = c_1(\|u\|_{H^2(\Omega)}, \|u^0\|_{H^2(\Omega)}, \alpha)$.

Proof.

$$\begin{aligned}\|u - J_h u\|_{H^1(\Omega\cap\Omega_h)} &= \left\|(u^0 - u_h^0) - (I_h(u-u^0) - (u-u^0))\right\|_{H^1(\Omega\cap\Omega_h)}\\ &\le \|u^0 - u_h^0\|_{H^1(\Omega\cap\Omega_h)} + ch|u-u^0|_{H^2(\Omega)}\\ &\le ch,\end{aligned}$$

where $c = c(\|u\|_{H^2}, \|u^0\|_{H^2}, \alpha)$. Since

$$\|u\|_{H^1(\Omega_h\setminus\Omega)} \le ch\|u\|_{H^2(\Omega_h\setminus\Omega)} \le ch\|u\|_{H^2(\Omega)}$$

by elementary estimates and (12), the result follows.

Lemma 12. *If* $\varphi_h \in X_{h0}$ *then*

$$|\langle D'_H(J_h u; \Omega_h), \varphi_h\rangle| \le c_2 h|\varphi_h|_{H^1(\Omega_h)},$$

where $c_2 = c_2(\|u\|_{H^2(\Omega)}, \|u^0\|_{H^2(\Omega)}, \alpha, H)$.

Proof.

$$\begin{aligned}&\langle D'_H(J_h u; \Omega_h), \varphi_h\rangle\\ &\quad = \left(\langle D'_H(J_h u; \Omega_h), \varphi_h\rangle - \langle D'_H(u; \Omega_h), \varphi_h\rangle\right) + \langle D'_H(u; \Omega_h), \varphi_h\rangle\\ &\quad =: A + B\end{aligned}$$

From the Taylor series expansion for $V'(\cdot\,;\Omega_h)$ and Remark 10

$$\begin{aligned}|A| &= \left|\int_{\Omega_h} \nabla(J_h u - u)\nabla\varphi_h + 2H\langle V'(J_h u;\Omega_h),\varphi_h\rangle - \langle V'(u;\Omega_h),\varphi_h\rangle\right| \\ &\le |J_h u - u|_{H^1(\Omega_h)}|\varphi_h|_{H^1(\Omega_h)} + 2|H||V''(u;\Omega_h)(J_h u - u,\varphi_h)| \\ &\quad + |H||V'''(u;\Omega_h)(J_h u - u, J_h u - u,\varphi_h)| \\ &\le |J_h u - u|_{H^1(\Omega_h)}|\varphi_h|_{H^1(\Omega_h)} \\ &\quad + c|H||u|_{H^1(\Omega_h)}|J_h u - u|_{H^1(\Omega_h)}|\varphi_h|_{H^1(\Omega_h)} \\ &\quad + c|H||J_h u - u|^2_{H^1(\Omega_h)}|\varphi_h|_{H^1(\Omega_h)} \\ &\le ch|\varphi_h|_{H^1(\Omega_h)},\end{aligned}$$

from Lemma 11 and (12), where $c = c(\|u\|_{H^2(\Omega)}, \|u^0\|_{H^2(\Omega)}, \alpha, H)$.

Also,

$$\begin{aligned}B &= \left|\int_{\Omega_h} \nabla u\nabla\varphi_h + 2H\int_{\Omega_h}[\varphi_h, u_x, u_y]\right| = \left|\int_{\Omega_h}(-\Delta u + 2Hu_x \times u_y)\cdot\varphi_h\right| \\ &= \left|\int_{\Omega_h\setminus\Omega}(-\Delta u + 2Hu_x \times u_y)\cdot\varphi_h\right| \le c\|\varphi_h\|_{L^2(\Omega_h\setminus\Omega)} \le ch\|\varphi_h\|_{H^1(\Omega_h)}\end{aligned}$$

where $c = c(\|u\|_{H^2}, H)$, as follows from (12), a Sobolev imbedding theorem, and elementary calculus.

The required result follows.

Remark 13 (A "Discrete Eigenspace" Decomposition). If $\varphi_h \in X_{h0}$ let φ_h also denote the zero extension to $\Omega \cup \Omega_h$. Note that $\varphi_h \notin H_0^1(\Omega)$ unless Ω is convex. For this reason *define* $P : X_{h0} \to H_0^1(\Omega)$ to be the $|\cdot|_{H^1}$ projection, i.e.

$$\int_\Omega \nabla(P\varphi_h)\nabla\varphi = \int_\Omega \nabla\varphi_h\nabla\varphi$$

for all $\varphi \in H_0^1(\Omega)$.

One has

$$|\varphi_h|_{H^1(\Omega_h\setminus\Omega)} \le ch^{1/2}|\varphi_h|_{H^1(\Omega_h)} \tag{22}$$

$$|P\varphi_h|_{H^1(\Omega)} \le |\varphi_h|_{H^1(\Omega_h)} \tag{23}$$

$$|\varphi_h - P\varphi_h|_{H^1(\Omega)} \le ch^{1/2}|\varphi_h|_{H^1(\Omega_h)} \tag{24}$$

To see (22) note that $\nabla\varphi_h$ is constant on any triangle $T \in \mathcal{T}_h$ and that $|T \cap (\Omega_h \setminus \Omega)| \le ch|T|$. Inequality (23) is immediate, since P is just $|\cdot|_{H^1}$ orthogonal projection onto $H_0^1(\Omega)$. For (24) first note that $|\varphi_h - P\varphi_h|_{H^1(\Omega)} \le |\varphi_h - \varphi|_{H^1(\Omega)}$ for any $\varphi \in H_0^1(\Omega)$, by orthogonality. Now choose φ by suitably deforming φ_h in a boundary strip.

Let $(P\varphi_h)^+, (P\varphi_h)^- \in H_0^1(\Omega)$ be the components of $P\varphi_h$ as in (9). Note that $(P\varphi_h)^-$ is smooth, and in particular

$$\|(P\varphi_h)^-\|_{H^2(\Omega)} \le \nu|\varphi_h|_{H^1(\Omega_h)} \tag{25}$$

since

$$\|(P\varphi_h)^-\|_{H^2(\Omega)} \le \nu|(P\varphi_h)^-|_{H^1(\Omega)} \le \nu|P\varphi_h|_{H^1(\Omega)} \le \nu|\varphi_h|_{H^1(\Omega_h)}$$

from (11), (9) and the $|\cdot|_{H^1}$-orthogonality of $(P\varphi_h)^+$ and $(P\varphi_h)^-$, and (23).

Define a discrete analogue of (9) by

$$\begin{aligned} \varphi_h^{(-)} = I_h(P\varphi_h)^- \in X_{h0}, \quad \varphi_h^{(+)} = \varphi_h - \varphi_h^{(-)}, \\ \varphi_h = \varphi_h^{(+)} + \varphi_h^{(-)}. \end{aligned} \tag{26}$$

Taking the zero extension of $(P\varphi_h)^-$ and $(P\varphi_h)^+$ to Ω_h, and of $\varphi^{(-)}$ and $\varphi^{(+)}$ to Ω, we claim

$$\begin{aligned} |(P\varphi_h)^- - \varphi_h^{(-)}|_{H^1(\Omega\cup\Omega_h)} \le ch|\varphi_h|_{H^1(\Omega_h)}, \\ |(P\varphi_h)^+ - \varphi_h^{(+)}|_{H^1(\Omega\cup\Omega_h)} \le ch^{1/2}|\varphi_h|_{H^1(\Omega_h)}, \end{aligned} \tag{27}$$

where $c = c(\nu)$

Proof of claim.

$$|(P\varphi_h)^- - \varphi_h^{(-)}|_{H^1(\Omega_h)} \le ch|(P\varphi_h)^-|_{H^2(\Omega)} \le ch\nu|\varphi_h|_{H^1(\Omega_h)}$$

from (25). Also

$$|(P\varphi_h)^-|_{H^1(\Omega\setminus\Omega_h)} \le ch|(P\varphi_h)^-|_{H^2(\Omega)} \le ch\nu|\varphi_h|_{H^1(\Omega_h)}.$$

This gives the first result.

For the second,

$$\begin{aligned} &|(P\varphi_h)^+ - \varphi_h^{(+)}|_{H^1(\Omega_h)} \\ &\quad \le |P\varphi_h - \varphi_h|_{H^1(\Omega_h)} + |(P\varphi_h)^- - \varphi_h^{(-)}|_{H^1(\Omega_h)} \le c(h^{1/2} + h\nu)|\varphi_h|_{H^1(\Omega_h)} \end{aligned}$$

from the first result and (24). On $\Omega \setminus \Omega_h$, $\varphi_h = \varphi_h^{(+)} = 0$ and so the required estimate now follows from (24). □

We also have

$$\begin{aligned} |\varphi_h^{(-)}|_{H^1(\Omega_h)} \le (1 + ch)|\varphi_h|_{H^1(\Omega_h)} \\ |\varphi_h^{(+)}|_{H^1(\Omega_h)} \le (1 + ch^{1/2})|\varphi_h|_{H^1(\Omega_h)} \end{aligned} \tag{28}$$

from (27), the orthogonal decomposition $P\varphi_h = (P\varphi_h)^- + (P\varphi_h)^+$ and (23).

Thus (26) is an "almost orthogonal" decomposition for small h.

Lemma 14. *If $\varphi_h \in X_{h0}$ then*

$$D_H''(J_h u; \Omega_h)(\varphi_h, \varphi_h^{(+)} - \varphi_h^{(-)}) \ge \frac{3\lambda}{4}|\varphi_h|^2_{H^1(\Omega_h)}$$

provided $h \le h_1$ where $h_1 = h_1(\|u\|_{H^2(\Omega)}, \|u^0\|_{H^2(\Omega)}, \alpha, \Omega, d, H, \lambda)$.

Proof. Since $V(\cdot\,; \Omega_h)$ is cubic, from (14)

$$D''_H(J_h u; \Omega_h)(\varphi_h, \varphi_h^{(+)} - \varphi_h^{(-)}) \\ = D_H''(u; \Omega_h)(\varphi_h, \varphi_h^{(+)} - \varphi_h^{(-)}) + 2HV'''(u; \Omega_h)(J_h u - u, \varphi_h, \varphi_h^{(+)} - \varphi_h^{(-)}).$$

But

$$\left|2HV'''(u; \Omega_h)(J_h u - u, \varphi_h, \varphi_h^{(+)} - \varphi_h^{(-)})\right| \\ \leq c|J_h u - u|_{H^1(\Omega_h)} |\varphi_h|_{H^1(\Omega_h)} |\varphi_h^{(+)} - \varphi_h^{(-)}|_{H^1(\Omega_h)} \leq ch|\varphi_h|^2_{H^1(\Omega_h)}$$

where $c = c(\|u\|_{H^2(\Omega)}, \|u^0\|_{H^2(\Omega)}, \alpha, \nu, H)$, from Remark 10, also Lemma 11 and (28).

Now

$$D_H''(u; \Omega_h)(\varphi_h, \varphi_h^{(+)} - \varphi_h^{(-)}) = D_H''(u; \Omega)(\varphi_h, \varphi_h^{(+)} - \varphi_h^{(-)}) + E_1$$

where

$$|E_1| \leq c(1 + \|u\|_{L^\infty}) |\varphi_h|_{H^1(\Omega_h \setminus \Omega)} |\varphi_h^{(+)} - \varphi_h^{(-)}|_{H^1(\Omega_h \setminus \Omega)} \leq ch|\varphi_h|^2_{H^1(\Omega_h)}$$

with $c = c(\|u\|_{H^2(\Omega)}, \nu, H)$, from (22) and (27), since $(P\varphi_h)^+ = (P\varphi_h)^- = 0$ in $\Omega_h \setminus \Omega$.

Also

$$\begin{aligned} D_H''(u; \Omega)&(\varphi_h, \varphi_h^{(+)} - \varphi_h^{(-)}) \\ &= D_H''(u; \Omega)(P\varphi_h, (P\varphi_h)^+ - (P\varphi_h)^-) \\ &\quad + D_H''(u; \Omega)(\varphi_h - P\varphi_h, (P\varphi_h)^+ - (P\varphi_h)^-) \\ &\quad\quad + D_H''(u; \Omega)(\varphi_h, (\varphi_h^{(+)} - (P\varphi_h)^+) - (\varphi_h^{(-)} - (P\varphi_h)^-)) \\ &\geq \lambda |P\varphi_h|^2_{H^1(\Omega)} + E_2 + E_3 \end{aligned}$$

from (10). But

$$|E_2|, |E_3| \leq c(1 + \|u\|_{L^\infty}) h^{1/2} |\varphi_h|^2_{H^1(\Omega_h)}$$

from (24) and (23), and (27) respectively.

It follows that

$$\begin{aligned} D''_H(J_h u; \Omega_h)(\varphi_h, \varphi_h^{(+)} - \varphi_h^{(-)}) &\geq \lambda |P\varphi_h|^2_{H^1(\Omega)} - ch^{1/2} |\varphi_h|^2_{H^1(\Omega_h)} \\ &\geq \frac{3\lambda}{4} |\varphi_h|^2_{H^1(\Omega_h)} \end{aligned}$$

from (24), for $h \leq h_1 = h_1(\|u\|_{H^2(\Omega)}, \|u^0\|_{H^2(\Omega)}, \alpha, \Omega, d, H, \lambda)$.

Lemma 15. *If* $v_h \in u_h^0 + X_{h0}$ *and* $\varphi_h, \psi_h \in X_{h0}$ *then*

$$|D''_H(v_h; \Omega_h)(\varphi_h, \psi_h) - D''_H(J_h u; \Omega_h)(\varphi_h, \psi_h)| \leq \frac{\lambda}{4} |\varphi_h|_{H^1(\Omega_h)} |\psi_h|_{H^1(\Omega_h)}$$

provided $|v_h - J_h u|_{H^1(\Omega_h)} \leq \varepsilon_1$ *where* $\varepsilon_1 = \varepsilon_1(H, \lambda)$.

Proof. This follows from

$$\begin{aligned} D''_H(v_h;\Omega_h)(\varphi_h,\psi_h) - D''_H(J_h u;\Omega_h)(\varphi_h,\psi_h) \\ = 2HV_h'''(J_h u;\Omega_h)(v_h - J_h u,\varphi_h,\psi_h), \end{aligned}$$

(21) and Lemma 11.

Completion of proof of Main Theorem. We use Lemma 9 with $\mathcal{X} = u_h^0 + X_{h0}$, $X = X_{h0}$, $Y = X_{h0}^*$, $x_0 = J_h u$, $f = D'_H(\cdot\,;\Omega_h)$. The norm on X_{h0} is $|\cdot|_{H^1(\Omega_h)}$ and on X_{h0}^* is the corresponding dual norm. Note that

$$D'_H(\cdot\,;\Omega_h)\colon u_h^0 + X_{h0} \to X_{h0}^*$$

with derivative

$$D''_H(\cdot\,;\Omega_h)\colon u_h^0 + X_{h0} \to L(X_{h0}, X_{h0}^*)$$

using standard identifications.

From Lemma 12

$$\|D'_H(J_h u;\Omega_h)\| \le c_2 h. \tag{29}$$

From Lemma 14, $D''_H(J_h u;\Omega_h)$ is invertible and

$$\left\|[D''_H(J_h u;\Omega_h)]^{-1}\right\| \le \left(\frac{3\lambda}{4}|\varphi_h|_{H^1(\Omega_h)}\Big/|\varphi_h^{(+)} - \varphi_h^{(-)}|_{H^1(\Omega_h)}\right)^{-1}$$

provided $h \le h_1$. But

$$\begin{aligned} |\varphi_h^{(+)} - \varphi_h^{(-)}|_{H^1(\Omega_h)} &\le |(P\varphi_h)^+ - (P\varphi_h)^-|_{H^1(\Omega_h)} + |\varphi_h^{(+)} - (P\varphi_h)^+|_{H^1(\Omega_h)} \\ &\quad + |\varphi_h^{(-)} - (P\varphi_h)^-|_{H^1(\Omega_h)} \\ &\le (1 + ch^{1/2})|\varphi_h|_{H^1(\Omega_h)} \end{aligned}$$

where $c = c(\nu)$, from (23) and (27). Hence

$$\left\|[D''_H(J_h u;\Omega_h)]^{-1}\right\| \le \left(\frac{\lambda}{2}\right)^{-1} \tag{30}$$

if $h \le h_3$ where $h_3 = h_3(\|u\|_{H^2(\Omega)}, \|u^0\|_{H^2(\Omega)}, \alpha, \Omega, d, H, \lambda)$.

Finally, from Lemma 15

$$\|D''_H(v_h;\Omega_h) - D''_H(J_h u;\Omega_h)\| \le \frac{\lambda}{4} \tag{31}$$

if $|v_h - J_h u|_{H^1(\Omega_h)} \le \varepsilon_1$ where $\varepsilon_1 = \varepsilon_1(H,\lambda)$.

Take $\delta = c_2 h$, $\alpha = \lambda/2$, $\beta = \lambda/4$ and $\varepsilon = \varepsilon_1$. Then from (29)–(31) the hypotheses of Lemma 9 are satisfied provided $h \le h_3$, $c_2 h \le \frac{\lambda}{4}\varepsilon_1$. This establishes the first (uniqueness) part of the main theorem with $\varepsilon_0 = \varepsilon_1$ and $h_0 = h_0(h_3, \varepsilon_1, \lambda, c_2) = h_0(\|u\|_{H^2}, \|u^0\|_{H^2}, \alpha, \nu, H, \lambda)$.

Taking $\delta = c_2 h$, $\alpha = \lambda/2$, $\beta = \lambda/4$ and $\varepsilon = \lambda^{-1} c_0 h$ the hypotheses of Lemma 9 are again satisfied from (29)–(31) provided $h \le h_3$, $c_2 h \le \frac{1}{4} c_0 h$. This establishes the second ($O(h)$ convergence) part of the main theorem with $h_0 = h_3$ and $c_0 = 4c_2$. □

4 Numerical Results

In Tables 1 and 2 we present the results of test computations for the explicitly known spherical solutions described in Example 2 with $H = 0.5$ and $\Omega = B_1(0)$. Denote by e_h the error between the continuous solution and the discrete solution in the chosen norm. For two successive grids with grid sizes h_1 and h_2 the experimental order of convergence is

$$eoc = \ln \frac{e_{h_1}}{e_{h_2}} \Big/ \ln \frac{h_1}{h_2}.$$

The test computations confirm the order 1 for the $H^1(\Omega)$-norm and additionally show the order 2 for the $L^2(\Omega)$-norm.

nodes	level	h	L^2-error	L^2-eoc	H^1-error	H^1-eoc
9	2	1.0000	1.0020e-1	-	0.2607	-
25	4	0.7368	3.9040e-2	3.09	0.1822	1.17
81	6	0.4203	1.0682e-2	2.31	9.6455e-2	1.13
289	8	0.2219	2.6916e-3	2.16	4.8223e-2	1.09
1089	10	0.1137	6.6871e-4	2.08	2.3909e-2	1.05
4225	12	0.05736	1.6621e-4	2.04	1.1876e-2	1.03
16641	14	0.02893	4.1401e-5	2.02	5.9160e-3	1.01

Table 1. Small solution, $H = 0.5$

nodes	level	h	L^2-error	L^2-eoc	H^1-error	H^1-eoc
81	6	0.4203	1.2292	-	6.1915	-
289	8	0.2219	0.4677	1.51	2.9080	1.18
1089	10	0.1137	0.1610	1.60	1.3131	1.19
4225	12	0.05736	0.04707	1.81	0.5870	1.18
16641	14	0.02893	0.01239	1.94	0.2772	1.09

Table 2. Large solution, $H = 0.5$

Figures 2 and 3 show computational results with $\Omega = B_1(0)$, $H = 0.5$ and boundary values $u(e^{i\varphi}) = (\cos(\varphi), \sin(\varphi), (2 + \sqrt{3})\cos(2\varphi) - 0.5\cos(6\varphi))$ on a grid with 8192 triangles. For better visibility the resulting surfaces are scaled, but the boundaries of the solution surfaces are the same.

Figure 4 shows a solution for the annular domain $\Omega = \{x \mid 1 < |x| < 2\}$ and boundary data which give knotted boundary curves.

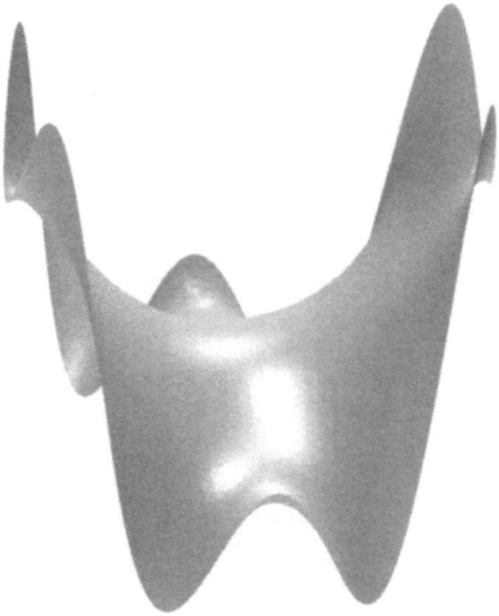

Fig. 2. Small solution, $H = 0.5$

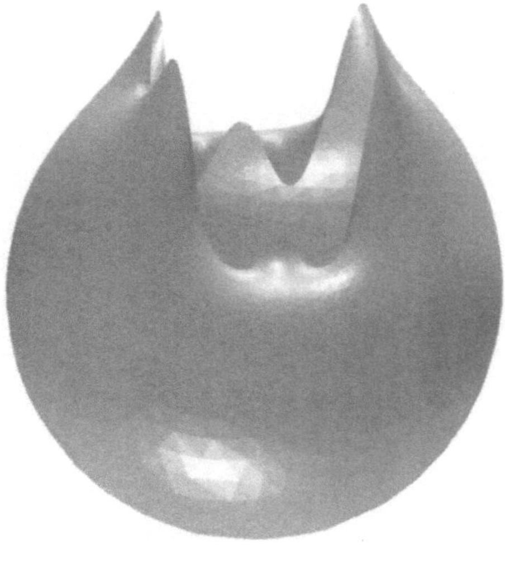

Fig. 3. Large solution, $H = 0.5$

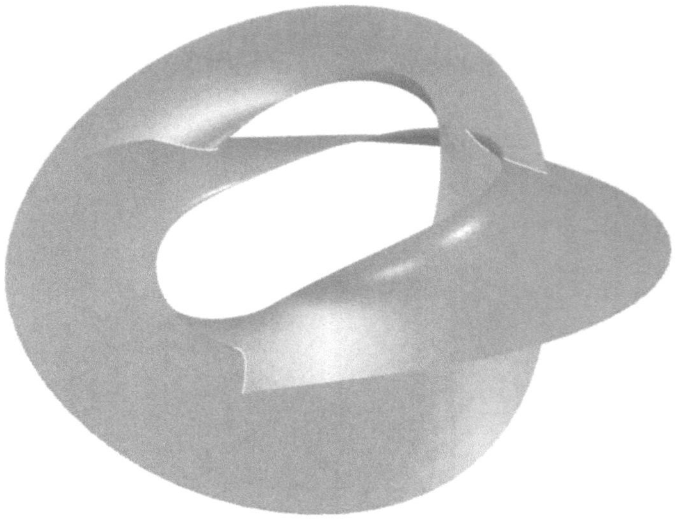

Fig. 4. Annulus, $H = 0.5$

References

1. M. BERGER, *Nonlinearity and Functional Analysis*, Academic Press, 1977.
2. H. BREZIS, J.-M. CORON, *Multiple solutions of H-systems and Rellich's conjecture*, Comm. Pure Appl. Math. *37*, 1984, 149–187.
3. P. G. CIARLET, *The Finite Element Methods for Elliptic Problems*, North Holland, 1978.
4. M. DOBROWOLSKI, R. RANNACHER *Finite element methods for nonlinear elliptic systems of second order*, Math. Nachr. *94*, 1974, 155–172
5. G. DZIUK, J.E. HUTCHINSON, *On the approximation of unstable parametric minimal surfaces*, Calc. Var. *4*, 1996, 27–58.
6. G. DZIUK, J.E. HUTCHINSON, *A finite element method for approximating minimal surfaces*, Preprint *4* Mathematische Fakultät Freiburg, 1996 CMA Math. Res. Rep. *5*, Australian National University, 1996.
7. G. DZIUK, J.E. HUTCHINSON, *A finite element method for approximating surfaces of prescribed mean curvature*, in preparation.
8. T. RADÓ, *The isoperimetric inequality and the Lebesgue definition of surface area*, Trans. Amer. Math. Soc. *61*, 1947, 530–555.
9. M. RUMPF, A. SCHMIDT ET AL, *GRAPE, Graphics Programming Environment*, Report 8, SFB 256, Bonn (1990).
10. M. STRUWE, *Plateau's Problem and the Calculus of Variations*, Princeton University Press, 1988.
11. M. STRUWE, *Variational Methods*, Springer Verlag, 1990.
12. H. C. WENTE, *An existence theorem for surfaces of constant mean curvature*, J. Math. Anal. Appl. *26*, 1969, 318–344.

Efficient Volume-Generation During the Simulation of NC-Milling

Georg Glaeser[1] and Eduard Gröller[2]

[1] University of Applied Arts, Austria
[2] Vienna University of Technology, Austria

Abstract. This paper presents an efficient and robust algorithm for the geometric determination of swept volumes during the simulation of numerical controlled (NC-) milling. The boundary Ψ of the volume swept by a cutter Φ is represented polygonally by using instantaneous helical motions to exactly determine the line of contact between Φ and Ψ.

Applying concepts of differential geometry allows a better and more efficient approximation of tool paths. Tool paths are explicitly calculated when a design surface Γ is to be milled along prescribed curves.

We also describe how to quickly determine a polygonized representation of the truncated material during the milling process by means of "Γ-buffering". This polygon-oriented algorithm is suitable for Boolean subtractions and error assessment.

1 Introduction

Numerically controlled (NC) milling technology is a production process where a rotating cutter is sequentially moved along prescribed tool paths in order to manufacture a free-form surface from raw stock.

Different approaches of the simulation of the process were introduced in the past years, each of which has advantages and drawbacks. Two major approaches can be developed: the exact, analytical approach and the approximation approach.

The main problem with the accurate approach – mostly done in *Constructive Solid Geometry* (CSG) – is its computational expense. The cost of simulation using the CSG approach is reported to be $O(N^4)$ [8], where N is the number of tool movements. A complex NC program might consist of ten thousand movements, thus making the computation intractable.

In order to increase efficiency, a number of approximate simulation methods have been devised. The computational cost of this methods simplifies to O(N) [8].

1.1 Accurate Techniques

In 1991, Kawashima et al. [10] used a special geometric modeling method called "Graftree" to speed up their solid modeling approach. A Graftree is a combination of an Octtree and CSG. At the leaf nodes of the Octtree CSG elements are "grafted" onto the tree, if the corresponding spatial partition contains a boundary of the object. Using CSG modeling, their approach allowed accurate and precise representation of the workpiece and the tool, while the Octtree helped to decrease the number of ray-intersection calculations in the rendering step.

Implementations of special Boolean set operations (addition, subtraction, difference) were developed, which worked on Graftrees. Using the Boolean difference operator, NC milling was simulated.

Another approach to find the exact representation of the envelope of a swept volume was chosen by Sourin et al. [15]. They describe the cutting tool analytically with the use of procedurally implemented time-dependent defining functions. The swept volume of the tool consists of the instances of the cutter in initial and final position together with the envelope of the moving cutter. Boolean set operations are defined by Rvachevs R-functions [13]. The milling process can be visualized by ray tracing the analytically defined scene, which is, however, a very time-consuming process.

1.2 Approximate Techniques

In 1986, van Hook [4] managed to get update rates of 10 cutting operations per second by using special hardware equipment. A solid model was milled by a cutting tool which followed an NC path. A real time shaded display of the solid model was achieved by using image-space Boolean set operations. The workpiece and the tool were converted into a quasi "dexel structure" which is an extended Z-Buffer. Dexel (depth elements) represent thereby rectangular solids aligned to the viewing direction. The Boolean subtraction operator could be easily applied to these quasi one-dimensional representations of solid models. It was not necessary to use the computational-expensive and time-consuming Boolean difference operation in 3-space.

This first approach by van Hook had two major problems: the z-direction vector of the dexel structure was limited to the viewing vector, and the whole simulation process had to be redone when changing the viewpoint.

These problems were solved by an extension of the method by Huang and Oliver in 1994 [6]. The dexel structure used had its own coordinate system, independent of the viewing vector. The view dependency of Hooks approach is also overcome by converting the dexel structure into a set of contour lines, which can be viewed from any direction. Huang and Oliver also introduced the possibility of error assessment to Hooks method. Deviations of the final workpiece from the design part are displayed with color coding.

In applications where fast response time is essential, evaluation of the tool-envelope surfaces in object space and then scan converting these surfaces to a dexel representation is undesirable. Scan converting is a time consuming process since it may require triangulating the surface and the scan converting all the created triangles.

If a dexel representation of the sweeping object already exists, it is possible to apply the sweeping operation in image space directly. No further scan conversion is required in this case. In 1994, Hui [5] has developed a method to speed up the creation of the swept volume by doing the sweep calculations in image space. In 1991 Saito and Takahashi also used an extension of the Z-Buffer to simulate NC milling [14]. A so-called G-Buffer is used to store various information per pixel: depth, normal vector, patch or object identifier, and other patch information. Such a G-Buffer can be produced by conventional rendering techniques. A tool path for NC machining can be generated by simply scanning the G-Buffer for height. This method can only be used for three-axis machining without pocketing. Error assessment is also impossible.

A completely different approach (called "lawn mower" method) was chosen by Jerard et al. in 1989 [8]. The design surface is approximated by a polygonal mesh, where vectors (e.g., normal vectors or z-axis aligned vectors) are stored during the simulation at the mesh points. These surface vectors are intersected with a polygonal approximation of the tool. The vectors are shortened to the amount of the over- or undercutting error when the tool moves over them. Finally the mesh points are displaced by the remaining parts of the surface vectors and color coded by the amount of the error. The resulting polygonal surface can be viewed from any direction using conventional Z-Buffer rendering techniques. This approach is very useful to assess the different kinds of errors which may evolve during simulation. These possibilities are investigated in an additional paper by Jerard et al. [7].

Yang and Lee [21] developed a method suitable for wire-EDM (wire cut electric discharge machining), where cutting is only done at the sides of the workpiece. They modified the dexel approach to a cylindrical R-map, where dexels may be accessed by the height at the cylinder axis and a rotation angle. Cutting errors are again color coded.

Some other related work is given in [2], [9], [11], [17], [18], and [19].

1.3 Our Approach via Differential Geometry

This paper presents an efficient and robust algorithm for the geometric determination of swept volumes during the simulation of NC-milling. This is essential for the optimization of cutter shapes and cutter paths. Both the three-axis machining and the five-axis machining are covered.

First, we introduce an efficient algorithm for the polygonal representation of the boundary Ψ of the volume swept by the boundary Φ of a cutter (which is the union of a surface of revolution and possible additional boundary circular disks). It takes advantage of the fact that in each moment any spatial motion can be interpreted as an instantaneous helical motion (the axis and the parameter of this motion are determined). This allows to very efficiently calculate the line of contact between Φ and Ψ (called the characteristic curve

	representation of workpiece	representation of tool	error assessment	tool path	tool shape	viewpoint dependent	swept volume generation	number of axes
"dexel approach" [4]	dexel				arbitrary	•	stamping in image space	3, 5 (slower)
"extended dexels" [6]	dexel		•	correction	arbitrary		stamping in dexel space	5
"lawn mowing" [8]	mesh with normal vectors	polygonal	•		ball or cylinder		explicit as polygons	3/5
"Graftree" [10]	Octtree / CSG		•		arbitrary		explicit as CSG	5
"G-Buffer" [14]	extended Z-Buffer		•	generation/ verification	arbitrary	•	scanning of G-Buffer	3
"wire EDM" [21]	R-Buffer		•		wire		explicit as polygons	4 (wire-EDM)
"moving solid" [15]	analytical				arbitrary		analytically, 4D functions	
our technique "Γ-Buffer"	Γ-Buffer	polygonal	•	generation verification	arbitrary		scanning of Γ-Buffer explicit as polygons	3/5

Table 1. Overview of some (previous) work on NC milling

c of Φ). Now Ψ can be interpreted as a surface swept by a curve c, which is much easier to polygonize.

The tool path is explicitly calculated when a design surface Γ is to be milled along a prescribed curve.

Second, we describe how to quickly determine a polygonized representation of the truncated material, during the milling process and after the cutter has been moved along all the prescribed paths. For this, the set of all the sweeps Ψ_k has to be 'subtracted' from the raw material. The polyhedra Ψ_k are of potentially complex shape and strongly tend to have self-intersections that are difficult to eliminate. For fast applications, they are not suitable for solid modeling.

Instead, we introduce a so-called "Γ-Buffer" (related to a G-Buffer). This polygon-oriented algorithm works in object space and is suitable for Boolean subtractions and error assessment. It creates a polygonized representation of the truncated material which can be displayed very efficiently.

To sum up, our approach has three major advantages:

- The consequent use of differential geometric results speed up the determination of the sweeps Ψ_k.
- The Γ-Buffer works efficiently and produces polygonized representations of the workpiece. Additionally, error assessment is done without further computational expense.
- There are no limitations to the shape of the cutting tool.

2 Swept Volumes

Sweeping is a technique that creates new solids which are not simple Boolean constructions and combinations of solid primitives like cylinders, spheres etc. In [15], a method is described how to treat both solids modeled with set-theoretic operations and solids modeled with sweeping on the general basis of the representation by a real function of three variables. This provides solutions for problems like the sweeping of CSG-solids, self-intersections, and set operations on sweeps. The method, however, is extremely computation-intensive.

It uses a result of differential geometry: Let Φ be the boundary surface of a solid undergoing a motion, given by the implicit equation $f(x, y, z, t) = 0$ (where t is a motion parameter). Then Φ and its envelope Ψ have a line of contact c in common, determined by the additional equation

$$\frac{\mathrm{d}f(x, y, z, t)}{\mathrm{d}t} = 0.$$

This line of contact is determined by a numerical algorithm. We now try to introduce some geometric considerations that help to determine c much faster, especially when Φ is a surface of revolution.

2.1 The Instantaneous Helical Motion

A major theorem of spatial kinematics is that in any moment an arbitrary spatial motion can be replaced by an infinitesimal helical motion [20]. A helical motion is a rotation around an axis plus a proportional translation along this axis. The proportion factor is called parameter. The path curves of space points are helical curves. Each spatial motion can thus be interpreted as an integral motion of an infinite number of helical motions. In general, the manifold of all instantaneous helical axes is a ruled surface.

In order to determine the corresponding helical motion $\mathfrak{J}$ (i.e., the helical axis and the helical parameter), we need two "neighboring" point-triples, i.e., two "snap shots" of a rigid triangle at the time t and $t + \mathrm{d}t$ [20].

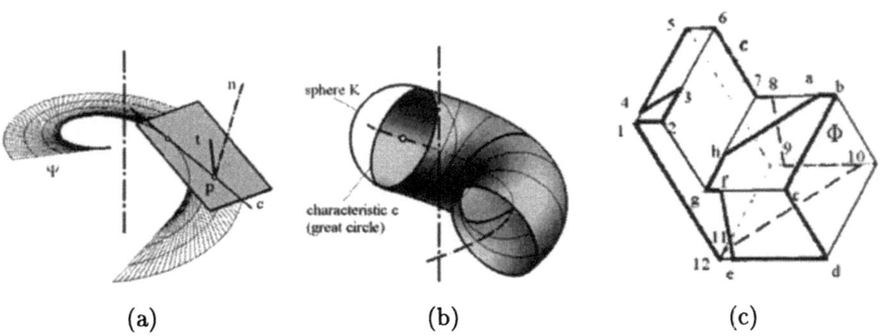

Fig. 1. Plane undergoing $\mathfrak{J}$ (a), sphere undergoing a helical motion (b), line of contact on a polyhedron (c).

As a well-known application, consider a plane undergoing $\mathfrak{J}$ (Figure 1(a)). It will in general envelope the tangent surface Ψ of a certain helix that is a path curve of $\mathfrak{J}$. The only tangent c of this helix in the plane is the so-called characteristic straight line c of the plane. For the points p of the straight line c we have the linear condition $\mathsf{t} \cdot \mathsf{n} = 0$ (where t is the tangent vector of a point p undergoing $\mathfrak{J}$ and n is the constant normal vector of the plane).

Another well-known example for a characteristic curve c occurs when a sphere is undergoing a helical motion (Figure 1(b)), sweeping a tubular helical surface. In this case, c is the great circle of the sphere the axis of which is the helical-tangent of its center.

2.2 The Line of Contact on a Polyhedron

Let Φ be a closed polyhedron (the polygonized boundary of a solid), undergoing a helical motion $\mathfrak{J}$. Let F be a polygonal face of Φ in some plane π. When the interior points of F are submitted to $\mathfrak{J}$, the dot product $\mathsf{t} \cdot \mathsf{n}$ of the corresponding helical tangent vector t and the normal vector n of π will vary linearly (which is easy to prove). We can distinguish three cases:

1. There are points inside the polygon with $\mathsf{t} \cdot \mathsf{n} = 0$. In this case, the straight characteristic line of π intersects the polygon. Because of the linearity of the condition, the two points of c_π on the sides of the polygon can be found by means of linear interpolation.
2. $\mathsf{t} \cdot \mathsf{n} < 0$ for all the points inside F. This is true for all interior points of the polygon, when it is true for all the vertices of the polygon.
3. $\mathsf{t} \cdot \mathsf{n} > 0$ for all the points inside F (same condition as in case 2).

The characteristic polygon c_Φ of Φ now consists of one or more closed polygonal branches, the edges of which are either parts of straight characteristic lines c_π or edges that faces of type 2 and type 3 have in common [16].

Figure 1(c) shows an example, where c_Φ consists of two branches $1, \ldots, 12$ and $a, \ldots, h$. As one can see, characteristic polygons can be quite tricky, especially when the helical axis intersects the polyhedron.

2.3 The Line of Contact on a Surface of Revolution

During the milling process, the cutting tool is rotating around its axis, thereby generating a surface of revolution Φ. (The rotation of the cutting tool around its axis can be neglected for geometrical considerations). Thus, it is sufficient to consider this surface, which we will call "the cutter" henceforth. The motion of Φ is subsequently interpreted as an infinite sequence of instantaneous helical motions. Furthermore, we will always talk about the *relative motion* of Φ in regard to the workpiece (regardless of the fact that the workpiece itself might be undergoing a motion).

When the cutter is moved along, it is sufficient to know about the path curve of one point on the cutter-axis and the corresponding direction of the axis. The axis positions determine a set of instantaneous helical motions $\mathfrak{J}$, as will be explained in Section 3.

Now we determine the characteristic curve of the surface of revolution. To do so, we have two possibilities:

1. We approximate the surface Φ by a polyhedron and then we apply the above mentioned algorithm for the determination of the characteristic polygon on polyhedra. In some cases, this is a quite practicable way.
2. We use the fact that a surface of revolution can always be interpreted as an envelope of a one parameter family of spheres K the centers of which lie on the rotational axis of Φ (which may degenerate to planes perpendicular to the axis).

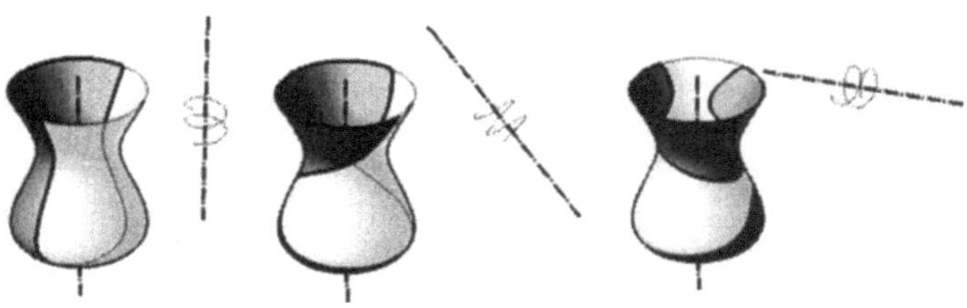

Fig. 2. Shape of the characteristic curve c on a surface of revolution Φ varies considerably with the position of the instantaneous helical axis.

Each sphere K touches Φ along a parallel circle k. When K is undergoing the helical motion , it sweeps a tubular surface Ψ_K. As we know, the line of contact c_K of K and Ψ_K is a great circle of K the axis of which is the helical tangent of the center of the sphere. The two possible intersection points s_1 and s_2 of k and the plane containing c_K are points on Φ that also belong to the swept volume Ψ of Φ, since the corresponding tangent planes of K and Φ are identical.

For a sufficient number of parallel circles on Φ, we can compute the points s_1 and s_2 very inexpensively. As long as each parallel circle carries two real (non-imaginary) points, the line of contact c consists of two separate branches. When the intersection points s_1 and s_2 are identical, the two branches of c "grow together". Since we deal with solids, parts of the bordering circles have to be included to c as well. Therefore, c consists of one or more closed branches with corner points at the bordering circles.

The shape of the characteristic curve c on a surface of revolution Φ varies considerably with the position of the instantaneous helical axis. When Φ is concave (Figure 2), c may consist out of several closed branches. Cutter shapes, however, are usually convex surfaces of revolution (Figure 3). In this case, c is always one closed curve.

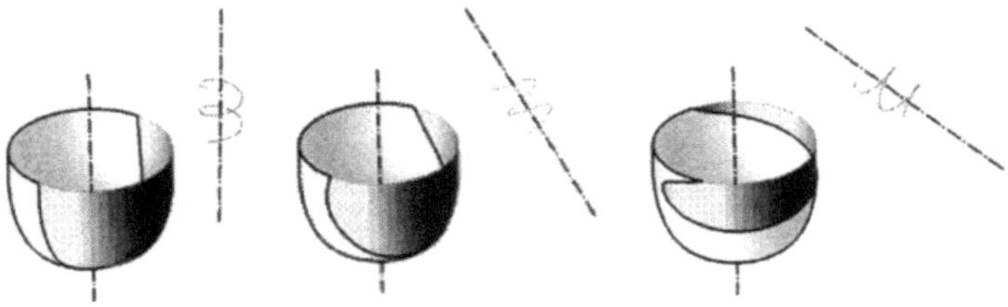

Fig. 3. Characteristic curve c of convex cutter shapes is always one closed curve.

2.4 The Swept Volume of the Cutter

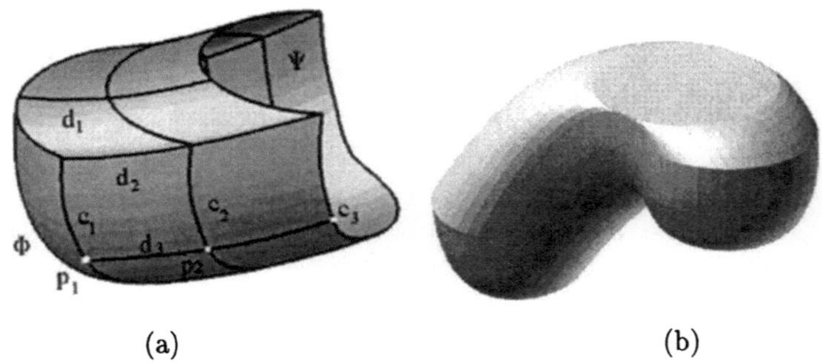

Fig. 4. Interpolation of a set of lines $\{c\}$ with a set of lines $\{d\}$ (a), resulting sweep object (b).

When we calculate the line of contact c for a sufficient number of instantaneous helical motions, we get a one-parametric set of lines $\{c\}$ on the boundary Ψ of the volume swept by Φ. The question is now: how can we interpolate a second set $\{d\}$ of lines on Ψ so that Ψ corresponds to the actual cutting surface as much as possible?

Surprisingly, corner points on c do *not* produce sharp edges on Ψ (though it looks like it in Figure 4(a)). This is because the plane spanned by the two tangents in a corner point is tangent plane of Φ. The only potential sharp edges of Ψ are produced by self-intersections and by parts of the bordering circles at the beginning and at the end of the sweeping (Figure 4(b)). Thus, a line d may also cross over the lines determined by all corner points.

To find a line d, we start from an arbitrary point p_1 on the first characteristic line c_1 (Figure 4(a)). The tangent t_1 of d lies in the corresponding helical tangent plane τ_1. As close as possible to t_1, we look for a new point $p_2 \in d$ on the second characteristic line c_2. The tangent t_2 of d in p_2 lies in the corresponding helical tangent plane τ_2. When t_2 coincides with the intersection point of t_1 and t_2, the line segment p_1p_2 of d can be approximated by a Bezier curve (e.g. [3]). Another approach (a rational model of the surface swept by a curve) is given in [9].

Figure 5 shows how two successive c-lines can have quite different behavior (details from Figure 4(b)). The image to the left shows that a self-intersection of the sweep Ψ occurs. The image to the right illustrates the not unusual case when the cutter's axis is parallel to the instantaneous helical axis and the cutter path reaches a relative maximum.

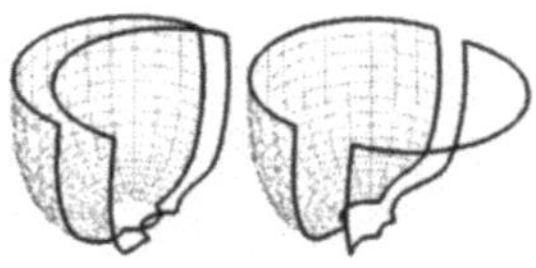

Fig. 5. Two successive c-lines can have a quite different behavior.

3 Tool Path Generation

Let Γ be the C^1-surface that has to be milled (the design surface). In each point $p \in \Gamma$ (position vector p), we have a surface normal n of Γ perpendicular to the tangent plane τ. The cutter Φ is considered to be a surface of revolution. When the direction vector d of the cutter axis a is given, it will in general be possible to find a point $p_0 \in \Phi$ where the normal of Φ has the direction of n. If not, a point p_0 on a bordering circle can be taken. We can move the cutter so that p_0 and p coincide. Now Φ and Γ touch in p. When Φ is convex, the correspondence $p \rightarrow p_0 = p_0(p, \mathsf{d})$ is well-defined.

Let α be the angle between d and n. It can also be interpreted as the angle between the tangent of the cutter's meridian and a base plane perpendicular to the cutter's axis (Figure 6). For further calculations, we parametrize the meridian of the cutter Φ by means of the angle α:

$$r = r(\alpha), \quad z = z(\alpha) \quad (0 \leq \alpha < \pi/2).$$

We now look for points on the cutter's axis, provided that Φ is convex and s is a regular point on the cutter's meridian.

According to Figure 6, the position vectors to the three points a_1, a_2, a_3 on the cutter's axis are given by the vector equations

$$\mathsf{a}_1 = p - \frac{r}{\sin\alpha\cos\alpha}\mathsf{d}^n, \quad \mathsf{a}_2 = p + \frac{r}{\sin\alpha}\mathsf{n}, \quad \mathsf{a}_3 = \mathsf{a}_2 - \frac{r}{\tan\alpha}\mathsf{d}.$$

(d^n is the normal projection of d on the tangent plane τ with the length $\sin\alpha$; n and d are normalized.) We can even describe a fixed reference point a ($\mathsf{a} = \mathsf{a}_3 - z\mathsf{d}$) on the cutter's axis:

$$\mathsf{a}(\mathsf{p}, \mathsf{d}) = \mathsf{p} + \frac{r(\alpha)}{\sin\alpha}\mathsf{n} - \frac{z(\alpha) + r(\alpha)}{\tan\alpha}\mathsf{d}.$$

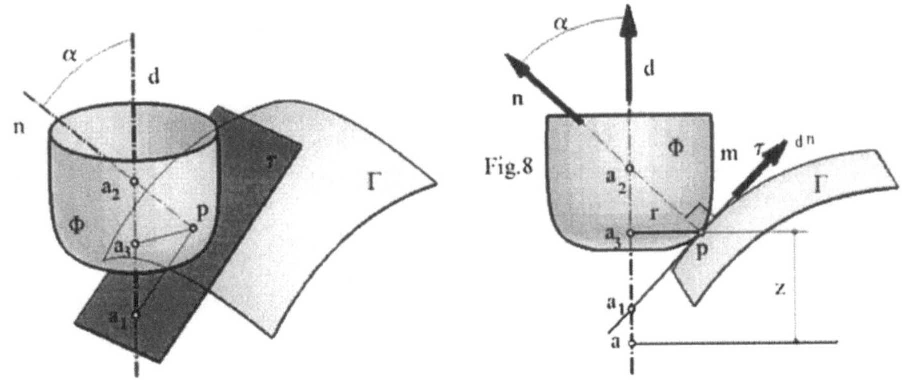

Fig. 6. Tool path generation.

The above equation describes a surface that carries the path curves of a during any milling process along the surface. In the case of 3-axis milling (where d is constant), such surfaces are called "general offset surfaces". They were introduced by Brechner [1] and more thoroughly investigated by Pottmann [12].

When we want to mill the surface Γ along an arbitrary line of contact $p(t) = \{p\}$, we can choose a set of corresponding axis directions $\mathsf{d}(t) = \{\mathsf{d}\}$. The parameter t is a motion (time) parameter. (For 3-axis milling, $\mathsf{d}(t)$ is constant). $p(t)$ and $\mathsf{d}(t)$ determine the motion of the cutter axis in a unique way. This allows to determine the instantaneous helical motion $\mathfrak{J}(t)$ as follows:

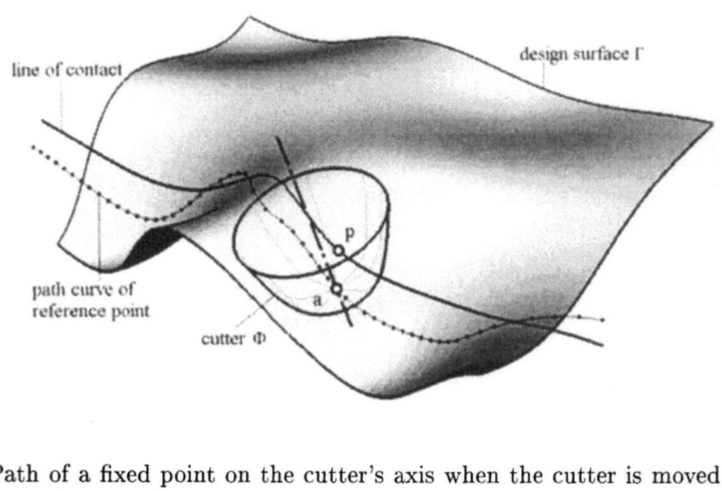

Fig. 7. Path of a fixed point on the cutter's axis when the cutter is moved along the design surface.

We consider three "neighboring" axes, corresponding to the times $t-\varepsilon$, t and $t+\varepsilon$ ($\varepsilon > 0$), where $a^{-\varepsilon}$, a and $a^{+\varepsilon}$ are the previously mentioned fixed points on these axes, and $d^{-\varepsilon}$, d and $d^{+\varepsilon}$ are the corresponding direction vectors. Let c be the circle determined by $a^{-\varepsilon}$, a and $a^{+\varepsilon}$. Then we can define three Cartesian systems $\Sigma^{-\varepsilon}$, Σ and $\Sigma^{+\varepsilon}$ as follows: The origin is the fixed point, the z-direction is the corresponding axis-direction, and the yz-plane contains the corresponding tangent of the circle c. Thus we get a helical $\mathfrak{J}^{-\varepsilon}$ that transforms $\Sigma^{-\varepsilon}$ into Σ and another one that transforms Σ into $\Sigma^{+\varepsilon}$. When ε converges to zero, the circle c converges to the osculating circle of the path of a, and the two helical motions converge to one helical motion which may be interpreted as the corresponding instantaneous helical motion $\mathfrak{J}$.

Note that in the case of three-axis milling (where the neighboring axes are parallel) the helical axis has the same main direction. The helical motion $\mathfrak{J}$, however, is in general not degenerated to a simple rotation or translation!

Figure 7 shows the path of a fixed point on the cutter's axis when the cutter is moved along the design surface.

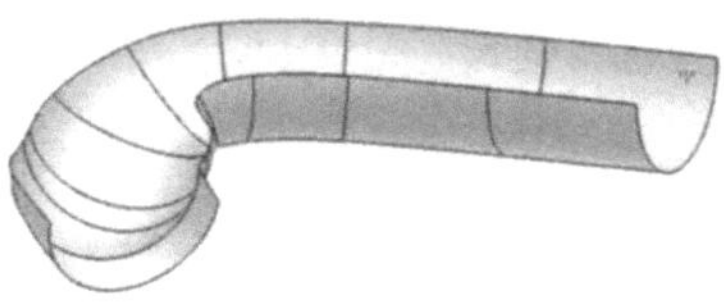

Fig. 8. Parameter lines on the surface of the swept volume Ψ with varying time differences Δt.

For a comparatively small number of instantaneous helical motions $\mathfrak{J}(t)$ ($t = t_1, t_2, \ldots, t_n$), we can now quickly determine a one-parametric set of characteristic curves $c(t)$. They are parameter lines on the surface of the swept volume Ψ. The time differences Δt should in general not be constant (Figure 8). As a rule of thumb, Δt can be increased when $\mathfrak{J}(t)$ is nearly a translation or rotation. For a pure translation, we only need the characteristic curve of the start position and of the final position.

4 The Γ-Buffer Representation of Surfaces

In [14], a method called "G-Buffering" for the representation of surfaces (function graphs) is introduced. A G-Buffer ("Geometry Buffer") is used to store various information per pixel: depth, normal vector, patch or object identifier, and other patch information. The method is viewpoint dependent and has some additional drawbacks, as mentioned in Section 1.

We now want to introduce a related object-space oriented Buffer that we will call Γ-Buffer. It will enable us to simulate the milling process, supporting Boolean subtractions very efficiently.

4.1 Γ-Representation of Function Graphs

Let the design surface be a function graph Γ, defined over a base rectangle $x \in [x_1, x_2]$, $y \in [y_1, y_2]$. The graph need not be represented by a mathematical function $z = f(x, y)$. It should rather be an arbitrary polyhedral graph. (Depending on the complexity, the number of polygons may vary from only a few up to tens of thousands). In a top view on the xy-plane, the polygons need not obey any pattern like regular grid arrangement.

We now define a Γ-Buffer. In terms of object-oriented programming, a Γ-Buffer is meant to be a class with the following members:

- A linear coordinate transformation from the rectangular base $\{(x, y)|x \in [x_1, x_2], y \in [y_1, y_2]\}$ to the rectangle $\{(u, v)|u \in [0, n_1], v \in [0, n_2]\}$ plus its inverse.
- A dynamic list of polygons (faces) with information about normal vector and vertices (in xy-coordinates) and physical properties.
- A large two-dimensional cell-structure ($n_1 \times n_2$ cells, $n_1, n_2 \approx 1000$), being placed as a fine grid over the base uv-rectangle. Each cell of the grid now contains an index that uniquely indicates the corresponding surface, two floating point z-values and a pointer to the polygon on the surface. The first z-value is used for the representation of the workpiece, the second one represents the design surface.
- Member functions for initializing, updating and displaying, read-write options for disk-storage and error-assessment.
- The most frequently used member function is to put a polygon into the polygon list and buffer its contents. It works as follows: The data of the polygon (vertices, normal etc.) are stored. Then the xy-coordinates of the vertices are transformed into the uv-base. The transformed polygon defines a plane γ. For each grid point inside the uv-projection of the polygon, the z-coordinate in γ is buffered in the *first* z-value of the corresponding cell (that means, it is stored when the existing z-value is larger than the current one). This can be done very quickly and is similar to Z-buffering (Z-buffering, however, is done in image space and is thus viewpoint dependent, whereas Γ-Buffering is not!). In addition to the z-values, we also store the pointer to the corresponding polygon and a unique index (> 0) for the surface.

The milling process can then be simulated by a Γ-Buffer in several steps:

1. The cells are initialized: The first z-value is initialized by a theoretical maximum number (e.g., $z = 10^{10}$). The surface indices are all initialized

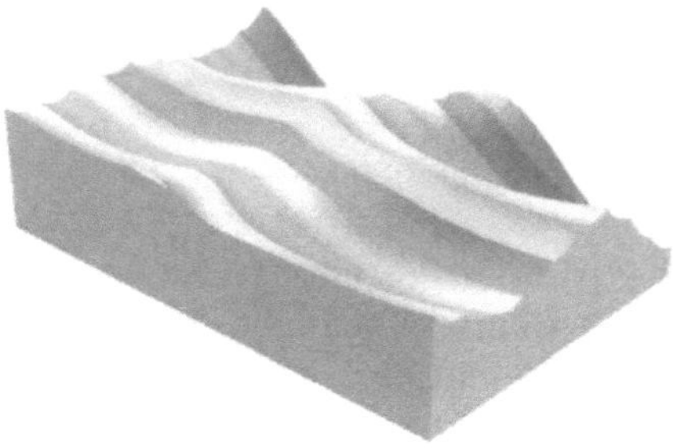

Fig. 9. Object with several color coded cutter paths Ψ_k.

by zero. For quick error assessment, the second z-values have to be initialized: All the polygons of design surface Γ are put into the buffer in a modified way: The z-values of the polygon's plane γ are stored in the *second* z-values.

2. We buffer the polygons of the raw stock's surface (this might be the top plane of a simple box or any other surface).
3. We update the buffer with the polygons of the sweeps Ψ_k (each Ψ_k has a distinct surface number). The non-constant first z-value should never be smaller than the constant second one, otherwise the cutter over-cuts the design surface Γ. For the time being, we only buffer those polygons of the sweeps Ψ_k that are "on the lower side". (In Section 4.3 we will see how we can also do "pocketing").
4. At any time, we can display the contents of the Γ-Buffer in an efficient way, using a recursive algorithm (see implementation in Section 4.2).
5. At the end of our simulation (or at any time in between) we can estimate with high accuracy the amount of material that still needs to be removed: For each cell of the grid, we subtract the second z-value from the first one. This z-difference is proportional to the volume above the design surface.

4.2 Implementation of the Γ-Buffer

Before we extend Γ-buffering to general surfaces, we briefly want to discuss the implementation of a Γ-Buffer.

The fine grid over the uv-rectangle ($n_1 \times n_2$ grid points) makes it necessary to allocate vast amounts of memory. In our representation, we used grid sizes like $n_1 = n_2 = 1024$. (In order to accelerate the display-function, n_1 and n_2 should be dividable by 16.)

Each cell of the grid contains a 2-byte integer (the index of the corresponding surface), two 4-byte floating point z-values and one 4-byte pointer to a polygon, which makes 14 bytes altogether.

Thus, the grid itself requires about 15 Mbytes RAM. Compared to this memory requirement, the polygons themselves need comparatively little space (approximately 1 Megabyte for 50,000 polygons).

The buffering process itself can be implemented very computation-inexpensively (on any platform that we used, even on a 200 MHz PC, thousands of polygons can be processed per second).
From time to time, the number of necessary polygons can be reduced by buffering all the polygons once again (this will not change the contents of the buffer). When the grid did not have to be updated by a polygon, this polygon can be removed from the G-buffer. In this way, even complicated buffers rarely exceed 20,000 polygons.

An important member function of the Γ-Buffer is the displaying of the contents.
The simplest - but unacceptably time-consuming - way is to display the $2 \times n_1 \times n_2$ triangles defined by the grid. This number can be reduced drastically without any loss of image quality, by using a recursive algorithm:
For a patch over for example 16×16 grid points, we test whether all these grid points belong to the same surface (surface index!). If that is so, we can display the surface over the patch quickly with two triangles. If not, we subdivide the patch into four squares of 8×8 grid points and repeat the process. The recursion ends when the square size is 1×1. In this case, we display the patch in any case, provided the surface index is not zero. The latter enables us to display surfaces with non-rectangular top view and/or holes (Figure 10(a)).

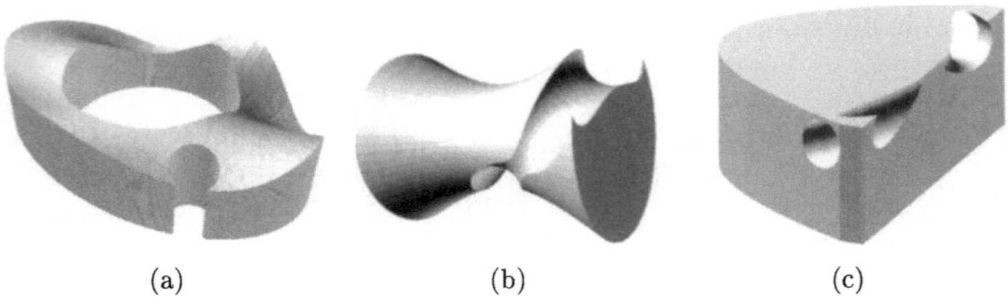

(a) (b) (c)

Fig. 10. Surfaces with non-rectangular top view and/or holes (a), surfaces with a maximum of two intersection points (b), workpiece with pocketing by means of a ball cutter (c).

This recursive display algorithm produces an equally good result as when we display all the 2-3 million triangles that the Buffer potentially provides, especially when we work with smooth shading. In Figure 9, the reasonable number of 20,000 polygons had to be displayed. On a 200 MHz Silicon Graphics workstation (Indigo 2), the surface can be displayed several times a second.

4.3 Γ-Representation of General Surfaces

So far we only discussed design surfaces Γ and sweeps Ψ_k that are function graphs (i.e., surfaces with only one z-value over a base point (x, y)).

With slight modifications, we can cover surfaces with a maximum of two intersection points on any z-parallel line (i.e., line perpendicular to the ground plane). Then we can split the surface into two function graphs, and apply two separate Γ-Buffers to the scene. Figure 10(b) shows an non-trivial example for this (the z-direction is chosen in such a way that we fulfill the restriction to the number of intersection points in that direction).

We have to provide a more general solution in order to be able to do "pocketing" and/or to mill general surfaces and/or to apply general 5-axis milling. For this purpose, we extend the definition of the Γ-Buffer:

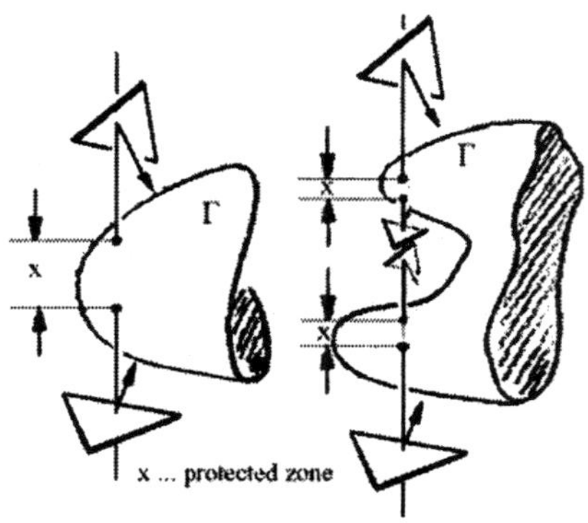

Fig. 11. Protected zone, where the cutter must not intersect the z-parallel through (x, y).

On a general surface, several z-values over a base point (x, y) are possible. Therefore, the grid cells must be extended. The idea is to store several "protected zones". Figure 11 shows what is meant with a protected zone: It is a zone, where the cutter must not intersect the z-parallel through (x, y). When m is the maximum number of possible intersection points of Γ with an arbitrary z-parallel ray, the contents of the extended cell is now

- the current number n of protected zones ($n \leq m/2$),
- the current number k of detected intersection points of the z-parallel ray with the polygons of the buffer ($k \leq m$),
- m floating point z-values z to these intersection points,
- m pointers to the corresponding intersecting polygons of the grid.
- For error assessment, we also store m floating point z-values z_0 (minima and maxima of the protected zones of the design surface Γ). They indicate the $2n$ intersection points of the vertical ray above the grid point with Γ.

(Each extended grid cell now requires $12m+2$ bytes altogether. For $m > 2$, this is quite memory intensive!)

The extended Γ-Buffer is filled and displayed in several steps:

1. The Buffer is allocated. The extended cells are initialized with $n = k = 0$.
2. The polygonized design surface Γ is stored. We process each polygon P of Γ as follows: We transform P into the (u, v)-base as usual. For each extended grid cell inside the (u, v)-projection of P, the corresponding z_0-value is stored. When the normal n is horizontal, i.e., when P is perpendicular to the (u, v)-plane, we only store points on the sides of P. – After all the polygons of Γ have been proceeded, the n detected z_0-values are sorted. Alternately, the z_0-values indicate the beginning and the end of a protected zone.
3. The surface of the raw stock is put into the buffer similar to how the design surface was stored. The normal vectors of the polygons have to be oriented to the *outside* of the surface. We store k z-values plus additional information (pointer to the intersecting face).
4. The sweeps Ψ_k are buffered (the Γ-Buffer is updated). The normal vectors of the polygons are oriented to the inside of the surface. When it comes to the buffering of a polygon, we determine for each cell the corresponding z-value ζ in the polygon's plane. This value must never be inside a protected zone, otherwise the sweep over-cuts Γ. Let the oriented normal vector be turned downwards. When ζ is greater than a buffered maximum and smaller than the following minimum, the maximum is updated (including polygon pointer). Analogously a minimum is updated when the oriented normal vector is turned upwards.
5. The contents of the extended Γ-Buffer can be displayed in a very similar way as the contents of a conventional Γ-Buffer. We start to check quadrangular patches with size 16×16. For all grid cells in this patch we first check whether the levels k are the same. If that is so, we can display all the k parts of the buffered object as described above (locally, we have k Γ-Buffers at our disposal). If not, we recursively split the patch into 4 smaller patches. The recursion ends when the minimum patch size 1×1 is reached. Then this patch is displayed, when the level of at least two of the 4 neighboring grid points are the same. (Otherwise this might build up unwanted "curtains" from lower points to higher ones). Figure 10(c) shows an example of a workpiece with pocketing by means of a ball cutter. In this case, we had a maximum of two protected zones ($m = 2$).

Although the computational expense and the request of memory increase for complex surfaces, Γ-buffering still is robust and works comparatively fast.

5 Conclusion and Future Work

The simulation of milling processes can be done effectively and without restrictions with the algorithms described in this paper. When a workpiece is stored in a Γ-Buffer, it can be shaded in very short time (several times a second on graphics workstations and - depending on the complexity of the surface - little more than one second on a 200 MHz Pentium PC with graphics hardware). The milling process itself can also be done very computational-inexpensively by making intensive use of the geometric results explained in this paper. The simulation is very accurate since it uses results from differential geometry.

Since these algorithms are very robust, it will now be possible to continue with further geometrical research towards improvements of NC-milling.

Here are two important questions, to which a final answer is not in sight so far (but to which the simulator can quite effectively verify or falsify various suppositions for a number of samples):

- How are the lines of contacts to be chosen in order to minimize the number of path-sequences? For example, it is conjectured that such lines might be the "relative principal curvature lines" of "generalized offset surfaces" ([12]).
- Which shape of the cutter is best for a specific surface? E.g., for a certain class of cutters (and 3-axis milling of a function graph) it is possible to prove the following theorem: If the cutter does not intersect the graph locally (in a small area around any point of contact), there will be no under-cuttings. To determine the optimal cutter among the possible ones is an important challenge.

In general, it can be said that 5-axis milling has so many degrees of freedom that it will require lots of research to get a better understanding of what is possible and what is not.

6 Acknowledgements

This research has been supported by the Austrian Science Foundation through project P11357-MAT. The authors thank Helmut Pottmann and Hellmuth Stachel for critical comments and Andreas König for his help with the introductory section.

References

1. E. Brechner, *General Offset Curves and Surfaces*, in R.E.Barnhill, ed., *Geometry Processing for Design and Manufacturing*. SIAM, Philadelphia, pp. 101-121 (1992).
2. I. T. Chappel, *The use of vectors to simulate material removed by numerically controlled milling*, Computer Aided Design, Vol. 15, No. 3, pp.156-158, (1983).
3. J. Foley van Damm, *Computer Graphics. Principles and Practice*, Addison-Wesley (1990).
4. T. van Hook, *Real Time Shaded NC Milling Display*, Computer Graphics, Vol. 20, No. 4, (Proc. SIGGRAPH '86), pp.15-20 (1986).
5. K.C., Hui, *Solid sweeping in image space- application in NC simulation*, The Visual Computer, Vol.10, pp.306-316 (1994).

6. Y. Huang and J.H. Oliver, *NC Milling Error Assessment and Tool Path Correction*, Computer Graphics Proceedings, Conference Proceedings July 24-19, 1994, (Proc. SIGGRAPH '94), pp. 287-294.

7. R.B. Jerard, R.L. Drysdale, K. Hauck, J. Magewick, and B. Schaudt, *Methods for Detecting Errors in Numerically Controlled Machining of Sculptured Surfaces*, IEEE Computer Graphics and Applications, 0272-1716/89, pp.26-39 (1989).

8. R.B Jerard, S. Z. Hussaini, R.L. Drysdale, and B. Schaudt, *Approximate methods for simulation and verification of numerically controlled machining programs*, The Visual Computer Vol. 5, pp.329-348 (1989).

9. J.K. Johnstone and J.P. Williams, *A rational model of the surface swept by a curve*, EUROGRAPHICS'95, Vol.14, pp.77-88 (1995).

10. Y. Kawashima, K. Itoh, T. Ishida, S. Nonaka, and K. Ejiri, *A flexible quantitative method for NC machining verification using a space-division based solid model*, The Visual Computer, Vol. 7, pp.149-157 (1991).

11. Ch. Liu and D.M. Esterling et al. *Dimensional Verification of NC Machining Profiles Using Extended Quadtrees*, Computer Aided Design, Vol. 28, No. 11, pp.845-842 (1996).

12. H. Pottmann, *General Offset Surfaces*, Neural, Parallel & Scientific Computations, 1997, to appear.

13. V.L. Rvachev, *Methods of Logic Algebra in Mathematical Physics*, Kiev: Naukova Dumka Publishers, p259, (1974).

14. T. Saito and T. Takahashi, *NC Machining with G-Buffer method*, Computer Graphics, Vol. 25, No. 4, (Proc. SIGGRAPH '91), pp.207-216 (1991).

15. A.I. Sourin and A.A. Pasko, *Function Representation for Sweeping by a Moving Solid*, IEEE Transactions on Visualization and Computer Graphics, Vol. 2, No. 2, March 1996, pp.11-18 (1996).

16. H. Stachel and A. Elsonbaty, *Generating solids by Sweeping Polyhedra*, 1996. Proc. 7th ICECGDG, Cracow 1996, Vol.I, pp. 245-249 (to appear in Journal for Geometry and Graphics 1 (1997)).

17. H.B. Voelker and K.K. Wang, *The Role of Solid Modeling for Swept Volume of Moving Solids*, SAE Technical Paper no. 810195, Feb. 1981.

18. S.W. Wang and A. Kaufmann, *Volume Sculpting*, ACM Symposium on Interactive 3D-Graphics, Monterey CA., pp.151-156.

19. W.P. Wang and K.K. Wang, *Geometric Modeling for Swept Volume of Moving Solids*, IEEE Computer Graphics and Applications, 6, 12, 1986, pp.8-17.

20. W. Wunderlich, *Darstellende Geometrie II*, BI-Hochschultaschenbuch, Mannheim 1967.

21. M. Yang and E. Lee, *NC verification for wire-EDM using an R-map*, Computer Aided Design, Vol. 28, No. 9, pp.733-740 (1996).

Constant Mean Curvature Surfaces with Cylindrical Ends

Karsten Große-Brauckmann[1], Robert B. Kusner[2], and John M. Sullivan[3]

[1] Mathematisches Institut, Universität Bonn, Germany
[2] School of Mathematics, Institute for Advanced Study, USA
[3] Mathematics Department, University of Illinois, Urbana, USA

Abstract. R. Schoen has asked whether the sphere and the cylinder are the only complete (almost) embedded constant mean curvature surfaces with finite absolute total curvature. We propose an infinite family of such surfaces. The existence of examples of this kind is supported by results of computer experiments we carried out using an algorithm developed by Oberknapp and Polthier.

The cylinder of radius $\frac{1}{2}$ is a surface with constant mean curvature 1 (a CMC surface for short). The cylinder has vanishing Gauss curvature K, and hence finite (indeed, zero) absolute total curvature $\int |K|\, dA$. It is the simplest example of an *unduloid.* These are the embedded CMC surfaces of revolution, described by Delaunay in 1841 [2] (see also [3]), which are simply periodic and have as generating curves the roulettes of ellipses (of major axis length 1). There is a one-parameter family of unduloids, depending on the eccentricity of the ellipse. For us it is more convenient to parameterize this family by the *necksize* $n \in (0, \pi]$, which is the length of the shortest closed geodesic. One extreme case of the family is the cylinder, whose necksize is π. At the other extreme, the necksize tends to 0 and the unduloids degenerate to a chain of unit spheres. Periodicity implies that every unduloid, aside from the cylinder, has infinite absolute total curvature.

The Delaunay unduloids play a significant role in the theory of embedded CMC surfaces with finite topology, that is, with finite genus g and a finite number of (necessarily annular) ends k. It is a result of Korevaar, Kusner, and Solomon [11] that each of the k ends is exponentially asymptotic to a Delaunay unduloid. Indeed, their results remain true for the slightly larger class of *almost embedded* surfaces, which are immersed surfaces whose immersion extends to the interior of the surface (see Section 1). We call any such CMC surface a *k-unduloid (of genus g).* For $k \leq 2$ the only k-unduloids are the sphere and the unduloids themselves [11,17].

More than a decade ago R. Schoen raised the question of whether there are any complete (almost) embedded CMC surfaces with finite absolute total curvature, besides the sphere and cylinder. Such a surface must have finite topology [1]. Thus, by the asymptotics theorem [11], the question is equivalent to the problem we address in the present paper:

Problem 1. *Can a k-unduloid have all of its ends cylindrical for $k \geq 3$?*

It is worth noting that simply or doubly periodic surfaces with (an infinite number of) cylindrical ends exist [4]. Of course, these all have infinite topology and infinite absolute total curvature, though the absolute total curvature of each (non-compact) fundamental domain is finite.

1 Immersed Examples and Almost Embeddedness

Interesting complete, non-compact immersed CMC surfaces of finite absolute total curvature are known: for example, Pinkall and Sterling depict a CMC surface with genus zero and two cylindrical ends [19]. It looks like a "two-lobed Wente torus" fused to a cylinder, and its existence was proven later (see [20], [4], and also [21] where similar surfaces can be constructed for arbitrary Delaunay ends). Since the ends of this surface are embedded they have again exponential decay to a true cylinder, so its absolute total curvature is finite. These (and many other) examples suggest that the class of all immersed CMC surfaces is too large to give much control on their geometry.

When studying minimal surfaces, embeddedness is often a natural condition to impose, especially for physically motivated problems. The maximum principle implies that, under continuous deformation, a complete minimal surface cannot suddenly stop being embedded unless self-intersections occur at the ends of the surface. The situation is different in the case of CMC surfaces, as seen in Figure 1: when we continuously deform an embedded CMC surface, embeddedness may be lost as bubbles start to overlap.

This leads us to concentrate on a natural class of immersed surfaces which arise when considering families of embedded CMC surfaces, the *almost embedded* surfaces, mentioned above. By definition, an immersed surface is almost embedded if it can be parametrized by an immersion $f : M \to \mathbb{R}^3$ which extends to an immersion $F : \Omega \to \mathbb{R}^3$, where Ω is a three-manifold with $\partial\Omega = M$. In fact for CMC surfaces of finite topology, the methods of [13] imply Ω can always be taken to be homeomorphic to a handlebody in $\mathbb{R}^3$.

The principal results on finite topology CMC surfaces are valid for this almost embedded class: for instance, ends are asymptotic to Delaunay unduloids [11], each k-unduloid remains a uniformly bounded distance from a k-ended piecewise-linear graph [10], and the moduli space of all these k-unduloids (near a surface with no L^2 Jacobi fields) is a real analytic manifold [14] of dimension $3k-6$.

2 Nonexistence Results for Cylindrical Ends

There is evidence that k-unduloids with only cylindrical ends are rare. For example, we have proven that there are no k-unduloids of genus zero with all ends cylindrical when there are only $k = 3$ ends [6]. More generally, when all the ends have their axes in one common plane — a case we call *coplanar* — there are at least two non-cylindrical ends provided $g = 0$ and k is odd.

Fig. 1. Two triunduloids indicate a continuous transition from embedded to non-embedded CMC surfaces. The second surface, whose bubbles overlap, is still almost embedded.

There is further evidence for the rarity of these examples from a different perspective. Gluing constructions were introduced by Kapouleas [8], and have become a powerful and general tool to produce examples of CMC surfaces. Two unduloids, for example, can be glued together by connecting them with a small, almost catenoidal neck. Here, as in general, the resulting surface will have slightly different axis directions of the ends, and it may be necessary to perturb the necksizes, too. These changes can be made arbitrarily small, however, when the connecting neck is small enough.

For Kapouleas' construction to be applicable, the two unduloids must have small necksize, and it is natural to ask if we can similarly glue two tangent cylinders together by a small neck. The two cylinders themselves form a degenerate surface which can be naturally regarded as lying in the boundary of the moduli space of 4-unduloids; in fact there is a one-parameter family of tangent cylinders, parameterized by the angle of their axes. We might expect to find examples with only cylinder ends in the interior of moduli space within a neighborhood of these boundary points. Kapouleas, Mazzeo, and Pollack have recently announced [16, p.7] a gluing construction (inserting a small catenoidal neck between the cylinders) which apparently yields all 4-unduloids in such a neighborhood. However, on each of the surfaces constructed this way, at least two of the ends necessarily decrease their necksize, and thus are no longer cylindrical. This change can be made arbitrarily small when the gluing neck is small, but the change is always present. Similarly, when h cylinders are glued to form a $2h$-unduloid, then at least h necksizes must change. Since this type of construction always changes some necksizes, there are no examples with all cylindrical ends near these boundary points of moduli space.

3 The Necksize Problem

Consider a k-unduloid of genus g. Let $n_1, \ldots, n_k \in (0, \pi]$ be the asymptotic necksizes of its ends (the lengths of the shortest closed geodesics on the limiting Delaunay unduloids), and let $a_1, \ldots, a_k \in \mathbb{S}^2$ be their (outward oriented) axis directions. These asymptotic quantities satisfy the *balancing formula* [11]

$$\sum_{i=1}^{k} n_i(2\pi - n_i)a_i = 0.$$

The balancing formula has the following physical interpretation: imagine the surface being made as a soap bubble. Due to surface tension and pressure, each end exerts a force on the surface which can be measured across any cap spanning a curve which separates the end. This force is independent of the particular cap chosen and is in fact equal to $n_i(2\pi - n_i)a_i$. Note that the force of a cylindrical end has maximal modulus, while zero force is approached in the spherical bead limit. The sum of these forces represents the net force on the remaining compact domain. If this net force were nonzero, the compact piece would tend to move in that direction; but the surface is in equilibrium so balancing must hold. We remark that the surfaces we consider are not stable, and therefore only small pieces can be realized as a soap bubble. Nevertheless, the physical argument can be made rigorous using the first-variation formula that characterizes the equilibrium of a CMC surface (see [11,10]).

Balancing gives a necessary condition on the asymptotic data of any k-unduloid. We can ask to what extent balancing is also a sufficient condition, leading to a question for which Problem 1 forms an extreme case.

Problem 2. *Given potential necksizes $n_1, \ldots, n_k \in (0, \pi]$ and axis directions $a_1, \ldots, a_k \in \mathbb{S}^2$ satisfying the balancing formula, is there a k-unduloid of genus g with this asymptotic data?*

Whenever all necksizes are sufficiently small, Kapouleas's construction [8] shows that the answer is yes. On the other hand, balancing and the requirement that all necksizes lie in $(0, \pi]$ are not sufficient to guarantee the existence of a k-unduloid. This follows for instance in the case of coplanar k-unduloids of genus zero with full dihedral symmetry: here the force balancing is automatically satisfied, but the maximal necksize reached is $2\pi/k$, so that the *necksize sum* $\sum n_i$ is at most 2π [4,6].

Also, for k odd, any coplanar k-unduloid of genus zero has necksize sum at most $(k-1)\pi$, and further constraints are known [6]. Conversely, we believe that coplanar k-unduloids of genus 0 exist with any necksize sum in the interval $(0, (k-1)\pi]$ for k odd, or $(0, k\pi)$ for k even. For example, a surface like the one pictured in Fig. 2 presumably exists with four cylindrical ends and two unduloid ends of arbitrary necksize $x \in (0, \pi)$. This figure was built from four congruent pieces; using instead $k-2$ pieces suggests a k-unduloid with all but two ends cylindrical. If k is even, the necksize sum is $(k-2)\pi + 2x$, while if k is odd, the noncylindrical ends have necksizes x and $\pi - x$ so the sum is $(k-1)\pi$. Any smaller necksize sum could be obtained in a similar fashion starting from pieces with no cylindrical ends. It is an interesting question whether the bound on the necksize sum for k odd extends to noncoplanar k-unduloids of genus zero; presumably it does not hold for higher genus surfaces.

For arbitrary genus, we can quantify the rarity of examples with all ends cylindrical by a parameter count. We already noted above that, for any genus, the moduli space of all k-unduloids has dimension $3k-6$. Generically, fixing the necksize of one end gives a subspace of codimension 1, but specifying an end to be cylindrical yields codimension 2, as we see in [6] for the case $k=3$. Thus one would expect the dimension of the space of k-unduloids with all ends cylindrical to be $k-6$.

Our methods produce examples with coplanar ends. Requiring the ends to be coplanar generically reduces dimension by $k-3$. Thus we would expect the subspace of coplanar k-unduloids with all ends cylindrical to have dimension $(k-6)-(k-3)=-3$. That means, to find any examples, we need some special condition (like symmetry) to make them nongeneric. Therefore it is natural to look for surfaces with a high degree of symmetry; this is precisely where our methods apply.

4 Numerical Examples

We have obtained good numerical evidence for the existence of CMC surfaces with only cylindrical ends and genus one. We carried out our computer experiments using an algorithm developed by Oberknapp and Polthier. This algorithm computes a polyhedral approximation to a minimal surface in $\mathbb{S}^3$, and then applies a discrete version of Lawson's conjugate surface construction [15,9]. It is well described in [18] so we refrain from further explanations here.

It is evident from the accompanying images that we had to search systematically for our surfaces. Indeed, as we will explain in the next section, the idea for our surfaces was found on the basis of theoretical work. The following result is illustrated in Color Plate 10, p. 371 and Figure 2.

Experimental result 3. *There exist almost embedded k-unduloids of genus one having only cylindrical ends, for $k=30$ and $k=72$.*

Let us describe these surfaces in more detail. First of all, they are coplanar, so that there is a *horizontal* symmetry plane of reflection, containing the axes of all ends. Moreover, each has a large group of symmetries around the vertical axis, namely a dihedral group with 15 (or 36) vertical mirror planes. The entire surface is generated from a fundamental piece depicted in Color Plate 10, top left, by reflection in its boundary planes. This fundamental piece contains half the cylindrical end, namely the portion above the horizontal plane; moreover, it contains two (half) bubbles of different sizes. In fact, 60

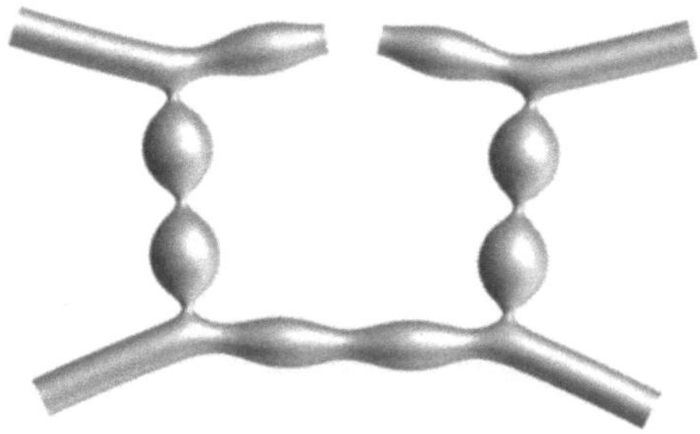

Fig. 2. A 72-unduloid of genus 1 with only cylindrical ends can be obtained from a fundamental domain with a 95° angle opposite a cylinder end. The picture shows four reflected fundamental domains leaving a 20° gap. Unlike the two segments at the top (across whose boundaries further reflection can be performed), the four cylindrical ends extend to infinity, and are truncated merely for the graphics.

(or 144) copies of the fundamental domain make up the entire surface. The asymptotic axes of the ends do not meet in the center of symmetry, but instead they all have the same nonzero distance to this point. Since each fundamental domain is embedded, an immersion of the closed solid torus with 30 (or 72) boundary punctures extends the immersion to the interior. Thus our surfaces are almost embedded, even though they are evidently not embedded.

5 The Fundamental Domains as Truncated Triunduloids

We think of the fundamental domains of our surfaces with cylindrical ends as truncated, and slightly deformed, triunduloids. Here, by *triunduloid* we mean specifically an almost embedded 3-unduloid of genus zero. In [6] we classify the triunduloids by triples of points in $\mathbb{S}^2$. This classification has the following consequence for the angles enclosed by the asymptotic axes of the ends (compare Figure 3).

Proposition 4. *A triunduloid can have at most one end with cylindrical necksize. Furthermore, the angle opposite the cylindrical end must be in the interval* $(\pi/2, 2\arccos(2/3)]$ *while the two angles adjacent to the cylinder end are in the range* $(\pi/2, \pi)$. *The latter two angles are equal only when the angle opposite the cylindrical end is exactly* $2\arccos(2/3)$.

For each possible angle opposite the cylinder end, in the allowed interval $(\pi/2, 2\arccos(2/3)]$, trigonometric formulas determine the remaining two angles exactly [6]. Currently we are working on an existence proof for triunduloids whose result would in particular imply the one-parameter family

Fig. 3. A one-parameter family of triunduloids with one cylindrical end. Cylinders with a perpendicular string of spheres attached form the degenerate limiting surfaces of the family. The pictures indicate half of the family running from a degenerate surface with axis graph ⊢, to a Y-shaped (or isosceles) surface, past which it can be continued with mirror images up to another degenerate ⊣.

suggested by Proposition 4. At present, however, the conditions stated in the proposition are only necessary conditions, and for sufficiency we take the computer experiments as support.

A triunduloid with one cylindrical end can be truncated (at some neck along each of the other two ends) to give a fundamental domain for a surface with only cylindrical ends. We want the curves along which we cut to become planar lines of curvature across which the surface can then be extended by (Schwarz) reflection. To achieve this, the surface must be slightly perturbed. In particular, the parameters of such a *truncated triunduloid*, namely the axis directions for given necksizes, must deviate from those of the complete triunduloid. While these parameters are known theoretically [6] for triunduloids, they must be determined experimentally in the truncated case in a way we will explain below. In fact, we even need to make the definition of these parameters precise in the truncated case. For the necksizes we simply take the length of the bounding curves. Since each of these curves is contained in a *vertical* plane, the axis direction can be defined to be the normal to this plane.

To see how much the parameters for truncated triunduloids differ from those of the complete surfaces, we look at the angle deviation for a given triple of necksizes. This point of view is natural since triunduloids and truncated triunduloids seem to exist for the same range of necksizes. Of course, if we truncate far out, leaving many bubbles on the truncated ends, then the angles for the truncated surface will approach those of the original triunduloid, as expected from the asymptotic convergence result [11].

Experimentally, we can determine the axis directions for truncated triunduloids by solving a *period problem*: since we need to reflect across three boundary curves in some horizontal plane, we require all three to be contained in the same plane. From the numerical construction of a fundamental domain, all we know is that they are contained in parallel planes. However, we can adjust the angle parameters to "kill" the periods, that is, to make the parallel planes coincide. As the end itself has no period, its two bounding arcs are contained in the same plane, so effectively we are left with only a single period problem. In our computer experiments, when we use (for the truncated triunduloid) the theoretically known parameters for a complete triunduloid, we find no observable period. Therefore we may conclude that the angle parameters (for a given triple of necksizes) are almost exactly the same for a truncated triunduloid as for the original triunduloid, even if only a small number of bubbles is left on each end. (This differs from our experimental observation for other problems, like the rhombic surfaces considered in [5,7]: there the deviations in the angle parameters between the truncated and complete surfaces are significant.)

Thus it is reasonable to consider the angle range of Proposition 4 as valid for the truncated case, too. On the grounds of this assumption, we now calculate the lowest number of ends possible. The angle φ opposite a cylindrical end lies in the interval between 90° and $2\arccos(2/3) \approx 96.4°$. The dihedral angle of the two truncating planes is $\pi - \varphi$. Hence, if successive reflection leads to a closed surface with k ends then $(\pi - \varphi)\cdot k = 2\pi l$ for some integer l. Note that the number of ends or fundamental domains k must be even, since according to Proposition 4, only for $\varphi = 2\arccos(2/3)$ itself does the truncated triunduloid have an extra mirror symmetry; in every other case the segments between the cylindrical ends alternate in necksize. Moreover, l gives the turning number of the k-gon one sees in the horizontal symmetry plane when the axes of the ends are deleted. It is not hard to determine the lowest value for even k when φ is in the given range: with $\varphi = 96°$ we obtain $k = 30$ and $l = 7$ as in Color Plate 10. Furthermore, every even $k \geq 30$ except for 32, 36, 40, 44, 48, 52 and 56 gives an example, while for every even $k \geq 102$ there is more than one example since different turning numbers are possible. (For $k = 72$, we have $\varphi = 95°$ and $l = 17$ as in Figure 2.)

This calculation, together with the experimental fact that truncation of a triunduloid does not significantly change the angle parameters, has the following consequence:

Experimental result 5. *Suppose that a coplanar k-unduloid of genus one has dihedral symmetry, acting transitively on its k cylindrical ends. Then k is even and $k \geq 30$ (and if k is a multiple of four, then $k \geq 60$).*

6 Conjectures

By the reasoning of Section 5 it is obvious that many further CMC surfaces with only cylinder ends must exist. First, we can increase the number of bubbles on each *segment*, that is, each annular portion of surface between successive cylindrical ends. Second, any other "rational" angle $\varphi \in (\pi/2, 2\arccos(2/3))$ will lead to a similar surface.

Conjecture 6. *For all but a finite number of even integers $k \geq 30$, there exist coplanar k-unduloids of genus 1 with only cylindrical ends and transitive dihedral symmetry on the set of ends. Moreover, for each such integer there is a countable family of these surfaces, distinguished by turning number and by the numbers of bubbles on the two distinct segments modulo symmetry.*

To find coplanar surfaces with fewer ends, we consider higher genus. For instance, we propose a 6-unduloid with genus 4 and dihedral symmetry transitive on the ends. The fundamental domain for this surface would be a coplanar 4-unduloid with one cylindrical end and two angles of 60° opposite this end. The axes of the segments then form an equiangular hexagon with all three of its diagonals, while the axes of the ends are six outward rays from the vertices. This surface fails to be embedded near the points where the diagonals cross. Because this surface is coplanar with simply connected fundamental domain, it is accessible by our experimental methods.

All the surfaces we have considered so far are almost embedded, but none are embedded. To find examples without self-intersection it seems we must drop the coplanarity assumption. We now propose an embedded 6-unduloid of genus 1. Consider the cycle consisting of the six edges on a cube which avoid the pair of opposite vertices along some body-diagonal direction; this cycle is a right-angled skew hexagon. Now compress the cube along the body-diagonal, until the equal angles in the hexagon increase to $2\arccos(2/3)$. Next, take an isosceles triunduloid with one cylindrical end and two truncated ends of necksize $\pi/2$. Place six copies of the triunduloid along the cycle, with the truncated ends along the edges, and a cylindrical end sticking outward at each vertex. With an appropriate edgelength for the original cube, these triunduloids should approximately match up at the truncations. The twist along each edge means there is no plane of mirror symmetry, so our present methods do not apply. Still, we believe that the truncated ends can be fused even in this case, producing an embedded example of a 6-unduloid with all ends cylindrical. In fact we suspect that this example is part of a one-parameter family of similar surfaces, using the other triunduloids from Proposition 4.

Recall that, generically, the space of k-unduloids with all ends cylindrical should have dimension $k-6$. This means that if the one-parameter family just mentioned does exist, it is nongeneric because of its symmetry. But similar symmetries fail to exist for $k = 5$, while other arguments rule out lower k. Thus the dimension count leads us to conjecture that k-unduloids with all ends cylindrical exist only for $k \geq 6$.

References

1. S. Cohn-Vossen, *Sur la courbure totale des surfaces ouvertes*, C. R. Acad. Sci. **197** (1933), 1165–1167.

2. C. Delaunay, *Sur la surface de révolution, dont la courbure moyenne est constante*, Journal de mathématiques **6** (1841), 309–320.
3. J. Eells, *The surfaces of Delaunay*, The Math. Intell. **9** (1987), 53–57.
4. K. Grosse-Brauckmann, *New surfaces of constant mean curvature*, Math. Zeit. **214** (1993), 527–565.
5. K. Grosse-Brauckmann, R. Kusner, *Moduli spaces of embedded constant mean curvature with few ends and special symmetry*, SFB 256 preprint **483**, Bonn, 1996.
6. K. Grosse-Brauckmann, R. Kusner, J. Sullivan, *Classification of embedded constant mean curvature surfaces with genus zero and three ends*, preprint, 1997.
7. K. Grosse-Brauckmann, K. Polthier, *Constant mean curvature surfaces derived from Delaunay's and Wente's examples*, p. 119–134 in: Visualization and Mathematics (H.-C. Hege, K. Polthier, eds.), Springer Verlag Heidelberg, 1997.
8. N. Kapouleas, *Complete constant mean curvature surfaces in Euclidean three-space*, Ann. of Math. **131** (1990), 239–330.
9. H. Karcher, *The triply periodic minimal surfaces of A. Schoen and their constant mean curvature companions*, Man. math. **64** (1989), 291–357.
10. N. Korevaar, R. Kusner, *The global structure of constant mean curvature surfaces*, Invent. Math. **114** (1993), 311–332.
11. N. Korevaar, R. Kusner, B. Solomon, *The structure of complete embedded surfaces with constant mean curvature*, J. Diff. Geom. **30** (1989), 465–503.
12. R. Kusner, *Bubbles, conservation laws, and balanced diagrams.* In: Geometric analysis and computer graphics (P. Concus, R. Finn, D. Hoffman, eds.), 103–108, Springer New York, 1990.
13. R. Kusner, *A maximum principle at infinity and the topology of complete embedded surfaces with constant mean curvature.* In: Global differential geometry and global analysis, 108–114, Lect. Notes Math. **1481**, Springer Berlin 1991.
14. R. Kusner, R. Mazzeo, D. Pollack, *The moduli space of complete embedded constant mean curvature surfaces*, Geom. Funct. Anal. **6** (1996), 120–137.
15. H.B. Lawson, *Complete minimal surfaces in* $\mathbb{S}^3$, Ann. of Math. **92** (1970), 335–374.
16. R. Mazzeo, D. Pollack, *Gluing and moduli for noncompact geometric problems*, preprint, 1996.
17. W.H. Meeks, *The topology and geometry of embedded surfaces of constant mean curvature*, J. Diff. Geom. **27** (1988), 539–552.
18. B. Oberknapp, K. Polthier, *An algorithm for discrete constant mean curvature surfaces*, p. 141–161 in: Visualization and Mathematics (H.-C. Hege, K. Polthier, eds.), Springer Verlag Heidelberg, 1997.
19. U. Pinkall, I. Sterling, *On the classification of constant mean curvature tori*, Ann. of Math. **130**, 407–451 (1989)
20. I. Sterling, H. C. Wente, *Existence and classification of constant mean curvature multibubbletons of finite and infinite type*, Indiana Univ. Math. J. **42** (1993), 1239–1266.
21. H. Wente, *Constant mean curvature immersions of Enneper type*, Memoirs of the AMS **478**, 1992.

Discrete Rotational CMC Surfaces and the Elliptic Billiard

Tim Hoffmann

Fachbereich Mathematik, Technische Universität Berlin, Germany

Abstract. It is a well known fact that the meridian curve of a rotational constant mean curvature (cmc) surface (which determines the surface completely) can be obtained as the trace of a focal point of an ellipse or hyperbola when rolling it on a straight line. In this paper, it will be shown that discrete rotational cmc surfaces can be obtained in a similar way. In fact, the meridian polygons are closely related to the elliptic (or hyberbolic) standard billiard: the discrete analogue of the ellipse or hyperbola will be the trace of a billiard in a continuous one.

1 Introduction

At the turn of the century the study of discrete objects often preceded continuous investigations (e.g. differential equations were viewed as limits of difference equations etc.). These discrete objects seemed to be lost for a while but due to the computer in our days they are in focus again. Nevertheless already in the early fifties mathematicians in Vienna like W. Wunderlich and R. Sauer started to study discrete analogues of smooth surfaces. These surfaces were discrete in the sense that they tried to discretize the geometric properties rather than to simply approximate smooth surfaces. In 1994, A. Bobenko and U. Pinkall benefited from this approach when they extended the definitions of Wunderlich [9] for discrete surfaces of negative Gaußian curvature (K-surfaces) and showed that they are equivalent to an integrable difference equation - the now famous discrete Sine Gordon equation [1]. Again, A. Bobenko and U. Pinkall found a discretization for surfaces of constant mean curvature (cmc) which leads in turn to an integrable discretization of the corresponding smooth integrable equation [2].

All these discretizations have in common that the (discrete) surfaces show the typical behavior of their smooth counterparts - even in very rough discretizations. They possess, for example, discrete versions of the Bäcklund transformations which are well-known for the continuous ones. Moreover, the construction can be done explicitly without numerically solving partial differential equations. For example, the construction of cmc surfaces is very difficult and general methods need a splitting in some loop group which can only be done approximatively for visualization purpose [5]. The discrete version, however, can be solved exactly [6].

Meanwhile, there exist not only discrete versions of K and cmc surfaces (see e.g. [4]). However, in the continuous case it is well-known that the rotational K and cmc surfaces are closely related and it turns out that this holds for the discrete case too: In fact, they both come from the same discrete difference equation. N. Kutz showed in [8] that rotational K-surfaces are equivalent to the standard billiard in an ellipse. The aim of this paper is to show that the billiard in turn gives discrete rotational cmc surfaces.

We start with a short review of the discrete surfaces we have to deal with:

2 Discrete Rotational Surfaces

It is a well-known fact that every (continuous) rotational surface allows isothermic parametrisation. A discrete analogue of isothermic surfaces is given in [2]. We will this definition:

Definition 1. A discrete isothermic surface in $\mathbb{R}^3$ is a map $F : \mathbb{Z}^2 \to \mathbb{R}^3$ for which the elementary quadrilaterals $[F_{n,m}, F_{n+1,m}, F_{n+1,m+1}, F_{n,m+1}]$ have crossratio[1] $= -1$[2].

A *dual* surface F^* of F is given by the equations:

$$\begin{aligned}(F^*_{m+1,n} - F^*_{m,n}) &= \lambda \tfrac{(F_{m+1,n} - F_{m,n})}{|F_{m+1,n} - F_{m,n}|^2}\\ (F^*_{m,n+1} - F^*_{m,n}) &= -\lambda \tfrac{(F_{m,n+1} - F_{m,n})}{|F_{m,n+1} - F_{m,n}|^2},\end{aligned} \tag{1}$$

$\lambda \in \mathbb{R} \setminus \{0\}$. The dual surface is itself isothermic again.

In general one can obtain discrete rotational surfaces by rotating a polygon $P_{n,m} \in \mathbb{R}^2$ around an axis, for example the x-axis: Choose $\varphi \in (0, \pi)$ and define $F : \mathbb{Z}^2 \to \mathbb{R}^3$ by

$$F(n,m) = F_{n,m} := ((P_m)_1, \cos(n\varphi)(P_m)_2, \sin(n\varphi)(P_m)_2) \tag{2}$$

The condition for this surface to be isothermic is the[3]:

$$||P_n - P_{n+1}||^2 = 4|cr| \sin^2 \frac{\varphi}{2} (P_n)_2 (P_{n+1})_2 \tag{3}$$

This is equivalent to the condition that the crossratio of P_n, P_{n+1} and their complex conjugates is constant. It can be interpreted as an arclength parametrization in the hyperbolic plane.

---

[1] The crossratio of four complex numbers a, b, c and d is given by $cr = \frac{(a-b)(c-d)}{(b-c)(c-a)}$. It is invariant under Möbius transformations (fractional linear transformations of the complex plane). The crossratio is real iff the four points lie on a circle. Therefore it is possible to demand a real crossratio for points in space: they must lie on a circle, and this defines a plane.

[2] In a more general version of this definition the crossratio is of the form $\frac{\alpha_n}{\beta_m}$.

[3] Choose $cr = -1$ for the narrow definition of discrete isothermic (crossratio = -1).

3 Unrolling Polygons and Discrete Rotational Surfaces

Let $Q : \mathbb{Z} \to \mathbb{R}^2$ be a polygon, $p \in \mathbb{R}^2$ a fixed point. Think of this as being a set of triangles $\Delta(Q_n, Q_{n+1}, p)$. Now take these triangles and place them with the edges $[Q_n, Q_{n+1}]$ on a straight line, for example the x-axis. The result is a sequence of points P_n. This new polygon P is the discrete trace of p when unrolling the polygon Q along the x-axis (Fig. 1).

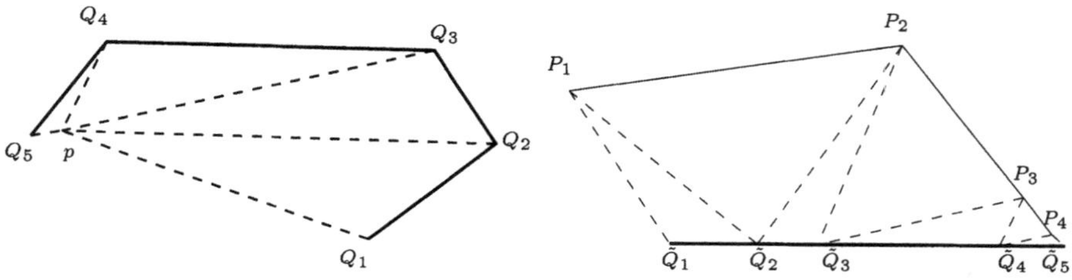

Fig. 1. Unrolling a polygon

We can rotate this polygon P around the x-axis to obtain a discrete rotational surface: Choose $\varphi \in (0, \pi)$ and define $F : \mathbb{Z}^2 \to \mathbb{R}^3$ by

$$F(n,m) = F_{n,m} := ((P_m)_1, \cos(n\varphi)(P_m)_2, \sin(n\varphi)(P_m)_2) \tag{4}$$

The condition (3) for F to be isothermic reads in terms of Q as follows:

$$\frac{1}{c} + 1 = \frac{1 + \cos(\alpha_n - \beta_n)}{1 + \cos(\alpha_n + \beta_n)} \tag{5}$$

where $\alpha_n = \angle(Q_{n-1}, Q_n, p)$, $\beta_n = \angle(p, Q_n, Q_{n+1})$ and $c = |cr| \sin^2 \frac{\varphi}{2}$. Or equivalently:

$$\tan \frac{\alpha_n}{2} \tan \frac{\beta_n}{2} = const. \tag{6}$$

4 The Standard Billiard in an Ellipse and Hyperbola

Let Ebe a given ellipse. Choose a starting point Q_0 on E and a starting direction in Q_0 pointing to the inner of E. Shooting a ball in that direction will give a new point Q_1 where it hits the ellipse again[4]. The new direction in Q_1 now is given by the usual reflection law: incoming angle = outgoing angle. This leads to a sequence $Q_0, Q_1, Q_2, \dots$ (see fig. 3 upper left). If one

[4] namely the point where the ball hits again the ellipse after moving on a straight line.

thinks of these points as the vertices of a polygon, it is again a well-known fact that the edges are tangential to either a confocal hyperbola or a confocal ellipse, depending whether the first shot goes between the two focii or not [7]. In a hyperbola the situation is similar with the one difference, that one must change the branch if the shooting line does not hit the first branch twice (Fig. 2).

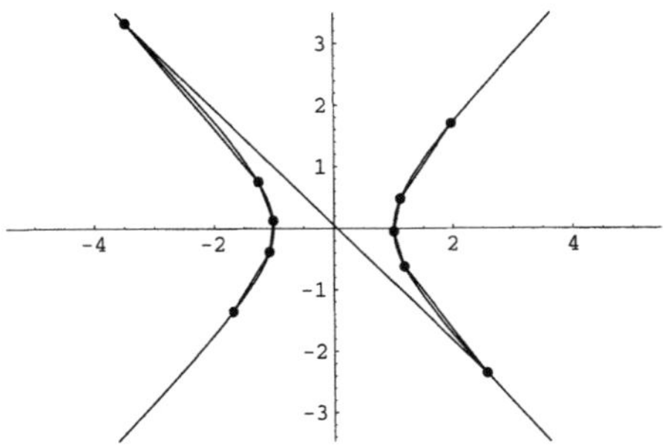

Fig. 2. Billiard in a Hyperbola

5 Discrete Rotational CMC Surfaces

Now we have to recall the notion of a discrete surface of constant mean curvature first defined in [3]. We will give it in the following way:

Definition 2. An isothermic surface F is a *cmc surface with mean curvature* $H \neq 0$ if there is a properly scaled and placed dual surface F^* of F in constant distance

$$||F_{n,m} - F^c_{n,m}|| = \frac{1}{H} \in \mathbb{R}.$$

Examples for discrete cmc surfaces and a general construction mechanism can be found in [6].

Now we will see that the polygon given by unrolling an elliptic or hyperbolic billiard trace is the meridian of a discrete rotational cmc surface. For shortness we will restrict ourselves to the elliptic case. First we show that it generates an isothermic surface:

Theorem 3. *The polygon obtained by playing the standard billiard in an ellipse together with one focus satisfies condition (5).*

Proof. Let us denote the two foci of the ellipse by $\tilde{p}$ and p. We will use the fact that the edges $[Q_n, Q_{n+1}]$ of the standard billiard in the ellipse are tangential

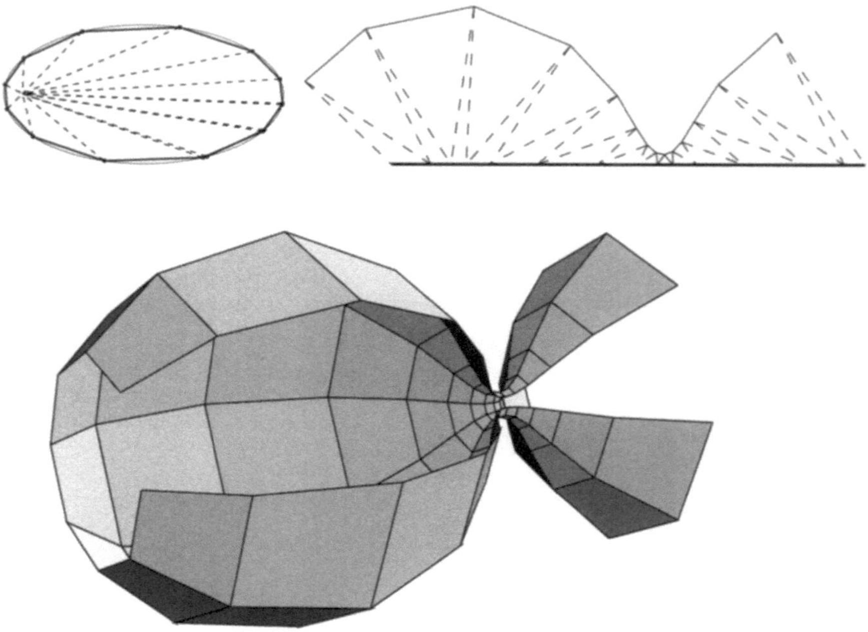

Fig. 3. Billiard in an ellipse, the meridian curve, and the resulting discrete cmc surface

to a confocal ellipse or hyperbola. We give the calculations for the case of the confocal quadric being an ellipse. The other case can be treated completely analogue.

If we look at the triangles made from $\tilde{p}, Q_n$ and p we see that the angle at Q_n is $\alpha_n - \beta_n$. For the edges $l_n = |p - Q_n|$ and $\tilde{l}_n = |\tilde{p} - Q_n|$ one has $l_n + \tilde{l}_n = const$. Set $c = |p - \tilde{p}|$ Now we have :

$$c^2 = l_n^2 + \tilde{l}_n^2 - 2 l_n \tilde{l}_n \cos(\alpha_n - \beta_n) \tag{7}$$

This gives:

$$1 + \cos(\alpha_{n+1} - \beta_{n+1}) = \frac{(l_{n+1} + \tilde{l}_{n+1})^2 - l_n^2 - \tilde{l}_n^2 + 2 l_n \tilde{l}_n \cos(\alpha_n - \beta_n)}{2 l_{n+1} \tilde{l}_{n+1}} \tag{8}$$

Denote the point obtained by mirroring $\tilde{p}$ at the straight line through Q_{n-1} and Q_n by $\hat{p}_n$ then $|p - \hat{p}_n| = \hat{c}$ is constant too since the intersection point of

$[Q_{n-1}, Q_n]$ and $[p, \hat{p}_n]$ is a point of the inner ellipse [5]. Since $|Q_n - \tilde{p}| = |Q_n - \hat{p}_n|$ and $\angle(\hat{p}_n, Q_n, p) = \alpha + \beta$ one has:

$$\hat{c}^2 = l_n^2 + \tilde{l}_n^2 - 2 l_n \tilde{l}_n \cos(\alpha_n + \beta_n) \tag{9}$$

And therefore:

$$1 + \cos(\alpha_{n+1} + \beta_{n+1}) = \frac{(l_{n+1} + \tilde{l}_{n+1})^2 - l_n^2 - \tilde{l}_n^2 + 2 l_n \tilde{l}_n \cos(\alpha_n + \beta_n)}{2 l_{n+1} \tilde{l}_{n+1}} \tag{10}$$

Combining equations (8) and (10) one gets:

$$\begin{aligned} \frac{1 + \cos(\alpha_{n+1} - \beta_{n+1})}{1 + \cos(\alpha_{n+1} + \beta_{n+1})} &= \frac{(l_n + \tilde{l}_n)^2 - l_n^2 - \tilde{l}_n^2 + 2 l_n \tilde{l}_n \cos(\alpha_n - \beta_n)}{(l_n + \tilde{l}_n)^2 - l_n^2 - \tilde{l}_n^2 + 2 l_n \tilde{l}_n \cos(\alpha_n + \beta_n)} \\ &= \frac{1 + \cos(\alpha_n - \beta_n)}{1 + \cos(\alpha_n + \beta_n)} \end{aligned} \tag{11}$$

showing the invariance of $\frac{1+\cos(\alpha_n - \beta_n)}{1+\cos(\alpha_n + \beta_n)}$.

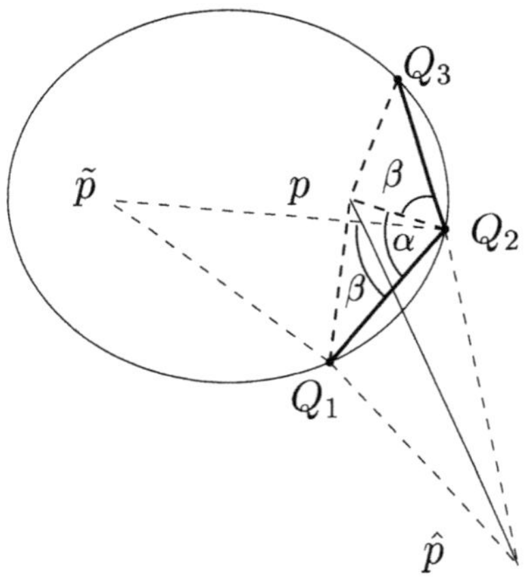

Fig. 4.

Discrete rotational cmc surfaces have been constructed in [6]. Here we present an alternative method based on above continuous construction. In the continuous case one can obtain rotational cmc surfaces if one rotates the trace of one focus when unrolling an ellipse or hyperbola. Here is the corresponding discrete

[5] One can see this from the fact that the normal of an ellipse in a point Q cuts the inner angle $\angle(p, Q, \tilde{p})$ into half.

Theorem 4. *The discrete rotational surface obtained by rotating the unrolled trace of a ball in an elliptic (or hyperbolic) billiard (see theorem 3) is a discrete rotational cmc surface.*

Proof. To proof this, one has to find a dual surface in constant distance. If we trace both foci when evolving the ellipse we get a second polygon $\hat{P}$. Now mirror it at the axis (Fig. 5). We already saw in the proof of theorem 3 that the distance $|P_n - \hat{P}_n| = \tilde{c}$ is constant. From the reflection law one gets that $|P_n - \hat{P}_{n+1}| = |P_{n+1} - \hat{P}_n| = c$ and therefore is constant. So $[P_n, P_{n+1}]$ and $[\hat{P}_n, \hat{P}_{n+1}]$ are parallel and

$$|P_n - P_{n+1}|^2 = \frac{2(c^2 - \tilde{c}^2)}{|\hat{P}_n - \hat{P}_{n+1}|^2}$$

This in turn leads to

$$|P_n - P_{n+1}| = \frac{\lambda}{|\hat{P}_n - \hat{P}_{n+1}|}$$

with $\lambda = \frac{c^2 - \tilde{c}^2}{4}$.

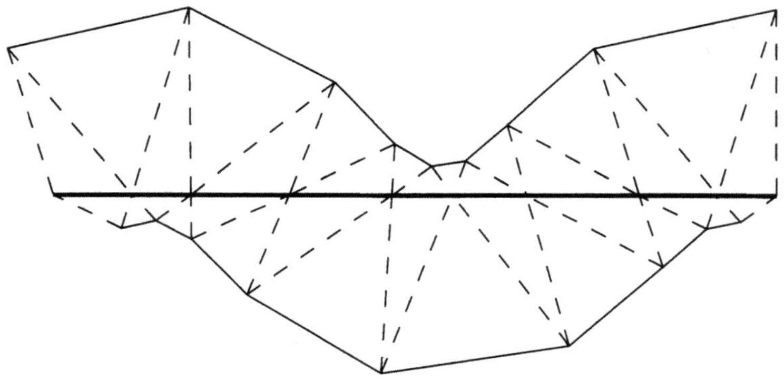

Fig. 5. A meridian and its dual

Acknowledgements

The author would like to thank N. Kutz and E.-H. Tjaden for helpfull discussions and remarks on the billiards. Special thanks go to G. Haak and I. Sterling who first brought up the idea of finding the discrete analogue of the smooth construction.

References

1. A. Bobenko and U. Pinkall, *Discrete surfaces with constant negative Gaussian curvature and the Hirota equation*, SFB288 Preprint TU Berlin.
2. ———, *Discrete isothermic surfaces*, J. reine angew. Math. **475** (1996), 178–208.
3. ———, *Discretization of surfaces and integrable systems*, to appear in Discrete Integrable Geometry and Physics (A. Bobenko and R. Seiler, eds.), 1997.
4. A. Bobenko and W. Schief, *Discrete affine spheres*, to appear in Discrete Integrable Geometry and Physics, GANG Preprint IV. 25 Amherst.
5. J. Dorfmeister, F. Pedit, and H. Wu, *Weierstrass type representation of harmonic maps into symmetric spaces*, G.A.N.G. Preprint III.25 Amherst (1994).
6. T. Hoffmann, *Discrete cmc surfaces and discrete holomorphic maps*, to appear in: Discrete Integrable Geometry and Physics (A. Bobenko and R. Seiler, eds.), 1997.
7. F. Klein, *Vorlesungen über höhere Geometrie*, Springer, 1926.
8. N. Kutz, *The doubly discrete sine-gordon equation in geometry and physics*, Ph.D. thesis, TU Berlin, 1996.
9. W. Wunderlich, *Zur Differenzengeometrie der Flächen konstanter negativer Krümmung*, Sitzungsber. Ak. Wiss. **160** (1951), 39–77.

Zonotope Dynamics in Numerical Quality Control

Wolfgang Kühn *

Konrad-Zuse-Zentrum für Informationstechnik Berlin (ZIB)

Abstract. Mathematical rigorous error bounds for the numerical approximation of dynamical systems have long been hindered by the wrapping effect. We present a new method which constructs high order zonotope (special polytopes) enclosures for the orbits of discrete dynamical systems. The wrapping effect can made arbitrarily small by controlling the order of the zonotopes. The method induces in the space of zonotopes a dynamical system of amazing geometrical complexity. We emphasis the visualization of the zonotopes to better understand the involved dynamics.

1 Discrete Dynamical Systems

The Cremona map is given by

$$z = \begin{pmatrix} x \\ y \end{pmatrix} \mapsto \begin{pmatrix} x\cos\lambda - (y - x^2)\sin\lambda \\ x\sin\lambda + (y - x^2)\cos\lambda \end{pmatrix}. \tag{1}$$

It is a planar, area-conserving map which arises via a reduction process in the study of celestial mechanics ([12]). For a given initial point $z_0 = (x_0, y_0)$, iterating the map yields the orbit $z_0 \mapsto z_1 \mapsto z_2 \mapsto \cdots$. Several orbits are shown in Fig. 1, which were created on a finite precision computer. The question to address is how well these computed orbits approximate the corresponding true orbits.

In this paper, we consider discrete dynamical systems in $\mathbf{R}^d$ defined by

$$z_n = f_n(z_{n-1}). \tag{2}$$

The initial value problem consist of finding the orbit $\{z_n\}_{n=0}^N$ for given initial point z_0. Discrete dynamical systems arise frequently in many areas of applied mathematics such as discrete evolution equations, the time discretization of continues dynamical systems (ordinary differential equation, hyperbolic and parabolic equations), recursion relations for special functions, the solution of sparse large linear systems etc.

System (2) can usually not be solved exactly, and one has to resort to approximation schemes executed on a computer. Practice shows that such schemes are easy to implement and generally yield accurate approximations.

* This work was done at School of Mathematics, Georgia Institute of Technology, Atlanta, GA

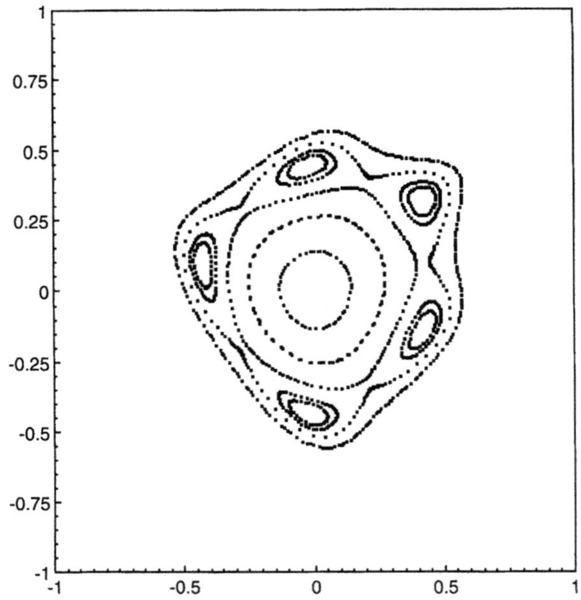

Fig. 1. Seven distinct orbits for the Cremona map.

However, to rigorously estimate the approximation error is a delicate problem which was until recently not solved in a efficient and effective way.

One possibility is to compute an approximate numerical orbit $\tilde{z}_n$ such that

$$\tilde{z}_n \approx f_n(\tilde{z}_{n-1}), \tag{3}$$

and to estimate the norm error

$$\varepsilon_n = ||z_n - \tilde{z}_n||. \tag{4}$$

Because the norm is a simple scalar quantity, it is often possible to carry out an a priori error analysis by hand, which is the main advantage of this approach. On the other hand, norm estimates tend to be very crude, overestimating the real error by many orders of magnitude.

A second and broader approach is to replace the original system (2) by the set valued dynamical system

$$\Omega_n = f_n(\Omega_{n-1}) \tag{5}$$

for given initial set Ω_0. Now sets, and not just singular points as in (2), get mapped forward in time . Again orbits for the system (5) cannot be computed exactly, and an approximation scheme now consist of finding an enclosing orbit $\tilde{\Omega}_n$ for (5) such that

$$\tilde{\Omega}_n \supseteq f_n(\tilde{\Omega}_{n-1}). \tag{6}$$

Note that the approximation error is already build into (6) because $\tilde{\Omega}_n$ contains Ω_n for all n. The orbit $\{\tilde{\Omega}_n\}$ should be chosen suitable for representation on a computer.

2 The Wrapping Effect

We measure the overestimation of the real orbit $\{\Omega_n\}$ by its approximation $\{\tilde{\Omega}_n\}$ with the Hausdorff distance

$$\text{dist}(\Omega, \tilde{\Omega}) = \text{ diameter of smallest ball } B \text{ s.t. } \tilde{\Omega} \subseteq \Omega + B. \tag{7}$$

The Hausdorff distance is measure of how close two sets are in visual terms, see Fig. 3. The difficulty in iterating (6) lies in the wrapping effect: At each stage the set $f_n(\tilde{\Omega}_{n-1})$ is wrapped into the superset $\tilde{\Omega}_n$ and is thereby overestimated proportionally to its size. If this process is iterated, one usually sees the overestimation growing at an devastating exponential rate. All efforts to curb the wrapping effect by choosing different shapes and construction schemes for the sets $\tilde{\Omega}_n$ have failed. Classes of sets that have been used are simplices, parallelepipeds ([7]), general polytopes ([1]) and ellipsoids ([3] and [4]). For a recent overview see [10].

In this paper we present a method which beats the wrapping effect effectively and efficiently by constructing high order zonotope enclosures of the orbit. Theorem 8 states that the proposed method indeed shows (arbitrarily slow) sub-exponential overestimations.

3 Zonotopes, Intervals and the Interval Hull

Definition 1. A zonotope Z of order m is the Minkowski sum

$$Z = P^1 + \cdots + P^m \tag{8}$$

of m parallelepipeds P^i, see Fig. 2.

This definition is biased towards our intended use of zonotopes. However, it is equivalent to the "sum of line segment" definition which prevails in the literature, see [13] and [11]. Recall that a parallelepiped P is the linear image $P = M\Xi$ (here we allow the map M to be singular) of the unit cube $\Xi = [-1, 1]^d$, and that the Minkowski sum of two sets X and Y (see Fig. 2) is defined by

$$X + Y = \{x + y : x \in X \text{ and } y \in Y\}. \tag{9}$$

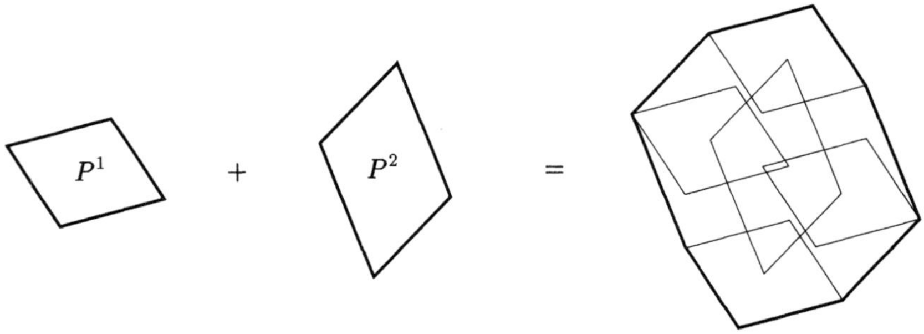

Fig. 2. The Minkowski sum of two parallelograms

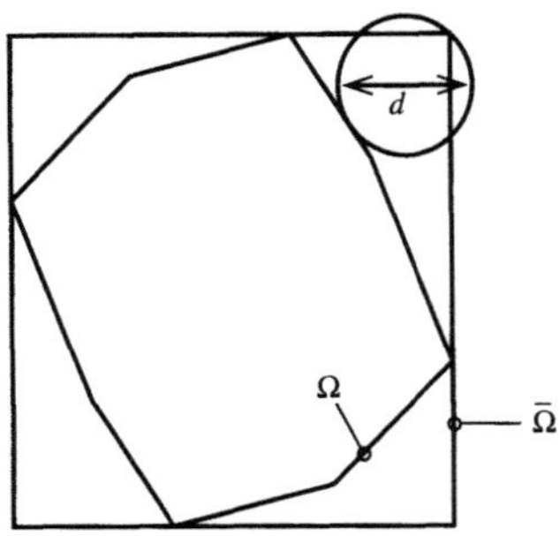

Fig. 3. The Hausdorff distance between Ω and $\tilde{\Omega}$.

The order m is a measure for the geometrical complexity of the zonotopes. It can be chosen freely and is a *performance parameter* for the method.

We should mention that zonotopes are relevant in such diverse areas as measure theory, computational geometry and convexity, but to the best of our knowledge have not yet been employed in numerical analysis.

Intervals are one of the most primitive sets and certainly the building blocks for numerical set-valued computation.

Definition 2. A (centered) interval E is a right rectangular cube aligned with the coordinate axes defined by the product

$$E = \prod_{i=1}^{d} [-e_i, e_i] = \left\{ x \in \mathbf{R}^d : |x_i| \leq e_i \right\}. \tag{10}$$

The interval hull $\square\, \Omega$ of a closed set Ω is the smallest interval that contains Ω, see Fig. 4.

The unit cube Ξ is an interval. Intervals are special for the fact they are the only sets whose section with respect to one coordinate is independent of all other coordinates. It is therefore relatively easy to compute the interval

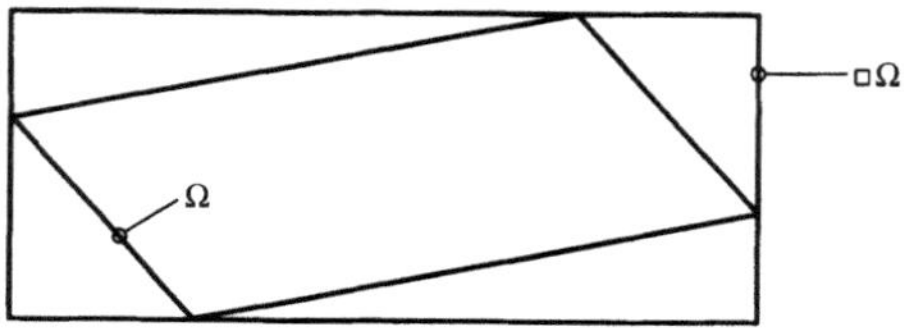

Fig. 4. The interval hull of a parallelogram and of a parallelepiped.

hull of a given set as in Lemma 3 for zonotopes and Lemma 4 for the range of functions. The branch of numerical analysis which replaces real machine numbers and vectors by intervals is called interval analysis, see [8] and [9].

Lemma 3. *Suppose $Z = P^1 + \cdots + P^k$ is a zonotope of order k with $P^l = M^l \Xi$. Then*

$$\square Z = \prod_i [-r_i, r_i] \qquad \textit{with} \qquad r_i = \sum_{l=1}^{k} \sum_{i,j=1}^{d} |M_{ij}^l|. \tag{11}$$

As a first order result we want to approximate a C^1 map $f : \mathbf{R}^d \to \mathbf{R}^d$ by a linear map plus an error term: for given set Ω, we compute an error set E such that $f(\Omega) \subseteq E + T\Omega$, see Fig. 5. The following lemma is a simple application of the Mean Value theorem to compute such an error set.

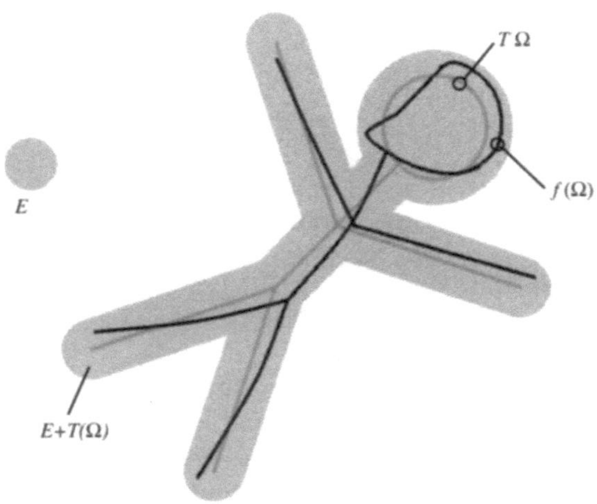

Fig. 5. The inflated linear image of a set encloses the non-linear image.

Lemma 4. *Suppose $f : \mathbf{R}^d \to \mathbf{R}^d$ is C^1 such that $f(0) = 0$. If T is a square matrix and $E = \prod_i [-e_i, e_i]$ is an (error) interval which satisfies, for all x*

and y in Ω,

$$e_i \geq \left| \sum_{j=1}^{d} (\mathrm{D}\, f_y - T)_{ij} x_j \right|, \tag{12}$$

then

$$f(\Omega) \subseteq E + T\Omega \tag{13}$$

holds.

4 Zonotope Dynamics

We can now give a concrete realization of the dynamical system (6) on the collection of zonotopes of order m. For given initial zonotope $\Omega_0 = Z_0$, consider the true orbit

$$\Omega_n = f_n(\Omega_{n-1}), \tag{14}$$

and the approximate orbit

$$Z_n = \mathcal{R}(E_n + T_n Z_{n-1}), \tag{15}$$

where T_n are square matrices and E_n are intervals such that $f_n(Z_{n-1}) \subseteq E_n + T_n Z_{n-1}$. In (15) we have introduced a reduction operator $\mathcal{R}$ which shall have the properties

- it maps zonotopes into zonotopes and
- it maps sets into supersets, i.e., $\Omega \subseteq \mathcal{R}\Omega$.

Before we discuss why $\mathcal{R} = I$ is not a practically feasible choice, let us state the following lemma (the easy proof is by induction and is omitted).

Lemma 5. *If the reduction $\mathcal{R}$ is such that $Z \subseteq \mathcal{R}Z$, then the orbit $\{\Omega_n\}$ is enclosed by $\{Z_n\}$, that is $\Omega_n \subseteq Z_n$ for all n.*

If Z_{n-1} in (15) is a zonotope of order m, then Z_n is a zonotope of order $m + 1$. This means that if (15) is iterated with the choice $\mathcal{R} = I$, then the order (complexity) of the resulting zonotopes increases linearly in the number of iterates n, and the computational cost increases quadratically in n. This is practically feasible only for n small.

5 The Cascade Reduction Algorithm

This section introduces a reduction from the space of zonotopes of order $m+1$ to the space of zonotopes of order m.

Definition 6 (Cascade Reduction). Let $Z = P^0 + P^1 + \cdots + P^m$ (here P^0 corresponds to the error term) be a $m+1$ zonotope and $1 < l \leq m$ be the largest integer such that

$$\operatorname{diam}(P^0 + P^1 + \cdots + P^{l-1}) \geq \operatorname{diam} P^l \tag{16}$$

or $l = 1$ otherwise. Then let

$$\mathcal{R}Z := \square(P^0 + P^1 + \cdots + P^l) + P^{l+1} + \cdots + P^m. \tag{17}$$

Remark 7.
1. The inclusion property $Z \subseteq \mathcal{R}Z$ indeed holds for the Cascade Reduction;
2. The zonotope $\mathcal{R}Z$ is of order $m - l + 1$ and an interval if $l = m$. In particular, the iterates in (15) are never of higher order than m.
3. The ordering relation (16) implies that $\operatorname{diam} P^k \leq \operatorname{diam} P^{k+1}$ for all $1 \leq k < m$. As a consequence, the interval hulls of small parallelepipeds are taken more frequently (l close to 1) than of big parallelepipeds (l close to m), resulting in the outstanding performance of the Cascade reduction. In [5], the rational and the implementation of the Cascade reduction is discussed in much more detail.

6 The Performance of the Cascade Reduction

We now state the sub-exponential overestimation property.

Theorem 8. *Suppose* $\operatorname{diam} E_n$ *and* $\|T_{n+k}T_{n+k-1}\cdots T_n\|$ *in (15) are uniformly bounded in both* k *and* n. *Then*

$$\operatorname{dist}(\Omega_n, Z_n) \leq c_1 c_2^{\sqrt[m]{n}} \tag{18}$$

for some constants c_1 *and* c_2.

The proof is rather technical and can be found in [5].

7 Example: The Cremona map

Let us now return to the Cremona map and use it to illustrate the dynamics of the Cascade Reduction in (15). Figure 6 shows an enclosed orbit of 2000 stages with initial value (0.0, 0.46). Zonotopes of order 5 and $\lambda = 1.3$ are used throughout this section.

The enclosures at five different stages are shown in Fig. 7. These stages were specifically chosen to demonstrate certain features. The first column contains both the zonotope enclosure $E_n + T_n Z_{n-1}$ of order 6 and its reduction $\mathcal{R}(E_n + T_n Z_{n-1})$. The second column shows the diameters of the 5 parallelogram summands on a logarithmic scale to base 10. As can be seen

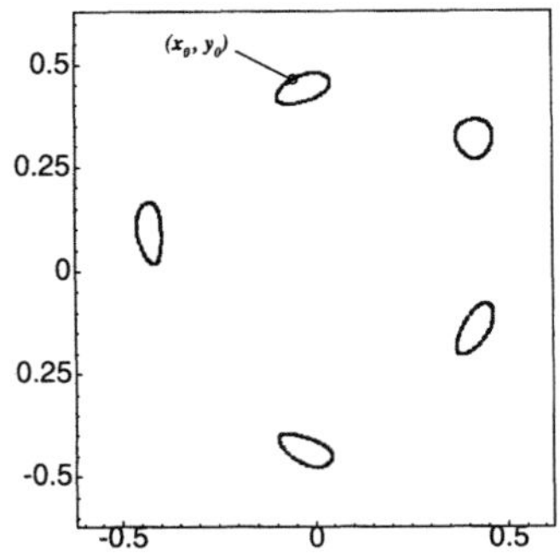

Fig. 6. Enclosed orbit with initial value (x_0, y_0).

from the second column, the values for l (by (16) equal to the number of zonotopes wrapped in that stage) are 3, 2, 4, 5 and 1 respectively. Big overestimation in (d) and (c) occur due to wrapping with zonotope of low order. In (d) the wrapper is an interval. No overestimation beyond the graphical resolution can be observed in (b) and (e). The zonotopes becomes more rounded if several summands have comparable magnitude as in (b). Case (e) is typical: there is one summand of dominating magnitude and a tail of smaller terms, thus giving the zonotope the shape of an parallelepiped.

8 Example: Langford's vector field

The following 3-d vector field is studied in [6] in connection with bifurcation to invariant tori.

$$x' = (\lambda - b)x - cy + x(z + d(1.0 - z^2)) \tag{19}$$

$$y' = cx + (\lambda - b)y + y(z + d(1.0 - z^2)) \tag{20}$$

$$z' = \lambda z - (x^2 + y^2 + z^2). \tag{21}$$

We set the parameters to $b = 3.0$, $c = 0.25$, $d = 0.2$, and $\lambda = 2.01$. The interval box centered at $(0.1, 0.1, 0.1)$ with radius 10^{-9} is chosen as initial set. Color plate 15, p. 375, in the Appendix shows 19 integrated forward in time with a total of 250 steps. The verifying initial value problem solver in [5] was used to compute the time-h map. We choose the stepsize $h = 0.02$, an order of integration of $p = 20$ and zonotopes of order $m = 5$. A modification of a *Mathematica* algorithm in [2] was implemented using the *Open Inventor 3D toolkit* to render the zonotopes on color plate 15, p. 375.

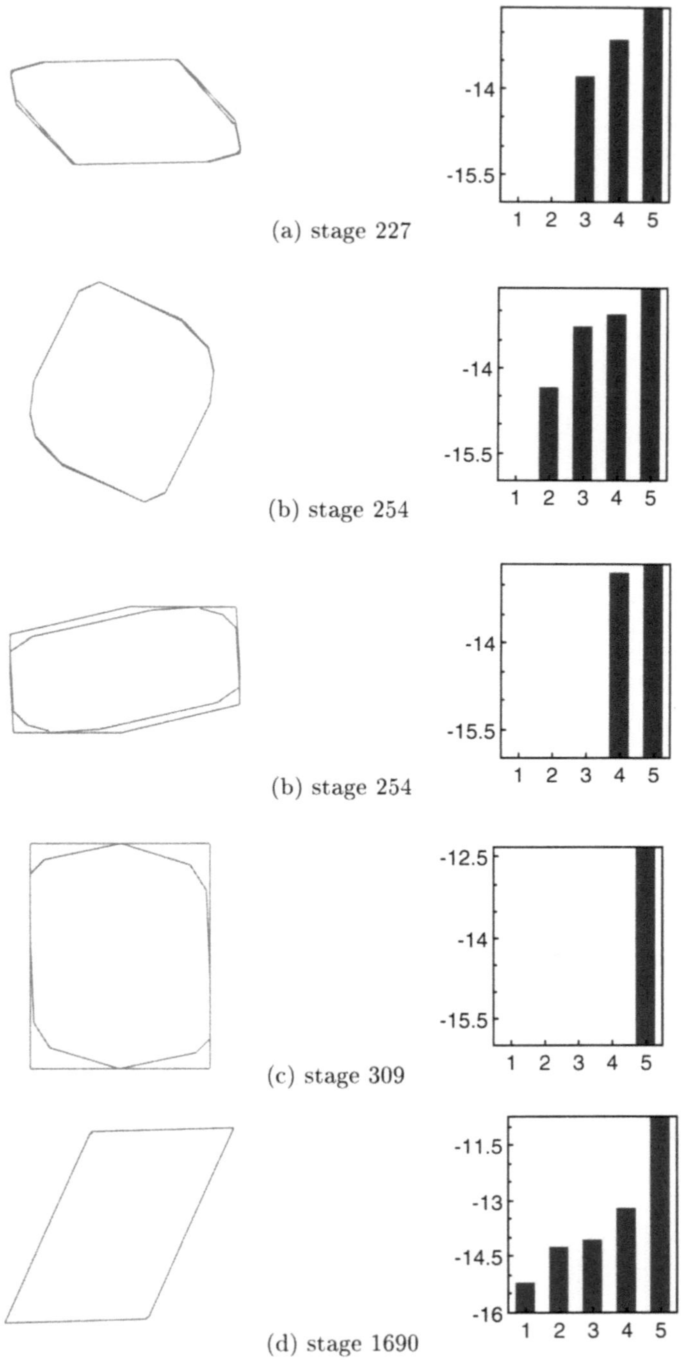

Fig. 7. Zonotope enclosure and corresponding radii for the orbit of the Cremona map at different stages.

References

1. D. P. DAVEY AND N. F. STEWART, *Guaranteed error bounds for the initial value problem using polytope arithmetic*, BIT **16** (1976), 257–268.
2. D. EPPSTEIN, *Zonohedra and zonotopes*, Tech. report, Dept. of Information & Computer Science U.C. Irvine, CA, 92717, 1995.
3. K. G. GUDERLEY AND C. L. KELLER, *A basic theorem in the computation of ellipsoidal error bounds*, Numer. Math. **19**:3 (1972), 218–229.
4. W. KAHAN, *A computable error bound for systems of ordinary differential equations*, Abstracts in SIAM Review **8** (1966), 568–569.
5. W. KUHN, *Rigorous and reasonable error bounds for the numerical solution of dynamical systems*, Ph.D. thesis, Georgia Institute of Technology, 1997.
6. W. LANGFORD, *Unfolding of degenerate bifurcations*, (P. FISHER AND W. SMITH, eds.), Marcel Dekker, 1985, pp. 87–103.
7. R. LOHNER, *Einschließung der Lösung gewöhnlicher Anfangs- und Randwertaufgaben und Anwendungen*, Ph.D. thesis, Univ. Karlsruhe, 1988.
8. R. E. MOORE, *Interval analysis*, Prentice-Hall, Englewood Cliffs, NJ, 1966.
9. A. NEUMAIER, *Interval methods for systems of equations*, publ Cambridge University Press, 1990.
10. R. RIHM, *Interval methods for initial value problems in odes*, (J. HERZBERGER, ed.), Elsevier Science B.V., 1994.
11. R. SCHNEIDER AND W. WEIL, *Zonoids and related topics*, Convexity and its Applications (P. M. GRUBER AND J. M. WILLS, eds.), Birkhäuser Verlag, Basel, 1983, pp. 296–317.
12. C. SIEGEL AND J. MOSER, *Lectures on celestial mechanics*, Springer-Verlag, 1971.
13. G. M. ZIEGLER, *Lectures on polytopes*, Springer-Verlag, 1995.

Straightest Geodesics on Polyhedral Surfaces

Konrad Polthier and Markus Schmies

Fachbereich Mathematik, Technische Universität Berlin, Germany

Abstract. Geodesic curves are the fundamental concept in geometry to generalize the idea of straight lines to curved surfaces and arbitrary manifolds. On polyhedral surfaces we introduce the notion of discrete geodesic curvature of curves and define straightest geodesics. This allows a unique solution of the initial value problem for geodesics, and therefore a unique movement in a given tangential direction, a property not available in the well-known concept of locally shortest geodesics.

An immediate application is the definition of parallel translation of vectors and a discrete Runge-Kutta method for the integration of vector fields on polyhedral surfaces. Our definitions only use intrinsic geometric properties of the polyhedral surface without reference to the underlying discrete triangulation of the surface or to an ambient space.

1 Introduction

Geodesics on smooth surfaces are the straightest and locally shortest curves. They generalize the concept of euclidean straight lines and play a fundamental role in the study of smoothly curved manifolds. Two basic properties are responsible for their importance: first, geodesics solve the initial value problem which states, that from any point of a manifold there starts a unique geodesic in any direction. Second, the length minimization property provides a solution of the boundary value problem of connecting two given points on a manifold with a locally shortest curve. On smooth surfaces geodesics possess both properties, in contrast to the situation on polyhedral surfaces.

The aim of this paper is to define *straightest curves* on two-dimensional polyhedral surfaces, as opposed to the concepts of locally shortest and quasi-geodesics. Such straightest geodesics will uniquely solve the initial value problem on polyhedral surfaces, and therefore allow to move uniquely on a polyhedral surface in a given direction along a straightest geodesic until the boundary is reached, a property not available for locally shortest geodesics. An application of straightest geodesics is the definition of parallel translation of vectors and higher order numerical integration methods for tangential vector fields. This allows the extension of Runge Kutta methods to polyhedral surfaces.

We consider polyhedral surfaces as two-dimensional simplicial complexes consisting of triangles. Each triangle has a flat metric and the common edge of two neighbouring triangles has the same length in both triangles. The definition of a metric on the polyhedral surface only requires the specification of edge lengths and does not refer to an immersion of the surface in an ambient space. This intrinsic approach allows the definition of straightest geodesics, discrete geodesic curvature, vector fields, and parallel translation of vectors in terms of the geometric data of the surface, such as edge lengths, triangle angles, and discrete curvature properties.

Geodesics on polyhedral surfaces were intensively studied using different definitions. The Russian school of A.D. Alexandrov [1] defines geodesics on polyhedral surfaces as locally shortest curves which leads to important implications in the study of non-regular and regular differential geometry. But shortest geodesics cannot be extended as shortest curves across a spherical vertex with positive Gauß curvature as, for example, the vertex of a cube. Beyond a hyperbolic vertex with negative Gauß curvature there even exists a continuum of extensions. Therefore, shortest geodesics fail to solve the initial value problem for geodesics at vertices of a polyhedral surface.

A.D. Alexandrov also introduced the concept of quasi-geodesics which are limit curves of geodesics on a family of converging smooth surfaces. They form a wider class than shortest geodesics and were amongst others studied by Pogorelov [7] on convex polyhedral surfaces. A quasi-geodesic through a spherical vertex is a curve with right and left angles both less than π, and therefore an inbound direction has multiple extensions.

Shortest geodesics appear in many practical applications. For example, the optimal movement of a robot should have minimal length in its parameter space. Such discrete minimization problems are studied in computational geometry, see for example Dijkstra [4], Sharir and Schorr [10], and Mitchell et.al. [6] for efficient algorithms on the computation of the shortest path in graphs and in polyhedral spaces.

Our paper starts in section 2 with a review of geodesics on smooth surfaces, especially since some of their properties differ from those of geodesics on polyhedral surfaces. In section 3 we will introduce polyhedral surfaces as metric spaces and recall basic facts. Straightest geodesics are defined in section 4 and discussed as solutions of the initial value problem. In section 5 we imbed the notion of straightest lines into the concept of discrete geodesic curvature of arbitrary curves on polyhedral surfaces. This general setting is more appropriate for our later discussions, and straightest geodesics turn out to be those class of curves with vanishing discrete geodesic curvature. As a validation of the definition we prove the Gauß-Bonnet theorem using our notion of discrete geodesic curvature. In section 6 we apply the concept to the definition of parallel translation of tangential vector fields and in section 7 we generalize Runge Kutta methods to the numerical integration of ordinary differential equations on polyhedral surfaces.

Applications of this paper are given in the video *Geodesics and Waves* [8]. The numerics were developed within the visualization environment OORANGE [5].

2 Review of Geodesics on Smooth Surfaces

Geodesics on smooth surfaces can be characterized by different equivalent properties. The generalized properties on polyhedral surfaces will no longer be equivalent and lead to different classes of discrete geodesics. The following material can be found in any introductory text book on differential geometry, see for example [3].

Let M be a smooth surface and $\gamma : I = [a, b] \to M$ a curve parametrized over an interval I. To avoid accelerations tangential to the curve we assume arc length parametrization, i.e. the tangent vector has constant length $|\gamma'| = 1$. A curve γ is called *locally shortest* if it is a critical point of the length functional $L(\gamma_{|[a,b]}) := length(\gamma_{|[a,b]})$ with respect to variations tangential to M which leave the endpoints fixed. Formally, if $\varphi : I \to T_\gamma M$ is a tangential vector field along γ with $\varphi(a) = 0$ and $\varphi(b) = 0$, then we have $\frac{\partial}{\partial\varepsilon} L(\gamma + \varepsilon\varphi)_{|\varepsilon=0} = 0$. A critical point of the length functional is usually not a global minimizer compared to curves with the same endpoints, see Color Plate 29 on page 382.

On smooth manifolds the length minimizing property of geodesics can be reformulated as an ordinary differential equation for γ, namely $\gamma''(s)^{\tan M} = 0$, the Euler-Lagrange equations of the variational problem.

The curvature $\kappa(s) = |\gamma''(s)|$ of a curve measures the infinitesimal turning of the tangent vector at every point $\gamma(s)$. For curves γ on surfaces $M \subset \mathbb{R}^3$, the curvature can be decomposed into the curve's bending in the normal direction n of the surface and its bending in the tangent space in direction of the binormal b. This decomposition leads to the definition of the geodesic curvature κ_g and the normal curvature κ_n of a curve:

$$\begin{aligned} \kappa^2(s) &= |\gamma''(s)|^2 \\ &= \left|\gamma''(s)^{\tan M}\right|^2 + \left|\gamma''(s)^{\text{nor } M}\right|^2 \\ &= \kappa_g^2(s) + \kappa_n^2(s). \end{aligned} \tag{1}$$

The geodesic curvature κ_g of a curve γ measures the tangential acceleration. If $\kappa_g = 0$ then the curve varies up to second order only in direction of the surface normal, therefore it is a *straightest curve* on the surface. The normal curvature κ_n is related with the bending of the surface itself and can be neglected from an intrinsic point of view.

Summarizing, one characterizes smooth geodesics as follows:

Definition 1. Let M be a smooth two-dimensional surface. A smooth curve $\gamma : I \to M$ with $|\gamma'| = 1$ is a *geodesic* if one of the equivalent properties holds:

1. γ is a locally shortest curve.
2. γ'' is parallel to the surface normal, i.e.

$$\gamma''(s)^{\tan M} = 0. \tag{2}$$

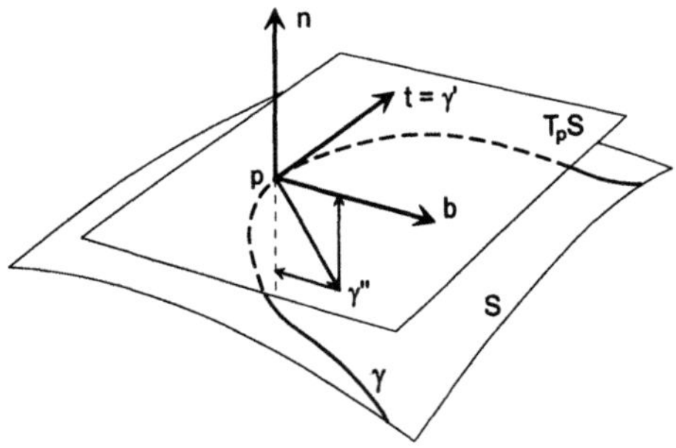

Fig. 1. Geodesic and normal curvature of a curve on a smooth surface.

3. γ has vanishing geodesic curvature $\kappa_g = 0$.

In section 4 we will consider geodesics on polyhedral surfaces and notice that the polygonal equivalents of the above properties lead to different notions of discrete geodesics.

The boundary value problem for geodesics has a solution in every homotopy class and is usually not unique. On the other hand, we have a unique solution for the initial value problem derived from equation (2):

Lemma 2. *Let M be a smooth manifold. Then for any point $p \in \overset{\circ}{M}$ in the interior of M and any tangent direction $v \in T_pM$ the initial value problem*

$$\begin{aligned} \gamma''(s)^{tan\ M} &= 0 \\ \gamma(0) &= p \\ \gamma'(0) &= v \end{aligned} \tag{3}$$

has a unique solution $\gamma : [0, \ell) \to M$, where ℓ is the length of the maximal interval of existence.

3 Curvature of Polyhedral Surfaces

In this section we review some facts on the geometry of polyhedral surfaces. Basic references are for example A.D. Alexandrov and Zalgaller [1] and Reshetnyak [9]. For simplification we restrict ourselves to two-dimensional surfaces consisting of planar triangles. A topological triangle f in a two-dimensional manifold S is a simple domain $f \subset S$ whose boundary is split by three vertices into three edges with no common interior points.

Definition 3. A *polyhedral surface* S is a two-dimensional manifold (with boundary) consisting of a finite or denumerable set F of topological triangles and an intrinsic metric $\rho(X, Y)$ such that

1. Any point $p \in S$ lies in at least one triangle $f \in F$.
2. Each point $p \in S$ has a neighbourhood that intersects only finitely many triangles $f \in F$.
3. The intersection of any two non-identical triangles g, $h \in F$ is either empty, or consists of a common vertex, or of a simple arc that is an edge of each of the two triangles.
4. The intrinsic metric ρ is flat on each triangle, i.e. each triangle is isometric to a triangle in $\mathbb{R}^2$.

Remark 4. Most of our considerations apply to a more general class of length spaces. Each face may have an arbitrary metric as long as the metrics of two adjacent faces are compatible, i.e. if the common edge has the same length in both faces, and the triangle inequality holds.

Let $\gamma \subset S$ be a curve whose segments on each face are rectifiable. Then the *length of* γ is well-defined and given by

$$Length(\gamma) = \sum_{f \in F} Length(\gamma_{|f}). \tag{4}$$

The neighbourhood of a vertex is isometric to a cone and is characterized by the total vertex angle:

Definition 5. Let S be a polyhedral surface and $v \in S$ a vertex. Let $F = \{f_1,..., f_m\}$ be the set of faces containing p as a vertex, and θ_i be the interior angle of the face f_i at the vertex p, compare figure 2. Then the *total vertex angle* $\theta(p)$ is given by

$$\theta(p) = \sum_{i=1}^{m} \theta_i(p). \tag{5}$$

Interior points p of a face or of an open edge have a neighbourhood which is isometric to a planar euclidean domain and we define $\theta(p) = 2\pi$.

All points of a polyhedral surface can be classified according to the sign of the *vertex angle excess* $2\pi - \theta(p)$:

Definition 6. A vertex p of a polyhedral surface S with total vertex angle $\theta(p)$ is called *euclidean*, *spherical*, or *hyperbolic* if its angle excess $2\pi - \theta(p)$ is $= 0$, > 0, or < 0. Respectively, interior points of a face or of an open edge are euclidean.

The neighbourhood of a vertex can be isometrically unfolded to a (partial or multiple) covering of a part of the euclidean plane. There exist three situations as shown in figure 2 which metrically characterize the vertex. For example, the tip of a convex cone is a spherical vertex and a saddle point is hyperbolic. On the other hand, a spherical vertex need not be the tip of

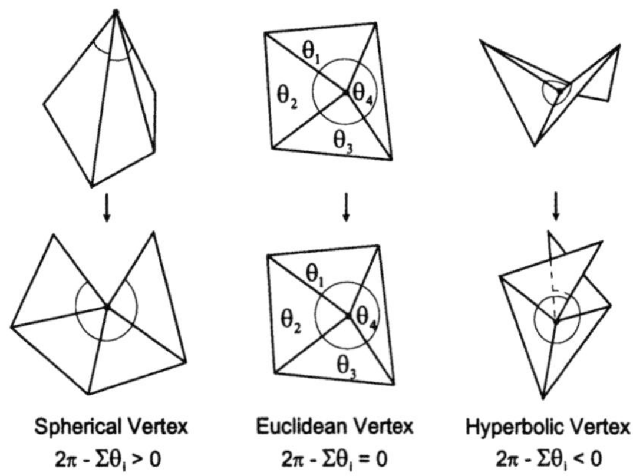

Fig. 2. Classification of vertices on a polygonal surface according to the excess of the vertex angle, and their unfolding in a planar domain.

a convex cone. The isometric unfolding of sets of a polyhedral surface is a common procedure to study the geometry.

The Gauß curvature of a general manifold is a central intrinsic property of the geometry and can be computed in terms of the metric. It influences, for example, the parallel translation of vectors along curves. The Gauß curvature of a piecewise linear surface is concentrated at the isolated vertices since all other points on the surface have a neighbourhood isometric to a planar euclidean domain with zero curvature. It is therefore more appropriate to work with the concept of total Gauß curvature. The Gauß curvature can be measured directly in metrical terms of the surface S:

Definition 7. The *(total) Gauß curvature* $K(p)$ of a vertex p on a polyhedral surface S is defined as the vertex angle excess

$$K(p) = 2\pi - \theta(p) = 2\pi - \sum_{i=1}^{m} \theta_i(p). \tag{6}$$

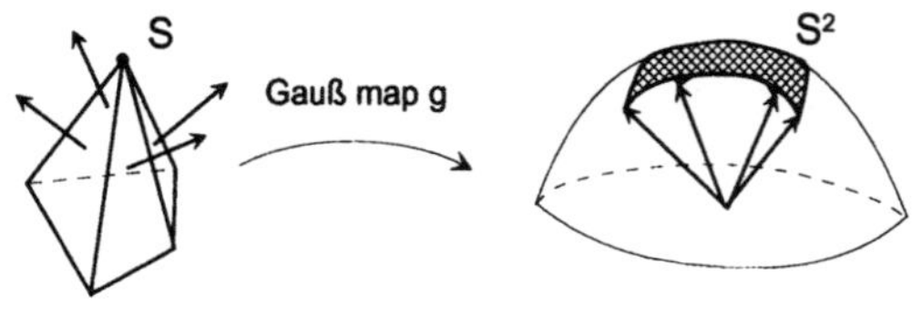

Fig. 3. The Gauß map assigns to each point $p \in S$ of a surface its normal vector $g(p) \in \mathbb{S}^2$. At edges and vertices of a polyhedral surface the Gauß map is the spherical convex hull of the normal vectors of adjacent faces.

An immediate consequence is that euclidean vertices have curvature $K = 0$, spherical vertices have $K > 0$, and hyperbolic vertices have $K < 0$. For example, the vertices of a cube each have Gauß curvature $\frac{\pi}{2}$.

For a smooth surface S imbedded into $\mathbb{R}^3$ the curvature measures the infinitesimal turn of the normal vector of the surface and can be defined via the *Gauß map* $g : S \to \mathbb{S}^2$ which assigns to each point p on a surface S its normal vector $g(p)$, see figure 3. The total Gauß curvature $K(\Omega)$ of a domain $\Omega \subset S$ is given by the area of its spherical image: $K(\Omega) = \text{area } g(\Omega)$. This relation also holds for the Gauß curvature of a vertex on a polyhedral surface.

4 Discrete Straightest Geodesics

Our approach to discrete geodesics on polyhedral surfaces concentrates on the property of a curve to be straightest rather than locally shortest. Both properties are equivalent for geodesics on smooth surfaces, as mentioned in section 2, but locally shortest curves on polygonal surfaces do not allow a unique extension, for example, beyond spherical vertices of the surface. The original motivation for our study was to define a unique way to move straight ahead in a given direction on a polyhedral surface. Applications are, for example, the tracing of moving particles restricted to flow along a polyhedral surface, the solution of initial value problems on polyhedral surfaces related with given tangential vector fields, and the intrinsic generalization of numerical algorithms for ordinary differential equations to polygonal surfaces.

The concept of shortest geodesics in graphs, polyhedral manifolds, and more general length spaces has been studied by a number of authors in different fields, see for example [4][6][1][2]. For our applications this concept misses a central property, namely, the initial value problem for geodesics is not uniquely solvable and in some cases has no solution: first, no shortest geodesics can be extended through a spherical vertex since it could be shortened by moving off the corner, and second, there exists a family of possible extensions of a geodesic as a shortest curve through a hyperbolic vertex: every extension with curve angles $\theta_l, \theta_r \in [\pi, \theta - \pi]$ is locally shortest where θ is the total vertex angle. See lemma 11 and figure 4.

Quasi-geodesics are a different approach introduced by A.D. Alexandrov (see the references to the original Russian literature in [1]) and investigated on convex surfaces by Pogorelov [7] and others. They appear as limit sets of smooth geodesics when smooth surfaces approximate, for example, a polyhedral surface. On polyhedral surfaces quasi-geodesics are characterized by their fulfillment of the inequality $|\pi - \theta_l| + |\pi - \theta_r| - |2\pi - \theta_l - \theta_r| \geq 0$ at each point, where θ_l and θ_r are the two angles of the curve, and $\theta_l + \theta_r = \theta$ is the total vertex angle of the point. Compare figure 4 for the notation. At hyperbolic vertices with $\theta > 2\pi$ the definition is identical to that for shortest geodesics, while at spherical vertices with $\theta < 2\pi$ curves with $\pi - \theta_l \geq 0$ and $\pi - \theta_r \geq 0$ are quasi-geodesics.

The following definition introduces straightest geodesics, a new class of discrete geodesics on polyhedral surfaces. This class has a non-empty intersection with the set of shortest geodesics and is a subset of quasi-geodesics.

Definition 8. Let S be a polyhedral surface and $\gamma \subset S$ a curve. Then γ is a *straightest geodesic* on S if for each point $p \in \gamma$ the left and right curve angles θ_l and θ_r at p are equal, see figure 4 and Color Plate 30 on page 382.

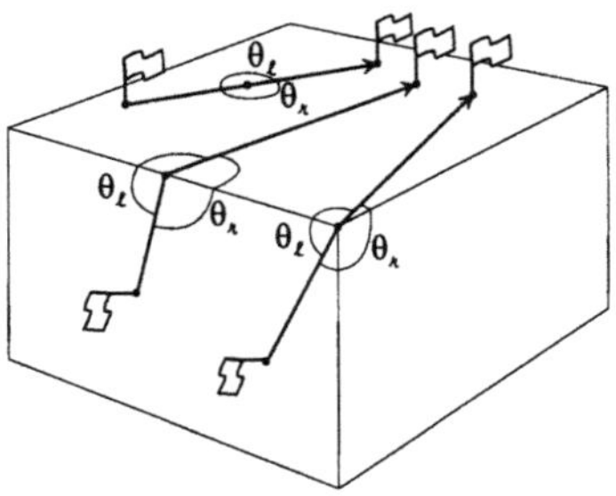

Fig. 4. Notion of left and right curve angles θ_l and θ_r with $\theta_l + \theta_r = \theta$, the total vertex angle.

A straightest geodesic in the interior of a face is locally a straight line, and across an edge it has equal angles on opposite sides. The definition of straightest geodesics on faces and through edges is identical to the concept of shortest geodesics but at vertices the concepts differ. Our definition fits into the more general discussion of discrete geodesic curvature of curves on a polyhedral surface. This will be discussed in section 5 in detail.

To state the initial value problem for straightest geodesics we start with the notion of a tangent vector on a polyhedral surface:

Definition 9. Let S be a polyhedral surface and $p \in S$ a point. A *polyhedral tangent vector* v with base point p lies in the plane of an adjacent face and locally points into the face. The polyhedral tangent space T_pS consists of all polyhedral tangent vectors at p.

We remark, that the polyhedral tangent bundle TS can be equipped with the structure of a topological vector bundle by introducing normalized angles as in definition 17, but we do not pursue this property in this paper.

Theorem 10 (Discrete Initial Value Problem). *Let S be a polyhedral surface and $p \in S$ a point with polyhedral tangent vector $v \in T_pS$. Then there exists a unique straightest geodesic γ with*

$$\begin{aligned} \gamma(0) &= p \\ \gamma'(0) &= v, \end{aligned} \tag{7}$$

and the geodesic extends to the boundary of S.

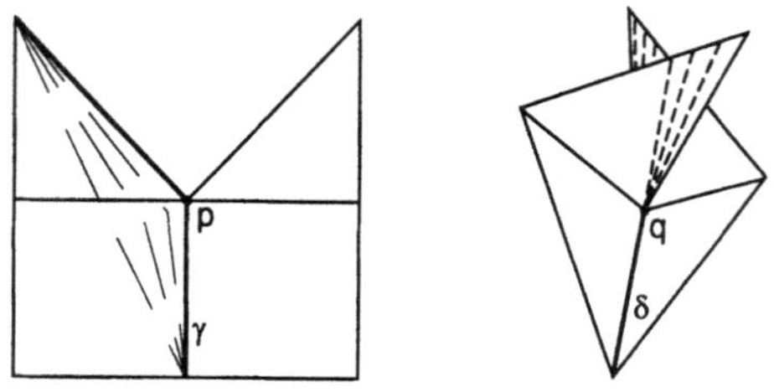

Fig. 5. Locally shortest geodesics cannot be extended through spherical vertices p, and provide multiple extensions at hyperbolic vertices q.

Proof. There exists a face f of S which contains the initial point p and, for a small number $\varepsilon > 0$, the straight line $\gamma(t) := p + tv$ with $t \in [0, \varepsilon)$. γ is a straightest geodesic and a solution of equation 7. If we extend γ beyond the small interval and γ reaches an edge or a vertex of S for larger values of t then definition 8 of straightest geodesics uniquely defines how to extend γ beyond the edge or vertex. That is to proceed in that direction for which the left and right curve angles of γ at the vertex are equal.◇

The concepts of straightest and shortest geodesics differ on polyhedral surfaces. For example, as shown in the following lemma, the theorem above does not hold for locally shortest geodesics approaching a spherical or hyperbolic vertex. As long as a geodesic γ does not meet a vertex of a polyhedral surface both concepts are equal and γ is both, straightest and locally shortest. The following lemma comprehends the differences:

Lemma 11. *On a polyhedral surface S the concepts of straightest and locally shortest geodesics differ in the following way (see figure 5):*

1. *A geodesic γ containing no surface vertex is both straightest and locally shortest.*
2. *A straightest geodesic γ through a spherical vertex is not locally shortest.*
3. *There exists a family of shortest geodesics γ_θ through a hyperbolic vertex with the same inbound direction. Only one of the shortest geodesics extends the inbound direction as straightest geodesic.*
4. *Straightest geodesics do not solve the boundary value problem for geodesics since there exist shadow regions in the neighbourhood of a hyperbolic vertex where two points cannot be joined by a straightest geodesic.*

Proof. Ad 1.) We unfold the faces met by the geodesic to an isometric strip of faces in the euclidean plane. The geodesic γ is unfolded to a euclidean straight line in the interior of the strip which is locally shortest and fulfills the angle condition of definition 8.

Ad 2.) Let γ be a straightest geodesic through a spherical vertex with curvature $K > 0$. We unfold the adjacent faces to a planar domain by cutting along the outbound direction of γ. The image of γ in the plane has a corner at the vertex with curve angle $\frac{\theta}{2} = \pi - \frac{K}{2} < \pi$ at both sides. Therefore, γ is not locally shortest since it can be shortened by smoothing the corner in either direction as shown on the left in figure 5.

Ad 3.) A hyperbolic vertex has curvature $K < 0$. Let γ_0 be the unique straightest geodesic though the vertex which extends the inbound direction. We unfold the adjacent faces to a planar domain by cutting along the outbound direction of γ_0, then γ_0 has a curve angle $\frac{\theta}{2} = \pi - \frac{K}{2} > \pi$ at both sides of the corner. Assume a curve with the same inbound but a different outbound direction. Whenever both angles between the inbound and outbound direction are bigger than or equal to π, we cannot locally shorten the curve. Therefore all such curves are locally shortest.◇

5 Discrete Geodesic Curvature

We define the notion of geodesic curvature of curves on piecewise linear surfaces with the later aim of defining parallel translation of vectors along arbitrary curves. Additionally, vanishing geodesic curvature should characterize straightest geodesics. The definition should comply with the known (total) curvature of polygons in the euclidean plane, and the Gauß-Bonnet equation should hold. In the following, we assume curves to be smooth on faces and to have well-defined polyhedral tangent directions at the edges and vertices of the surface. Similar to the discrete Gauß curvature for surfaces, the discrete geodesic curvature is the equivalent of the total geodesic curvature of smooth surfaces.

Definition 12. Let γ be a curve on a polyhedral surface S. Let θ be the total vertex angle and β one of the two curve angles of γ at p. Then the *discrete geodesic curvature* κ_g of γ at p is given by

$$\kappa_g = \frac{2\pi}{\theta}\left(\frac{\theta}{2} - \beta\right). \tag{8}$$

Choosing the other curve angle $\beta' = \theta - \beta$ changes the sign of κ_g.

Using the notion of discrete geodesic curvature we obtain a new characterization of straightest geodesics since they bisect the total vertex angle θ, i.e. $\beta = \frac{\theta}{2}$:

Lemma 13. *Let S be a polyhedral surface and $\gamma \subset S$ a curve. Then γ is a straightest geodesic if and only if γ has vanishing discrete geodesic curvature.*

Remark 14. 1.) Let γ be a polygon in the euclidean plane S and $p \in \gamma$ be a vertex with curve angle β. Then the discrete geodesic curvature equals the total curvature of γ at p defined by the spherical image of its normal vectors.
2.) Let S be a polyhedral surface and let γ touch a vertex $p \in S$, i.e. $\beta = 0$. Then the geodesic curvature of γ at p is $\kappa_g = \pi$, i.e. it can be measured in

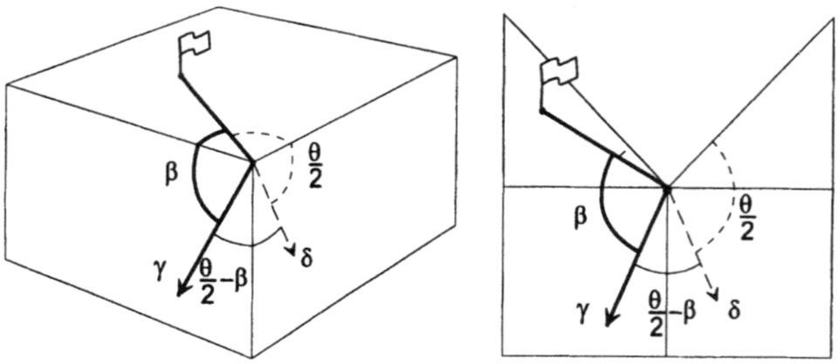

Fig. 6. The discrete geodesic curvature of a curve γ is the normalized angle between γ and a discrete straightest geodesic δ.

the euclidean face and without influence of the vertex angle θ at p.
3.) Shortest geodesics through a hyperbolic vertex with vertex angle $\theta > 2\pi$ have geodesic curvatures κ_g in the interval $\left[-\pi(1-\frac{2\pi}{\theta}), \pi(1-\frac{2\pi}{\theta})\right]$.

Straightest geodesics are natural generalizations of straight lines in euclidean space. For example, geodesic triangles on surfaces can be defined as simply connected regions bounded by three straightest segments, and geodesic polygons as piecewise straightest curves.

The Gauß-Bonnet theorem relates the topology and geometry of surfaces. It is a remarkable consequence of the definition of discrete geodesic curvature that this fundamental theorem still holds. In fact, one can even reverse the arguments and derive our formula for geodesic curvature from the requirement that the equation of Gauß-Bonnet should hold.

There have been different formulations of the Gauß-Bonnet theorem on polyhedral surfaces, each expressing the Euler characteristic $\chi(\Omega)$ of a domain Ω using different curvature terms. For example, Reshetnyak [9] only uses the Gauß curvature of interior vertices and defines the curvature of the boundary curve by $\kappa = \pi - \beta$, where β is the inner curve angle of the boundary. We refine this approach and split his definition of boundary curvature in two components, a geodesic curvature of the boundary curve and a partial Gauß curvature, where the vertices $p \in \partial\Omega$ contribute to the total Gauß curvature of Ω. The following natural definition determines the contribution of boundary vertices to the total Gauß curvature of Ω.

Definition 15. Let $\Omega \subset S$ be a domain on a polyhedral surface with boundary $\Gamma = \partial\Omega$. If $\theta(p)$ is the total vertex angle and $\beta(p)$ the inner curve angle at a vertex $p \in \Gamma$, then the *partial Gauß curvature* $K_{|\Omega}$ of Ω at p is proportional to β:

$$K_{|\Omega}(p) = \frac{\beta}{\theta} K(p). \tag{9}$$

If $\beta = 0$ then the vertex has no partial Gauß curvature, and $\beta = \theta$ leads to a full contribution of the total Gauß curvature $K = 2\pi - \theta$ to Ω. In the following we simplify the notation by omitting the subindex $_{|\Omega}$.

Theorem 16 (Discrete Gauss-Bonnet). *Let S be a polyhedral surface and $\Omega \subset S$ a domain with boundary curve Γ and Euler characteristic $\chi(\Omega)$. Then the equation*

$$\sum_{p \in \Omega} K(p) + \kappa_g(\Gamma) = 2\pi\chi(\Omega) \tag{10}$$

holds where the total Gauß curvature of Ω includes the partial Gauß curvature at boundary points. If Γ is piecewise straightest then the total geodesic curvature is the sum of the geodesic curvature at the vertices of Γ.

Proof. For the proof we use the version

$$\sum_{p \in \overset{\circ}{\Omega}} K(p) + \sum_{p \in \Gamma} (\pi - \beta(p)) = 2\pi\chi(\Omega)$$

proved by Reshetnyak [9] where only interior vertices of Ω contribute to the total Gauß curvature. Let $p \in \Gamma$ be a boundary vertex, then we have the splitting

$$K_{|\Omega}(p) - \kappa_g(p) = \pi - \beta(p)$$

which proves the assumption.◇

6 Parallel Translation of Vectors

Numerical methods for the integration of ordinary differential equations rely on the possibility for parallel translation of vectors in the euclidean plane. For example, higher order Runge-Kutta methods do several trial shots in a single integration step to compute the final shooting direction and translate direction vectors to their current positions. When transferring such integration methods to surfaces, which are not described by local charts, it is necessary to compare vectors with different base points on the curved surface.

We use the notion of polyhedral tangent vectors formulated in definition 9 and define an intrinsic version of parallel translation of vectors which uses no ambient space as reference. We start with two definitions of angles:

Definition 17. Let S be a polyhedral surface and $p \in S$ a point with total vertex angle θ. The *euclidean angle* $\measuredangle(v,w)$ between tangent vectors $v, w \in T_pS$ is the angle between corresponding vectors in the unfolded neighbourhood of p measured in $\mathbb{R}^2$, i.e. $\measuredangle(v,w) \in \left[-\frac{\theta}{2}, \frac{\theta}{2}\right]$. The *normalized angle* $\alpha(v,w)$ is obtained by scaling:

$$\alpha(v,w) := \frac{2\pi}{\theta} \measuredangle(v,w). \tag{11}$$

The normalized and euclidean angles are identical at points which are not vertices of the surface. In practical applications one measures the euclidean angle at first, and then uses the normalized angle to avoid case distinctions at vertices of the surface as seen, for example, in the following lemma:

Lemma 18. *Let Δ be a geodesic triangle on a polyhedral surface S whose edges are straightest segments. If α_1, α_2, and α_3 are the normalized angles of Δ then we have*

$$\alpha_1 + \alpha_2 + \alpha_3 - \pi = \int_{\Delta} K. \tag{12}$$

Proof. Denote the euclidean angles of Δ with β_i and the vertex angles with θ_i . Then the geodesic curvature of the boundary of Δ at one of its vertices is given by

$$\kappa_g = \frac{2\pi}{\theta}\left(\frac{\theta}{2} - \beta\right) = \pi - \alpha \tag{13}$$

and the assumption follows directly from the discrete Gauß-Bonnet equation (10).◇

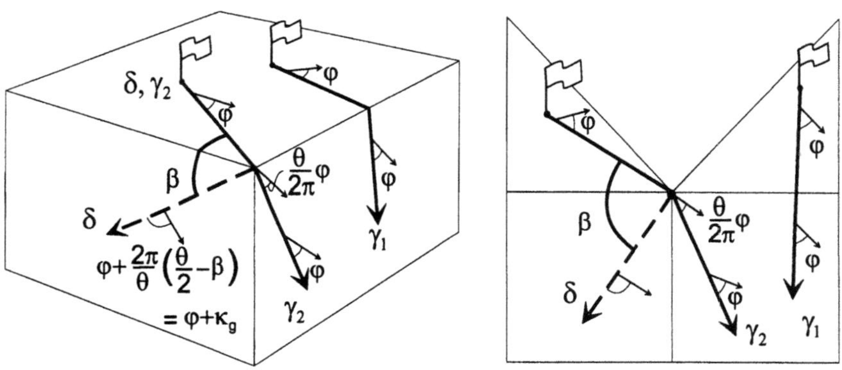

Fig. 7. Parallel translation of vectors along straightest geodesics γ_1,γ_2 and an arbitrary curve δ.

On polyhedral surfaces we can use the concept of straightest geodesics and normalized angles to define parallel translation along geodesics and arbitrary curves similar to the smooth case:

Definition 19. Let $\gamma : I \to S$ be a parametrized straightest geodesic on a polyhedral surface S. A tangential vector field $v : I \to TS$ with $v(s) \in T_{\gamma(s)}S$ is a *parallel vector field* along γ if the normalized angle $\alpha(v(s), \gamma'(s))$ is constant.

Definition 20. Let κ_g be the geodesic curvature of a curve $\gamma : I \to S$ with $\gamma(0) = p$ and let $v_0 \in T_pS$ be a tangent vector with normalized angle $\alpha(0) := \frac{2\pi}{\theta(p)} \measuredangle(v_0, \gamma'(0))$. Then v_0 uniquely extends to a parallel vector field v with $v(s) \in T_{\gamma(s)}S$ along γ with $v(0) = v_0$. $v(s)$ is defined by the normalized angle $\alpha(s)$ it encloses with $\gamma'(s)$:

$$\alpha(s) = \alpha(0) + \int_0^s \kappa_g(t)\, dt. \tag{14}$$

The formula is well-known for curves on smooth surfaces. In the discrete situation we have made direct use of the definition of discrete geodesic curvature and the notion of normalized angles at vertices.

7 Runge Kutta on Polyhedral Surfaces

The tracing of particles on a surface by integrating a given vector field with Euler or Runge Kutta methods requires an additional effort to keep the trace on the surface. For example, one may use local coordinate charts of a surface to transform the integration to the planar euclidean domain. Here the metrical distortion between surface and euclidean domain must be respected and a preprocessing step to generate the charts and transitions between neighbouring charts is required.

If the vector field is given on a curved surface in an ambient space, say $\mathbb{R}^3$, then a usual tangent vector "points into the ambient space", leading the numerical particle trace off the surface.

The concepts of straightest geodesics and polyhedral tangent vectors offer an intrinsic tool to solve these problems. In euclidean methods, the vector $v|_{\gamma(s)}$ is interpreted as tangent vector to the particle trace $\gamma(s)$, and the straight line through $\gamma(s)$ with direction $v|_{\gamma(s)}$ is the first order approximation of γ. The idea on surfaces is to use polyhedral tangent vectors defined in definition 9 and to replace the straight line with a straightest geodesic through $\gamma(s)$ with initial direction $v|_{\gamma(s)}$:

Definition 21 (Geodesic Euler Method). Let S be a polyhedral surface with a polyhedral tangential vector field v on S, let $y_0 \in S$ be an initial point, and let $h > 0$ a (possibly varying) stepsize. For each point $p \in S$ let $\delta(t, p, v(p))$ denote the unique straightest geodesic through p with initial direction $v(p)$ and independent parameter t. The iteration

$$y_{i+1} := \delta(h, y_i, v(y_i)). \tag{15}$$

produces a sequence of points $\{y_0, y_1, ...\} \subset S$ and connecting straightest geodesic segments of length h. For each $i \in \{0, 1, ...\}$ we define

$$\gamma(ih + t) := \delta(t, y_i, v(y_i)),\ t \in [0, h] \tag{16}$$

and obtain a piecewise straightest, continuous curve $\gamma : [0, \ell) \to S$ of some length ℓ such that each segment $\gamma_{|[ih,\ (i+1)h]}$ is a straightest geodesics.

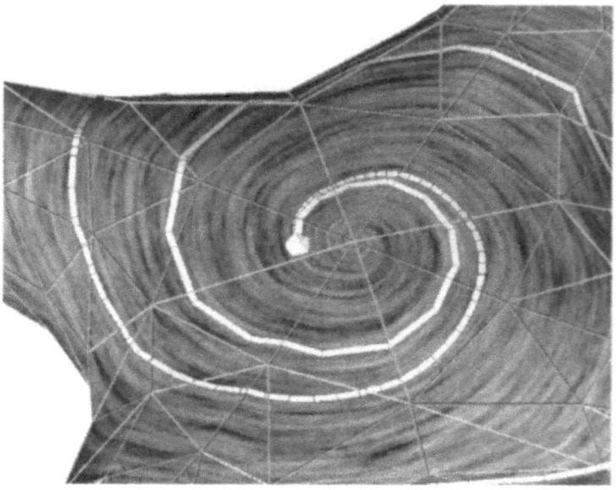

Fig. 8. The two piecewise straightest geodesics are solutions computed with the geodesic Euler method (outer curve, stepsize h) and 4th order Runge Kutta method (inner curve, stepsize $4h$). Note, that the geodesic segments extend across triangle edges and vertices. Also, a comparison with the underlying flow shows the expected better approximation quality of the geodesic Runge-Kutta method.

The definition of the geodesic Euler method is intrinsic and no projection of the tangent vectors or tangent directions onto the surface is required during integration. If the original vector field is not a polyhedral tangential field then an initial generation of a polyhedral tangential vector field is required in a preprocessing step, however, this step is part of the formulation of the numerical problem and not of the integration method. See figure 8 and Color Plates 31 and 32 on page 383 in the Appendix for applications.

Using the concept of parallel translation it is straight forward to define higher order integration methods in a similar intrinsic way. For simplicity, we restrict to a 4-th order geodesic Runge Kutta method:

Definition 22 (Geodesic Runge-Kutta Method). Let S be a polyhedral surface with a polyhedral tangential vector field v on S, let $y_0 \in S$ be an initial point, and let $h > 0$ a (possibly varying) stepsize. For each point $p \in S$ let $\delta(t, p, v(p))$ denote the unique straightest geodesic through p with initial direction $v(p)$ and independent parameter t. A single iteration step of the *geodesic Runge Kutta method* is given by

$$y_{i+1} := \delta(h, y_i, v_i) \tag{17}$$

where the direction v_i is a polyhedral tangent vector at y_i obtained as follows: we denote the parallel translation of vectors along a geodesic δ to $\delta(0)$ by $\pi_{|\delta}$ and iteratively define

$$\begin{aligned} v_i^1 &:= v(y_i) \\ v_i^2 &:= \pi_{|\delta_1} \circ v(\delta_1(\frac{h}{2}, y_i, v_i^1)) \end{aligned} \tag{18}$$

$$v_i^3 := \pi_{|\delta_2} \circ v(\delta_2(\frac{h}{2}, y_i, v_i^2))$$
$$v_i^4 := \pi_{|\delta_3} \circ v(\delta_3(h, y_i, v_i^3))$$

and

$$v_i := \frac{1}{6}(v_i^1 + 2v_i^2 + 2v_i^3 + v_i^4) \quad (19)$$

where the curves δ_i are straightest geodesics through y_i with initial direction v_i^j for $j \in \{1, 2, 3\}$.

8 Conclusion

On polyhedral surfaces we introduced the concept of straightest geodesics and discrete geodesic curvature of curves. We applied the concept to define the parallel translation of tangential vectors and generalized Runge Kutta methods to polyhedral surfaces. These concepts allow a uniform and intrinsic description of geometric and numerical properties on polyhedral surfaces.

References

1. A. D. Aleksandrov and V. A. Zalgaller, *Intrinsic geometry of surfaces*, Translation of Mathematical Monographs, vol. 15, AMS, 1967.
2. S. B. Alexander and R. L. Bishop, *Comparison theorems for curves of bounded geodesic curvature in metric spaces of curvature bounded above*, Diff. Geom. Appl. 6:1 (1996), 67–86.
3. M. P. Do Carmo, *Differential geometry of curves and surfaces*, Prentice-Hall, 1976.
4. E. Dijkstra, *A note on two problems in connection with graphs*, Numer. Math. 1 (1959), 269–271.
5. C. Gunn, A. Ortmann, U. Pinkall, K. Polthier, and U. Schwarz, *Oorange - a visualization environment for mathematical experiments*, Visualization and Mathematics (H.-C. Hege and K. Polthier, eds.), Springer Verlag, 1997.
6. J. S. B. Mitchell, D. M. Mount, and C. H. Papadimitriou, *The discrete geodesic problem*, SIAM J. Comput. 16:4 (1987), 647–668.
7. A. V. Pogorelov, *Quasigeodesic lines on a convex surface*, Amer. Math. Soc. Transl. I. Ser. 6:72 (1952), 430–473.
8. K. Polthier, M. Schmies, M. Steffens, and C. Teitzel, *Geodesics and Waves*, ACM Siggraph Video Review 120 (1997).
9. Y. G. Reshetnyak, Geometry IV, Encyclopaedia of Math. Sci., vol. 70, ch. 1. Two-Dimensional Manifolds of Bounded Curvature, Springer Verlag, 1993, pp. 3–164.
10. M. Sharir and A. Schorr, *On shortest paths in polyhedral space*, SIAM J. Comput. 15:1 (1986), 193–215.

Part III

Graphics Algorithms and Implementations

Support of Explicit Time and Event Flows in the Object-Oriented Visualization Toolkit MAM/VRS

Jürgen Döllner and Klaus Hinrichs

Institut für Informatik, Universität Münster, Germany

Abstract. We present an object-oriented architecture of a visualization toolkit which integrates geometric modeling and behavioral modeling. It is based on sharing graphics objects between geometrical scene descriptions and descriptions for the flow of time and events. We discuss the properties graphics objects should possess in such a system so that they can be used by different rendering toolkits and can be linked to time- and event-dependent processes. We introduce a new methodology for the symmetric modeling of geometry and behavior based on geometry graphs and behavior graphs.

1 Architectural Limitations of Visualization Software

During the last few years much progress has been made in the development of object-oriented 3D graphics and visualization techniques and toolkits (e.g. [4] [5] [10] [14] [19] [22]). However, the development of animated and interactive 3D visualization applications still represents a complex challenge for several reasons:

Most visualization toolkits provide support for geometric modeling, but do not integrate an explicit time management. Therefore, animation must be integrated at a low-level and is in general not supported by object-oriented animation concepts. For example, VRML [10] provides time sensors which generate events as time passes. However, the sampling frequency for continuous changes is implementation dependent because VRML makes no assumption how often a time sensor will generate events. Furthermore, the time management is based on these elementary building blocks but no high-level modeling of time is supported, e.g. distributing time accoding to time layouts.

Interaction components are typically provided as complex black-boxes (e.g. [19]) which makes it difficult to specialize these components, to reuse parts of them, and to integrate animation in interactive behaviors [9]. Often interaction and animation components are attached to geometric components. However, this is difficult to achieve for complex animations and in teractions because their relationships are more of a temporal than a spatial nature [1].

Furthermore, geometric modeling is most often based on one specific rendering package, but does not possess a rendering-toolkit independent architecture which would allow the developer to exchange rendering packages without further recoding. Different rendering packages are necessary to satisfy the needs of different visualization tasks, e.g., interactive visualization requires real-time rendering, post production of a movie requires high-quality rendering.

We present an approach for interactive, animated 3D visualization based on a new methodology for the symmetric modeling of geometry and behavior. Both geometry and behavior are represented by first class objects, the *geometry nodes* and *behavior nodes.* These nodes are organized in two types of directed acyclic graphs, the *geometry graphs* and *behavior graphs.* A geometry graph is a renderer-independent hierarchical scene description, whereas a behavior graph manages the flow of time and events, and is responsible for time- and event-dependent constraints. Geometry graphs and behavior graphs are associated with *shareable graphics objects* which are visualized by geometry nodes and animated by behavior nodes.

2 Graphics Objects: Basic Visualization Entities

Visualization is based on the manipulation of graphical entities, typically represented as instances of graphics classes, called *graphics objects* (e.g., triangle strips, NURBS, colors, textures, ...). We can identify several properties which have to be fulfilled by graphics objects so that they are as general as possible and usable for both geometric and behavioral modeling:

- *Integration of new graphics classes.* A visualization system has to provide built-in graphics classes, but this collection can never be complete, because new visualization applications may require new graphics types. Therefore, the visualization system should allow the developer to build new graphics classes which have the same "rights" as the built-in graph ics classes.
- *Decoupling of graphics objects and their rendering.* If graphics objects provide built-in rendering methods, the rendering technique cannot be exchanged easily without recoding these methods. Therefore, the rendering commands which depend on the rendering toolkit used in the visualization system should be decoupled from the (rendering toolkit independent) graphics class properties.
- *Full encapsulation by a language-independent interface* which allows the visualization system the automatic conversion to different application programming interfaces. For example, a graphics object might be used by a C++ API, accessed through a CORBA interface, or exported to an interpretative language such as Tcl [15].
- *Export of variables by parameter objects.* Graphics objects which possess a time-dependent nature can be controlled by time-dependent constraints. If constraints are modelled by constraint objects, we have to access the data fields of graphics objects in an object-oriented way. To build generic

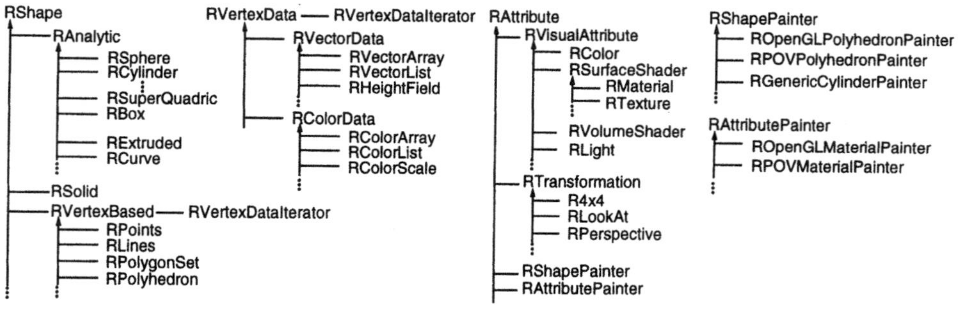

Fig. 1. Part of the VRS graphics class hierarchy.

constraint classes, constraint objects have to be based on (mostly elementary) parameter types (e.g., vectors, colors, floats, ...). Therefore, graphics objects must provide a way to export their parameters as parameter objects.

2.1 Graphics Classes: Shapes, Attributes, Rendering Engines.

In our approach, we provide an extendible collection of graphics classes. Instances of these classes can be processed by different rendering toolkits and can be attached to generic constraint objects.
Graphics classes can be characterized by their flyweight design [3] which ensures that their instances are as minimal and as small as possible, and that they can be used in large numbers and implemented efficiently [13]. They do not include any context information and do not make any assumptions about their visualization. A sphere, for example, stores its radius and midpoint, but does not include a transformation matrix, a color, a rendering context or a surface approximation.

We distinguish between two main categories of graphics classes, *shape classes* and *attribute classes*. Shape classes represent 2D and 3D geometric objects. Attribute classes represent the shape-related or scene-related options and modifications, e.g. surface shaders. Part of the class hierarchy is shown in Fig. 1. *Rendering engines* are abstract state machines. Their main features are:

- *Association of shapes with attributes.* Rendering engines store attribute objects in attribute stacks – one for each class. The top-most object of an attribute stack is the currently active attribute object of its category. Shape objects can be sent to a rendering engine in order to evaluate the shape's geometry, and attributes can be pushed (respectively popped) in order to modify the state of the rendering engine (Fig. 2).
- *Mapping of shapes and attributes to low-level rendering calls.* Rendering engines convert geometric descriptions and attributes into appropriate calls and structures of the 3D rendering toolkit actually used for image synthesis.

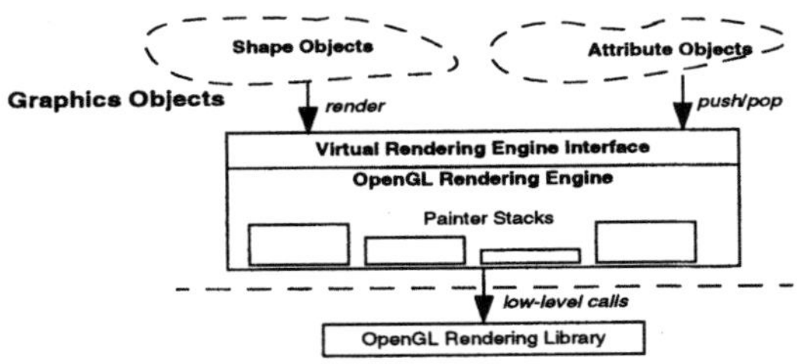

Fig. 2. Processing graphics objects for an OpenGL rendering engine.

- *Homogeneous rendering interface.* The interface of rendering engines consists of (mainly) three commands: *render, push,* and *pop.* Since all rendering engines have the same homogeneous interface, we can use different rendering toolkits for image synthesis without having to recode the application.

2.2 Painter Concept

The rendering commands used to map the shape description to an underlying 3D rendering toolkit are encapsulated in *shape painter classes.* For each shape class and for each rendering engine type, a shape painter class specifies the way a shape object is visualized. For non-elementary shape types, we provide generic shape painter classes which transform such a shape object into a set of lower-level shape objects. E.g. the generic cylinder painter transforms a cylinder object into a quad mesh object. Therefore, most high-level shape types can be used immediately for a new rendering engine type once the most elementary shape painters are defined for it.

In analogy to shape painters, *attribute painter classes* specify how attribute objects are evaluated by the underlying 3D rendering toolkit. Painter objects are bound to shapes at run time and form part of the state of a rendering engine. The painter concept implies the following advantages:

- The application code has not to be rewritten when the application wants to use a new low-level rendering library because the responsible rendering engine provides its specific shape and attribute painters. Fig. 3, for example, shows two images of Coxeter's polyhedron $\{4,6|3\}$ with hidden symmetries; the edges of the polyhedron are projected onto the canvas. Both images are derived from the same set of graphics objects using an OpenGL rendering engine and a POVRay rendering engine.
- Different views of a shape (e.g. a triangle mesh visualized by its surface and by its silhouette edges) can be easily modeled because the shape painter is just an attribute of a shape (and not a method of the shape

Fig. 3. Visualization of Coxeter's $\{4,6|3\}$. Rendered by OpenGL and POVRay.

class). In particular, this allows us to handle rendering algorithms by the generic attribute management.
- Applications can take advantage of an a-priori know nature of their data by providing specialized shape painters which implement an optimized rendering algorithm without having to touch the original shape class.

2.3 Geometry Nodes and Geometry Graphs

Graphics objects can be organized into graphs which facilitate their composition. Geometry graphs are declarative and rendering-library independent descriptions of 3D scenes. They consists of geometry nodes which are associated with graphics objects. Geometry graphs are similar to the OpenInventor scene composition technique [19]. However, they are rendering-toolkit independent and do only associate graphics objects with a 3D scene; the associated graphics objects may be used in other parts of the visualization system as well (e.g. embedded in behavior graphs which constrain them). We use four categories of geometry nodes to compose 3D scenes: *shape nodes*, *attribute nodes*, *geometry groups*, and *image controllers*. Shape nodes embed associated shape graphics objects (e.g. spheres, boxes, 3D characters) in a 3D scene. Attribute nodes apply associated attribute graphics objects (e.g. surface shaders, transformations) to all shapes in their subgraph. Geometry groups are container nodes which arrange the shapes of its subgraphs based on a layout strategy, e.g. the BSpline-aligned layout adjusts shapes along a BSpline. Image controllers manage the image synthesis process, e.g. a camera node provides a virtual rendering device, a lens node blends a scene in another scene.

3 Behavior Graphs: Time and Event Flows

Behavior graphs specify the flow of time and events. They simplify the design of an animated, interactive visualization because they provide a hierarchical

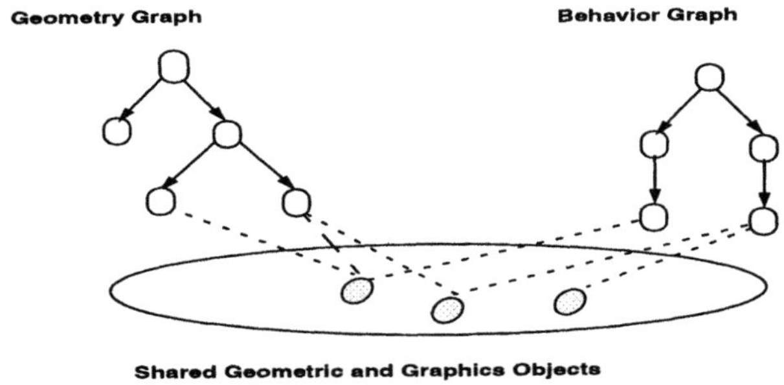

Fig. 4. Object sharing between geometry graph and behavior graph.

organization scheme for time-dependent and event-dependent processes which treats geometry descriptions separate from behavior descriptions. Basically, an application specifies two graphs, a geometry graph and a behavior graph, and associates graphics objects with geometry nodes (to anchor them in the 3D world) and with behavior nodes (to constrain them) (Fig. 4). Graphics objects play a central role, because they are shared between 3D worlds and time-dependent and event-dependent processes.

Geometry descriptions require a hierarchical modeling technique such as the directed acyclic graph; scene objects are positioned in different, nested coordinate systems. In analogy, animation descriptions require an organization scheme which models time coordinate systems. Current approaches (e.g., VRML [10]) do mix both description categories which leads to a simplier scheme for small models (the animation can be attached directly to a shape), but which tends to produce difficult animation and interaction specifications when they become large because time-related model elements are distributed across the geometric objects and are not inherently connected by hierarchical, nested time coordinate systems.

3.1 Temporal Abstract Data Types

We introduce the following temporal abstract data types to simplify the time management:

- *Time*: a float value, measured in milliseconds.
- *TimeRequirement*: describes the time demands of a behavior node. It consists of the natural (i.e. desired, optimal), minimal and maximal duration. Furthermore, it defines an alignment used to position time intervals within time intervals.
- *Moment*: represents a point in time in a time interval. A moment assigned to a behavior node determines the node's life time and the current point in time within this interval. Moments are essential for behavior nodes which specify animations. Based on the knowledge about their life time, they can plan and distribute their activity. Moments provide a local model time for behavior nodes.

- *TimeProcess*: a process which sends synchronization events to a target behavior node during a given moment and at a given frequency. It is typically implemented as a separate thread.

3.2 Behavior Nodes

Behavior nodes are categorized in time modifiers, behavior groups, and interactions (Fig. 5).

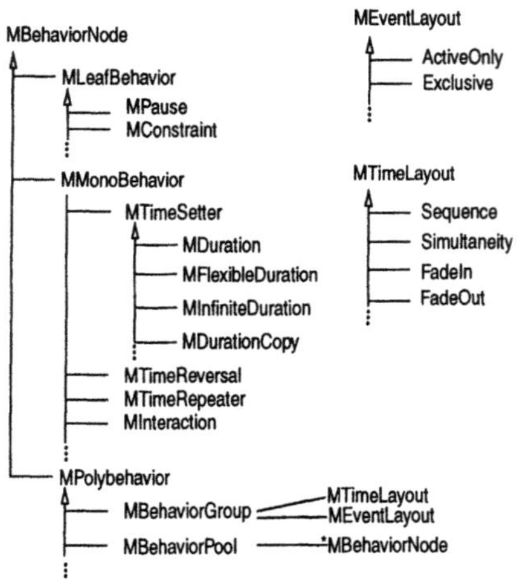

Fig. 5. Time and event related classes.

Time Modifiers Time modifier nodes define and transform time requirements for their sub graphs. Time modifier classes include:

- *Time setter*: A time setter node specifies the time requirement for its body. It can define time stretchability and time shrinkability (*MFlexibleDuration*), natural durations (*MDuration*), and infinite durations (*MInfiniteDuration*). It can also use the time requirement of another behavior node for its body (*MDurationCopy*).
- *Time repeater*: A *MTimeRepeater* node maps a moment modulo another moment, and passes the resulting moment to its body.
- *Time reversal*: A *MTimeReversal* node inverts incoming moments and delegates them to its body. Thus, for the body the direction of the time progress is inverted. Time reversal nodes are useful to model retrograde actions.

The introduction of time requirements and time setters was motivated by the space requirements and space setters developed in InterViews [11], and by the TeX [11] boxes and TeX glue model. Applied to time, it allows us to specify animations at an abstract level and makes it unnecessary to calculate in advance the absolut time requirements of an action in an animation.

Behavior Groups Behavior groups provide an automatic time negotiation mechanism based on time layouts and event layouts. A *time layout calculates* the individual life times of the child nodes based on their time requirements. If a behavior group is synchronized to a new time, the time layout checks which child nodes have to be activated or deactivated. It synchronizes all active child nodes to the new time, and assigns the calculated moments to them. Examples for time layouts are:

- *Sequence*: It distributes time in sequential, disjoint moments to its child nodes. Only one child node is alive at any given time during the duration of the sequence.
- *Simultaneity*: It aligns all child nodes to its own life time. If a child node defines a shorter (or longer) natural life time, the layout tries to stretch (or shrink) the child nodes' life time.
- *FadeIn and FadeOut*: FadeIn layouts assign moments to their child nodes which start in cascading order like in the case of the sequence layout, but all child nodes remain activated until the last child node is deactivated. FadeOut layouts are reversed FadeIn layouts.

Event layouts determine how events are dispatched to child nodes. For example, the ActiveOnly event layout dispatches an event to active child nodes only, whereas an Exclusive event layout dispatches an event to all child nodes until one child node has consumed the event.

Constraints Constraints form a main category of behavior nodes. A constraint node is as sociated with parameter objects (exported by graphics objects), and starts to constrain the object at the beginning of its life time, and releases the parameter objects when its life time ends. Constraint networks are anchored in behavior graphs and connected to the flow of time and events by behavior nodes.

In general, constraint nodes apply time-dependent functions to parameters of the graphics objects. These functions are encapsulated in mapping objects. A mapping represents a function which maps a given moment to a numerical value of a generic type. If the constraint node is synchronized, the mapping calculates the new parameter values and applies them to the rendering object. For example, a *MSlowInOut*<Vector> mapping interpolates two given vectors during the time interval defined by the moment. The interpolation proceeds slower at the beginning and at the end. A collection of mapping classes is shown in Fig. 6.

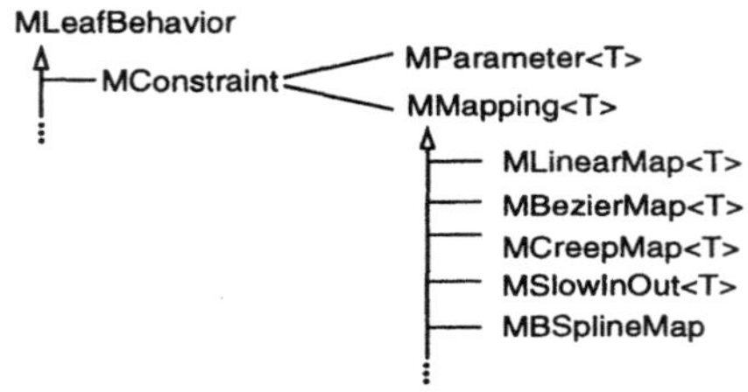

Fig. 6. Constraint and mapping classes.

3.3 Execution of Behavior Graphs

The execution of behavior graphs determines how fast time passes and how often behavior nodes are synchronized. There are two execution modes:

- *Real-time execution.* Behavior nodes are synchronized to the (scaled) real time. At each time step, the frame rendering time is used to calculate the next point in time. The real-time mode is used for simulations and interactions.
- *Model-time execution.* Behavior nodes are synchronized to the model time, i.e. time passes in application-defined time steps. This execution mode allows animations to produce a desired sampling frequency. The frame rendering time does not affect the time progress.

4 Example: An Animated, Interactive 3D Viewer

We sketch in this example the geometry graph and behavior graph of a simple 3D viewer which shows and animates a 3D geometric model (Fig. 7). The object viewer visualizes a polygonal graphics object and its convex hull. A rectangular area is used as a magic lens [2] which exhibits the facet normals of that object. The lens can be moved interactively across the view plane.

4.1 Object Viewer Geometry Graph

The object viewer's geometry graph is shown in Fig. 7.

The *MAdjuster* nodes fit their subgraphs in the volume *vol* given by a *RBox* graphics object. The graphics object rot_E applies a rotation matrix to the subgraph of its attribute node. The *MShape* node, the *MConvexHull* node and the *MNormalViewer* node refer to the same graphics object obj_E of type *RVertexBased* (e.g. triangle set, quad mesh, facets). A transmission graphics object tm_{CH} is applied to the convex hull. The *MNormalViewer* represents the normals of its associated shape object by glyphs (e.g. small arrows). The geometry graph is connected to an *MCamera* node. To generate an image, parts of the graphs are traversed twice.

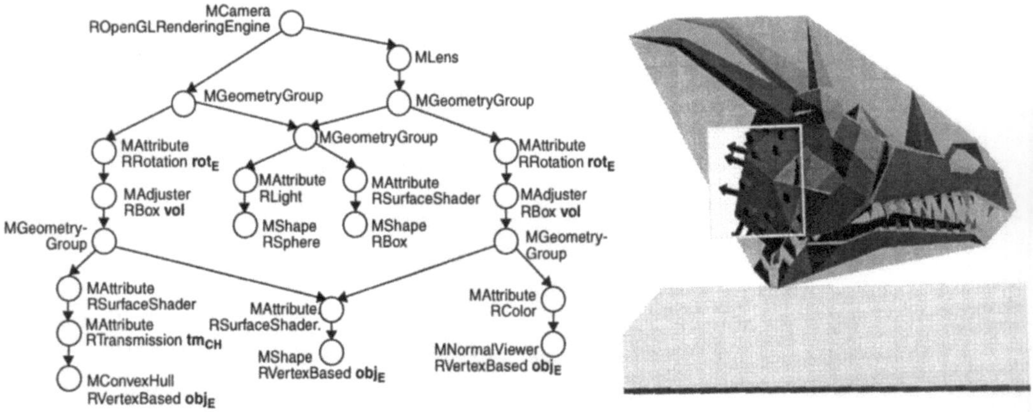

Fig. 7. Geometry graph of the object viewer.

4.2 Behavior Graph

The following animation fades in a new object loaded by the object viewer. In the beginning the object is hidden in its convex hull. The object's volume is enlarged to its default size. Then, the convex hull becomes transparent. To fade out the object, we reverse the fade-in behavior. Additionally, we define an automatic rotation for the object which can be activated and deactivated. The behavior graphs are shown in Fig. 8.

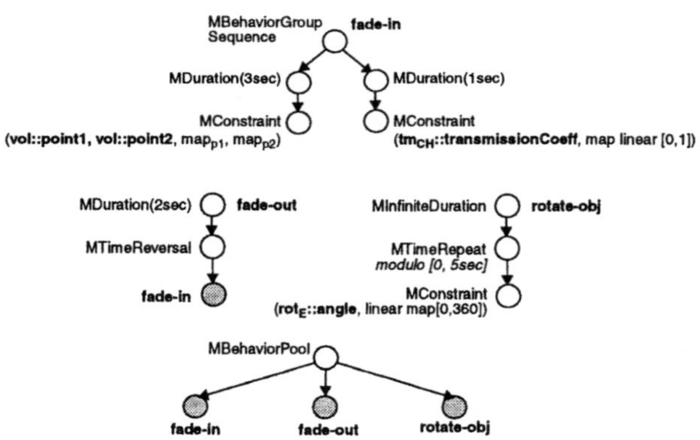

Fig. 8. Behavior graph of the object viewer.

The fade-in behavior consists of two sequential animations. The first animation constrains the bounding box *vol* of the object during 3 seconds. It uses two time-dependent functions, called mappings, which are applied to the minimal point and maximal point of the bounding box. The second animation constrains the transmission coefficient of tm_{CH} during the next 2 seconds. The fade-out behavior consists of a *MTimeReversal* node linked together with the fade-in behavior. Moments sent to the fade-out behavior node are reversed, i.e. we get the retrograde animation. The $rotate - obj$ behavior constrains the graphics object rot_E and has an infinite time requirement. Infinite moments are mapped to the moment $[0, 5sec]$. The constraint maps incoming moments to the numeric range $[0, 360]$, i.e. a full rotation is performed in 5 seconds. In the example, the fade-in behavior requires a natural time of 4 seconds. The actual time assigned to the sequential behavior group may differ but is distributed in a 3:1 proportion to both child nodes. Color plate 3, p. 367 in the appendix shows an object during the fade-in animation.

4.3 Interaction

Several research tools have been developed which explore new interaction techniques and interface styles. However, they are limited with respect to robustness, completeness, and portability [9]. Most 3D toolkits (e.g. [5] [19]) provide low-level interaction techniques, but their reusability is limited due to bindings to concrete geometry types and their coarse-grained object oriented design. In our approach, interaction can be specified by specialized behavior nodes, called *interaction nodes.* Interaction nodes can be used to specify complex in teractions and multi-state augmented transition networks [8]. The base class *MInteraction* defines four states: starting, processing, terminating, and canceled. The state changes if start, end, termination, or cancel conditions are satisfied. Interaction nodes can be linked together to build complex interactions. State changes are propagated to child interaction nodes. Since interaction nodes use their own event types for communication, other behavior nodes can be inserted between them. Basic interaction nodes are:

- Drag nodes associate events with transformation graphics objects. For example, *MRotationDrag* nodes, *MScalingDrag* nodes, and *MSliderDrag* nodes map mouse movements to transformations.
- Command nodes integrate application-specific callbacks in the behavior graph. For example, the *MAction* node executes an action if a given event condition is satisfied, and the *MCommandIA* node infiltrates actions in interaction processes.

4.4 Example: A Transformation Box

We construct a geometry graph and a behavior graph which build an interactive transformation box (Fig. 9) with handles to scale and to rotate a shape. These graphs could be added to the object viewer application. The geometry graph arranges 6 side scale handles, 8 corner resize handles, and 12 rotation handles around a wire-frame bounding box (Fig. 10). The behavior graph

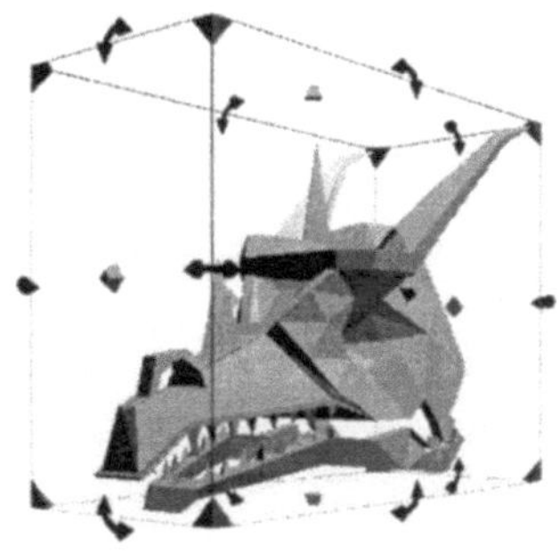

Fig. 9. Transformer box applied to a geometric object.

for the transformer consists of a drag node for each handle. The drag nodes constrain the associated *RRotation rottf* and *RScaling scltf* graphics objects which are shared with the geometry graph of the transformable shape.

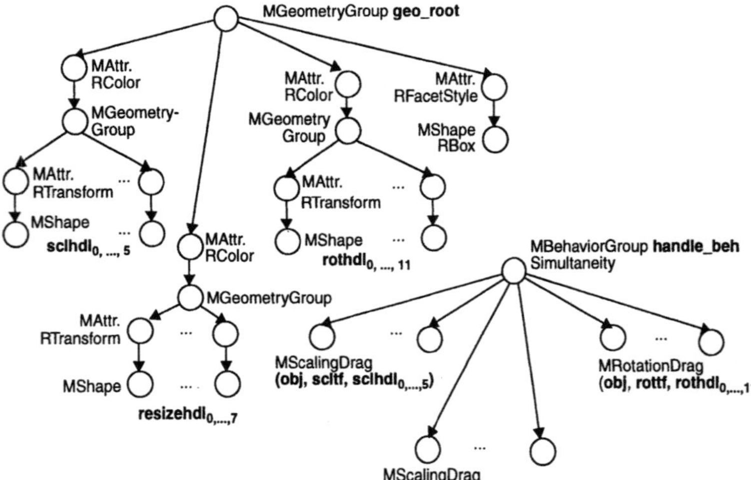

Fig. 10. Geometry graph and behavior graph for the object transformer.

5 Implementation

Our approach has been implemented in the visualization system MAM/VRS. The software architecture consists of two layers, MAM [6] as modeling layer,

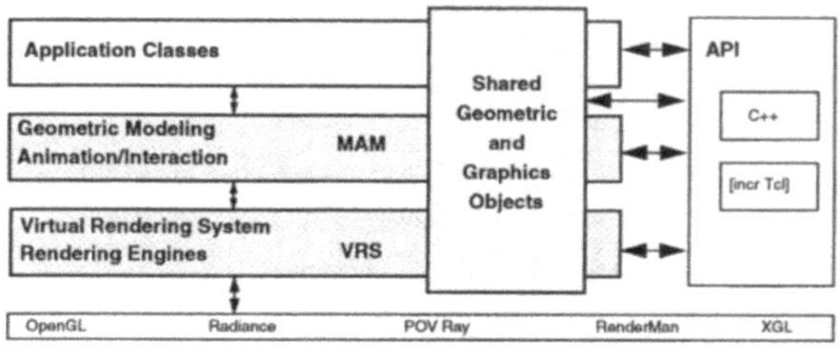

Fig. 11. MAM/VRS system architecture.

and VRS [7] as rendering layer (Fig. 11). The *Modeling and Animation Machine* is responsible for the geometric modeling and the modeling of animation and interaction, and implements geometry node classes and behavior node classes. The *Virtual Rendering System* is a rendering meta system which implements rendering engines and a collection of graphics objects. Currently, we provide rendering engines for OpenGL [17], Pixar's RenderMan [20], POV Ray [18], Radiance [21], and SUN's XGL.

The object model of MAM/VRS is language independent. MAM/VRS objects can be created and manipulated by different APIs. For example, MAM/VRS can be used as a C++ library or as an extension to the [incr Tcl] language [16], an object-oriented extension to Tcl/Tk. A native Java port of MAM/VRS is under development. Documentation and more details about MAM/VRS can be found at our WWW site:
`http://wwwmath.uni-muenster.de/~mam`.

6 Conclusions and Future Work

The fine-grained object-oriented architecture allows us to adapt the system efficiently and quickly to different visualization needs in industry and science. The integration of time and event control in an object-oriented fashion together with the shared objects which are used in all application layers results in a transparent software system which can be extended and maintained easily. The presented approach is currently used as a software platform in several visualization projects in medicine, chemistry, physics and geo-science. A visual language for the presented object model is under development.

Acknowledgements. The VRS project is supported by the Ministry of Science and Research, Northrhine-Westphalia, Germany.

References

1. M. Baker, *An Object-Oriented Approach to Animation Control*, In: Computer Graphics Using Object-Oriented Programming, S. Cunningham et al. (eds.), Wiley & Sons, 1992, pp. 187–212.

2. E. Bier, M. Stone, K. Pier, W. Buxton, T. DeRose, *Toolglasses and magic lenses: the see-through interface,* Proceedings of SIGGRAPH'93, pp. 73–80.
3. P. Calder, M. Linton, *Glyphs: Flyweight Objects for User Interfaces,* Proceedings of the ACM SIGGRAPH Third Annual Symposium on User Interface Software and Technology, 1990.
4. P. Egbert, W. Kubitz, *Application Graphics Modeling Support Through Object–Orientation,* Computer, October 1992, pp. 84–91.
5. C. Elliot, G. Schechter, R. Yeung, S. Abi-Ezzi SunSoft, Inc., *TBAG: A High Level Framework for Interactive, Animated 3D Graphics Applications,* Proceedings of SIGGRAPH '94, pp. 421–434.
6. J. Döllner, K. Hinrichs, *Object-Oriented 3D Modelling, Animation and Interaction,* The Journal of Visualization and Computer Animation, Vol. 8(1), pp. 33–64, 1997.
7. J. Döllner, K. Hinrichs, *The Design of a 3D Meta System,* 6th Eurographics Workshop on Programming Paradigms in Graphics, Budapest, 1997, to appear.
8. M. Green, *A Survey of Three Dialogue Models,* ACM Transactions of Graphics, Vol. 5, No. 3, July 1986, pp. 244–275.
9. K. Herndon, A. van Dam, M. Gleicher, *Workshop of the Challenges of 3D Interaction,* SIGCHI Bulletin, October 1994, Vol. 26, No. 4.
10. ISO/IEC DIS 14772-1, *The Virtual Reality Modeling Language,* April 1997, http://www.vrml.org/VRML97/DIS.
11. D. Knuth, *The TEXbook,* Addison-Wesley, Reading, MA, 1984.
12. L. Koved, W. Wooten, *GROOP: An Object-Oriented Toolkit for Animated 3D Graphics,* ACM SIGPLAN NOTICES OOPSLA'93, Vol. 28, No. 10, October 1993, pp. 309–325.
13. M. Linton, J. Vlissides, and P. Calder, *Composing user interfaces with InterViews,* IEEE Computer, pp. 8–22, February 1989.
14. M. Najork, M. Brown, *Obliq-3D: A High-Level, Fast-Turnaround 3D Animation System,* IEEE Transactions on Visualization and Computer Graphics, Vol 1, No. 2, Juni 1995, pp. 175–192
15. J. Ousterhout, *Tcl and the Tk Toolkit,* Addison-Wesley, 1994.
16. M. McLennan, *[incr Tcl]: Object-Oriented Programming in Tcl,* Proceedings of the Tcl/Tk Workshop 1993, University of California, http://www.wn.com/biz/itcl.
17. J. Neider, T. Davis, M. Woo, *OpenGL Programming Guide,* Addision-Wesley, 1993.
18. POV Team, *Persistency of Vision Ray Tracer (POV-Ray),* Version 1.0, Technical Report, 1991.
19. P. Strauss, R. Carey, *An object-oriented 3D graphics toolkit,* SIGGRAPH'92 Proceedings, Vol. 26, No. 2, pp. 341–349.
20. S. Upstill, *The RenderMan Companion. A Programmer's Guide to Realistic Computer Graphics,* Addison-Wesley, 1990.
21. G. Ward, *The RADIANCE Lighting Simulation and Rendering System,* Proceedings of SIGGRAPH' 94, pp. 459–472.
22. R. Zeleznik, D. Conner, M. Wloka, D. Aliaga, N. Huang, P. Hubbard, B. Knep, H. Kaufman, J. Hughes, A. van Dam, *An object-oriented framework for the integration of interactive animation techniques,* Proceedings of SIGGRAPH' 91, Vol. 25, No. 4, pp. 105–112.

A Survey of Parallel Coordinates

Alfred Inselberg

Multidimensional Graphs Ltd, Raanana, Israel

Abstract. Visualization provides *insight through images* [14], and can be considered as a collection of application specific mappings: *problem domain* $\longrightarrow$ *visual range.* For the visualization of multivariate problems a multidimensional system of *Parallel Coordinates* is studied which provides a one-to-one mapping between subsets of N-space and subsets of 2-space. The result is a systematic and rigorous way of doing and *seeing* analytic and synthetic N-dimensional geometry. Lines in N-space are represented by N-1 indexed points. In fact all p-flats (planes of dimension p in N-space) are represented by indexed points where the number of indices depends on p and N. The representations are generalized to enable the visualization of polytopes and certain kinds of hypersurfaces as well as recognition of convexity. Several algorithms for constructing and *displaying* intersections, proximities and points interior/exterior/or on a hypersurface have been obtained. The methodology has been applied to visual data mining, process control, medicine, finance, retailing, collision avoidance algorithms for air traffic control, optimization and others.

"Let no one ignorant of geometry enter" .... At entrance to Plato's academy

1 In the Spirit of Descartes

The fascination with dimensionality may predate Aristotle and Ptolemy who argued that space had only three dimensions. By the nineteenth century Riemann, Lobachevsky and Gauss unshackled the imagination, inhibited as it was by our 3-dimensional habitat, and higher-dimensional as well as non-Euclidian geometries came into their own. Over the years various multivariate methodologies came about (see [15] and [16] for non-technical reviews and [11] for a partial list of references). In 1959 Prof. S.S.Cairns, in his popular topology course at the University of Illinois, challenged his students to visualize higher dimensional geometries. In geometry parallelism, which does not require a notion of angle, rather than orthogonality is the more fundamental concept. This reason, and the fact that orthogonality "uses-up" the plane very fast, was my inspiration for proposing *Parallel* Coordinates. In spite of the professor's encouragement (and perhaps because of it ...but that's another story), I did not pursue the idea until 1977 when, while giving a course of Linear Algebra I was in turn, challenged by the students to *"show"* higher dimensional spaces. Parallel coordinates won the day and the systematic development got started [7,8,10–12,9].

In the Euclidean plane R^2 with xy-Cartesian coordinates, N copies of the real line labeled $\bar{X}_1, \bar{X}_2, \ldots, \bar{X}_N$ are placed equidistant (e.g. one unit apart)

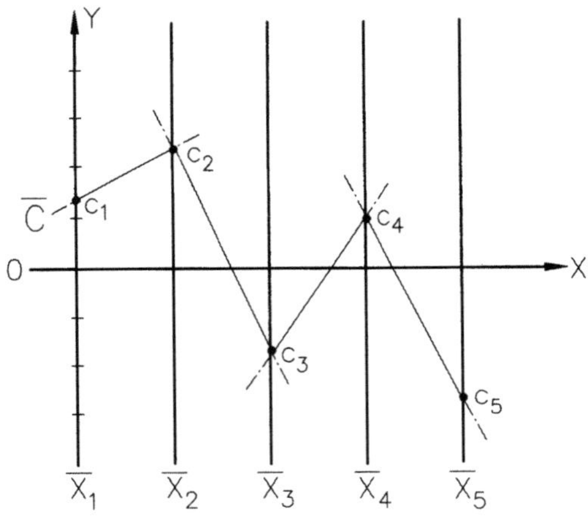

Fig. 1. Polygonal line $\bar{C}$ represents the point $C = (c_1, c_2, c_3, c_4, c_5)$.

and perpendicular to the x-axis. They are the axes of the parallel coordinate system for R^N all having the same positive orientation as the y-axis. A point C with coordinates (c_1,c_2,...,c_N) is represented by the complete polygonal line $\bar{C}$ (i.e. the lines of which only the local segments are usually shown) whose N vertices are at $(i-1, c_i)$ on the $\bar{X}_i$-axis for $i = 1, \ldots, N$ as shown in Fig. 1. In this way, a 1-1 correspondence between points in R^N and planar polygonal lines with vertices on the parallel axes is established (see [11] and [12] for more details as well as extensive bibliography). The image, also called the *representation*, of an object F in parallel coordinates is denoted by $\bar{F}$.

So far, the most widespread application of Parallel Coordinates (abbr. ||-coords) has been in visual data mining (knowledge discovery) [1] of multivariate datasets in finance, process control, retailing, medicine, and many other areas. There are other applications in air traffic control (collision avoidance algorithms), computer vision (edge and surface recognition, line and p-flat topologies), nonlinear modeling and optimization [3] (see [12] for a partial list of references on the applications).

2 Duality in 2-D

$\mathcal{P}$oints on the plane are represented by segments between the x_1 and x_2 axis and, in fact, by the *line* containing the segment. In Fig. 2, the distance between the parallel axes is d. The line

$$l : x_2 = mx_1 + b, \tag{1}$$

is a collection of points A. They are represented by the infinite collection of lines $\bar{A}$ on the xy plane which, when $m \neq 1$, intersect at the *point* given in

xy-coordinates:

$$\bar{l} : (\frac{d}{1-m}, \frac{b}{1-m}). \tag{2}$$

The reason for representing a point P by the *whole* line $\bar{P}$, rather than

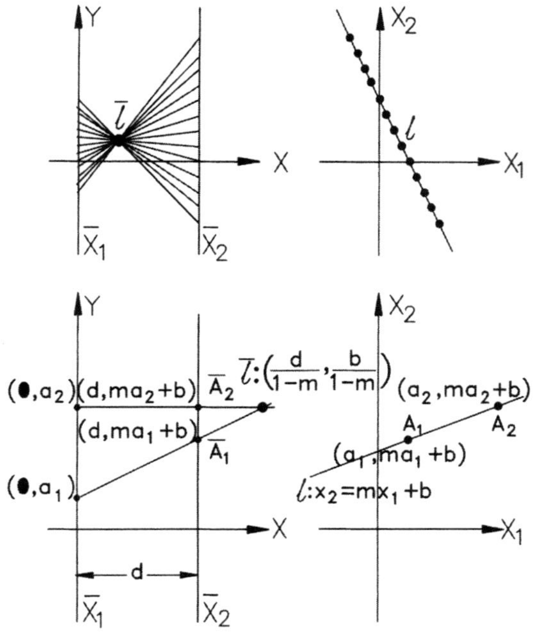

Fig. 2. In the plane parallel coordinates induce a point $\longleftrightarrow$ line duality.

just the segment between the parallel axes, is that $\bar{l}$ may lie outside the strip between the axes. Note that the x-coordinate of $\bar{l}$ depends only on m. Hence parallel lines are represented by points on the same *vertical* line; a property which provides some useful applications. For lines with $m = 1$, we consider xy and x_1x_2 as two copies of the *projective plane* so that such an l of the x_1x_2-plane corresponds to the *ideal point* $\bar{l}$ of the xy-plane with slope b/d. It is best to express this relation in homogeneous coordinates with triples within [...] and triples within (...) denoting line and point coordinates respectively, and yielding the linear transformation:

$$l : [m, -1, b] \longrightarrow \bar{l} : (d, b, 1-m). \tag{3}$$

It is easy to see that the point to line correspondence in 2-D is given by:

$$P : (p_1, p_2, p_3) \longrightarrow$$
$$\bar{P} : [(p_1 - p_2), dp_3, -dp_1]. \tag{4}$$

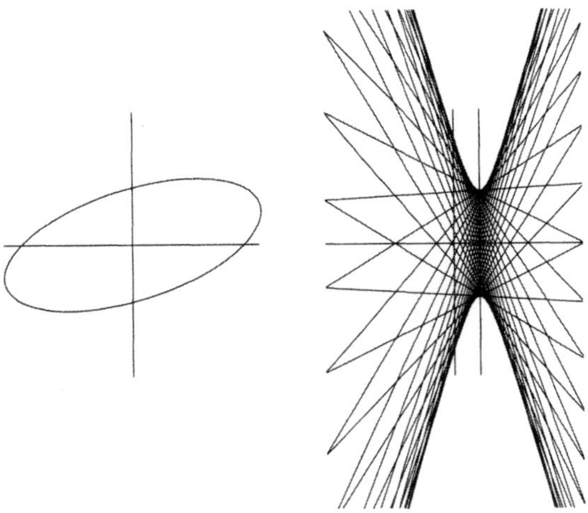

Fig. 3. Ellipse (point-curve) $\rightarrow$ Hyperbola (line-curve).

So, parallel coordinates *induce* a fundamental *Point* $\rightleftharpoons$ *Line* duality, formal-

Conic, *Gconic* $\rightarrow$	*Image*
ellipse, *estar*	hyperbola (Fig. 3), *hstar*
parabola, *pstar*	parabola,*pstar* or hyperbola,*hstar* with vertical assymptotes
hyperbola, *hstar*	ellipse (Fig. 4), *estar*, parabola, *pstar*, hyperbola, *hstar*

Table 1. Point curves, conics & *gconics* $\rightarrow$ line curves, conics & *gconics*

ized by Equations (3) and (4). The image (representation) $\bar{r}$ of a curve r can be obtained as the *envelope* of the lines representing the points of r. This is distinguished by referring to the original curves as *point-curves* and their images as *line-curves.* It can be shown that conics(point-curves) map into conics(line-curves) in 6 cases the orientation of the parabola or hyperbola determining the choice of its image. In fact, this a special case of a more general result involving generalized conics (abbr. gconics) which are sections of a double cone whose bases are two copies of a bounded convex set. For brevity, referring to bounded convex sets as *estars*, unbounded convex sets as *pstars* and the sections with two unbounded branches as *hstars*, the mapping has analogous associations (see table 1). This provides a new duality for convex sets from which some optimal duality algorithms were obtained (see [13]). There is also a nice *inflection point* $\rightleftharpoons$ *cusp* duality of interest in CAD/CAM

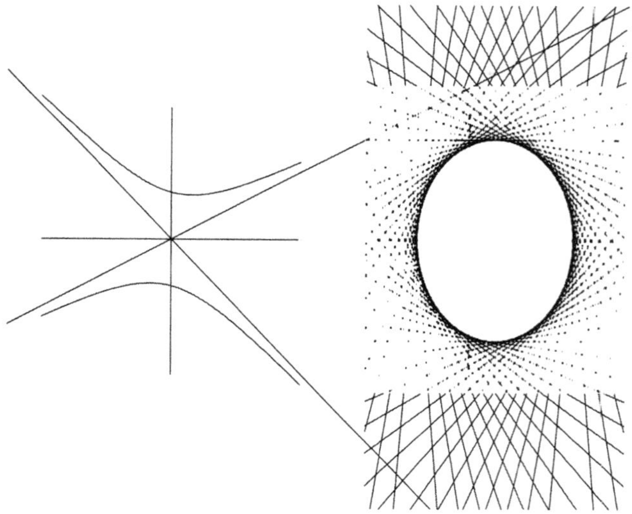

Fig. 4. Hyperbola(point-curve) $\rightarrow$ ellipse(line-curve) – image depends on orientation of hyperbola.

and computer vision. Still in 2-D, it turns out, that the rotation of a line l about one of it's points P is the dual of the point $\bar{l}$ translated along the line $\bar{P}$.

Law of attraction of unfortunate events:
"Unfortunate events tend to attract others of their kind."...

3 Lines, p-flats and Polytopes in R^N

In projective N-space, P^N, the general duality is *point* $\rightleftharpoons$ *hyperplane*, ergo the *point* $\rightleftharpoons$ *line* duality for N=2. For this reason the *line* $\longrightarrow$ *point* correspondence of 2-D does not typically generalize into a *line* $\longrightarrow$ *"something nice"* relation. For ||-coords, however, some useful and intuitively satisfying generalizations are possible.

Consider a line l in R^N described by

$$l_{i-1,i} : x_i = m_i x_{i-1} + b_i, \tag{5}$$

for i=*2,...,N*. In the $x_{i-1}x_i$-plane the relation labeled $l_{i-1,i}$ is a line and by the duality, Eq. (3), is represented in homogeneous coordinates by the point:

$$\bar{l}_{i-1,i} : \tag{6}$$
$$((i-2)(1-m_i)+1, b_i, 1-m_i).$$

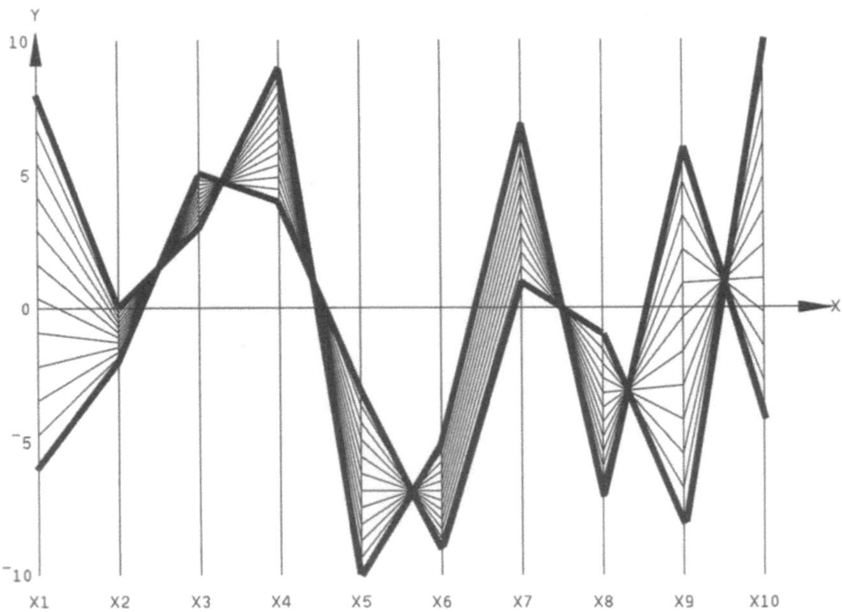

Fig. 5. Line interval in 10-D. Heavier polygonal lines represent end-points.

Hence, there are $N-1$ such points which represent the N-D line l with the indexing being an essential part of the representation. For Eq. (6) the distance between the axes is taken as 1. A polygonal line passing through all the $N-1$ points given by Eq. (6) necessarily represents a point on the line l since the pair of y-coordinates a_{i-1}, a_i of its vertices simultaneously satisfy Eq. (5) for $i = 2,...,N$. Such is the case in Fig. 5 where several polygonal lines representing points on an interval of a line in 10-D are shown. A number of construction algorithms based on this representation have been found including one for obtaining the minimum distance between two lines. This representation was used to obtain a collision avoidance algorithm in conjunction with the new air traffic control system in the USA (Patent # 4,823,272, # 5,058,024 and # 5,173,861).

The points given by Eq. (5) have a striking and very useful property. For $i \neq j \neq k \in [1,2,...,N]$ the three points $\bar{l}_{i,j}, \bar{l}_{j,k}, \bar{l}_{i,k}$ are always collinear. This can be seen by considering two points $P_r = (p_{1,r}, \ldots, p_{N,r})$, $r = 1,2$ on l and taking their projections on the x_i, x_j, x_k three-space as shown in Fig. 6. The projected portions of the points form two triangles and the collinearity

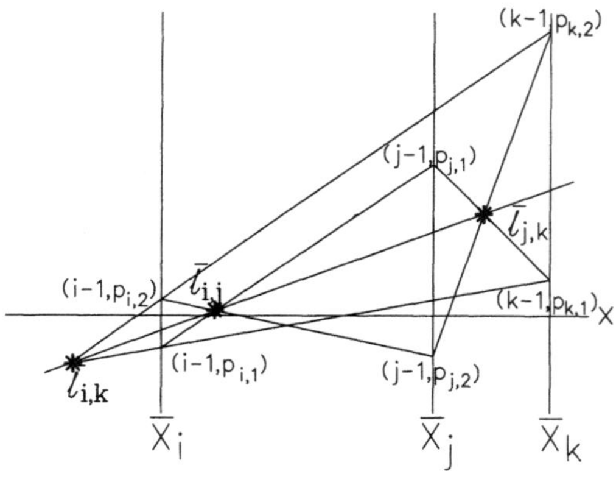

Fig. 6. The collinearity for the 3 points $\bar{l}_{i,j}, \bar{l}_{j,k}, \bar{l}_{i,k}$. The two triangles are in perspective with respect to the ideal point in vertical direction. The y-axis is offscale.

of $\bar{l}_{i,j}, \bar{l}_{j,k}, \bar{l}_{i,k}$ is obtained as a consequence of Desargue's Theorem[1]. This property is crucially important in the construction of the representation of higher dimensional p-flats in N-space (that is planes of dimension $2 \leq p \leq N-1$) and we will refer to it as the *the 3-point collinearity property.*

We saw in the previous section that parallel lines are represented by points on a vertical line whose x-coordinate is determined by the lines' slope. On a p-flat π in R^N, a coordinate system consisting of a grid formed by p families of parallel lines can be constructed. In turn, each family of lines can be represented by a vertical line. Altogether there are p distinct vertical lines. But the same grid can represent any p-flat parallel to π. A specific plane is then distinguished by choosing a point on it which is represented by a polygonal line. Hence p vertical lines together with a polygonal line represent a p-flat. With this representation a number of construction algorithms are obtained including finding whether a point is on a given p-flat (if not on what side it is), intersections as well as an extension of the duality between rotations about a line and translations of a point along a line. Also simple polytopes can be easily constructed and displayed as shown in in Fig. 8.

The vertical line representation of p-flats requires points on a regular grid of the p-flat. But such points are not always available. Eickemeyer [4] overcame this difficulty by describing a p-flat π_p in terms of its (p-1)-flats π_{p-1} (i.e. $\pi_{p-1} \subset \pi_p$). In fact, this gives rise to a *recursive* construction of p-flats where the recursion is on the dimension p. To illustrate, consider a plane π_2 in R^3 given by:

$$\pi_2 : c_1x_1 + c_2x_2 + c_3x_3 = c_0. \tag{7}$$

[1] Two triangles in perspective from a point are in perspective from a line.

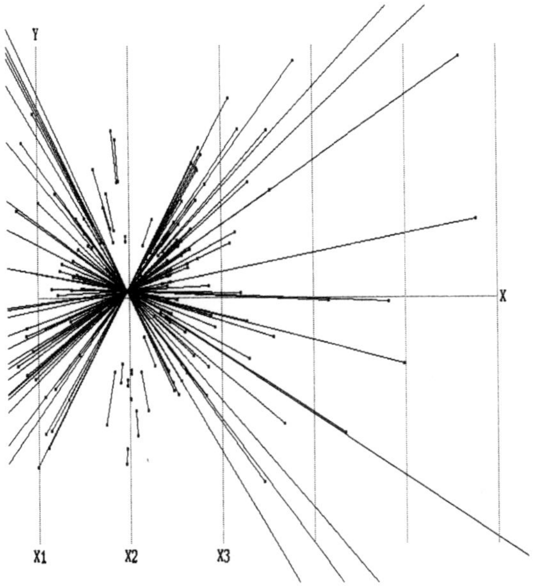

Fig. 7. Coplanarity. From a collection of lines l in 3-D another collection of lines $\bar{L}$ is constructed from the $\bar{l}$'s using the 3pt collinearity property. The lines $\bar{L}$ intersect at a point $\Longleftrightarrow$ the original lines l are coplanar.

From the representation of a line $l \subset \pi_2$ the points $\bar{l}_{1,2}, \bar{l}_{2,3}, \bar{l}_{1,3}$. are obtained and by the 3pt collinearity property they yield a line $\bar{L}$. Taking another line $l' \subset \pi_2$ in the same way a corresponding line $\bar{L}'$ is obtained. The point:

$$\bar{\pi}_{1,2,3} = \bar{L} \cap \bar{L}' : \qquad (8)$$
$$(c_2 + 2c_3,\ c_0,\ c_1 + c_2 + c_3)$$

has the remarkable property that for any other line $l'' \subset \pi_2$ the corresponding line $\bar{L''}$ also lies on $\bar{\pi}_{1,2,3}$. This property can also characterize coplanar points. Starting with a randomly chosen set $\mathcal{S}$ of points, a set $\mathcal{L}$ of lines is constructed from pairs of points. Next the corresponding $\bar{L}$'s are constructed from the 3pt collinearity property. It turns out that the set of points $\mathcal{S}$ is coplanar $\Longleftrightarrow$ all the $\bar{L}$'s intersect at a single point. Such a situation is shown in Fig. 7. Basically, Eq. (8) identifies an infinite *family* of 2-flats intersecting at a line and containing π_2. To distinguish π_2 more information is needed. For this reason the axis $\bar{X}_1$ is translated to a new position $\bar{X}_1'$ one unit to the right of $\bar{X}_3$. This corresponds to rotating the original x_1-axis to the equivalent x_1'-axis position. With the new coordinates x_1', x_2, x_3 another point $\bar{\pi}_{2,3,1'}$ is computed in the same way as the intersection of two $\bar{L}$'s. These two points completely identify π_2. Note that there are now *3 indices* to distinguish that these two points represent a 2-flat in R^3 rather than a line which is represented by two points but with *2 indices*; the reason being that a line and

$\bar{\pi}_2$ are described by equations involving *two* and *three* variables respectively. To recoup, we started with points (which are 0-flats i.e. 0-dimensional) on a 2-flat π_2 then constructed lines (1-flats) on π_2 and from them obtained the representation of the 2-flat still in terms of points. This outlines the recursive construction we alluded to earlier. In fact, the 3pt collinearity property generalizes to triples of points representing higher dimensional flats; the procedure is outlined later. The result is that a p-flat in R^N specified by $(N-p)$ linearly independent equations of the form:

$$\pi_{i_1 \ldots i_{(p+1)}} : \sum_{k=1}^{p+1} c_{i_k} x_{i_k} = c_0 \tag{9}$$

is represented by the $(N-p)p$ points:

$$\bar{\pi}_{i_1^* \ldots i_{(p+1)}^*} : (\sum_{k=1}^{p+1} d_{i_k}^* c_{i_k}, c_0, \sum_{k+1}^{p+1} c_{i_k}) \tag{10}$$

for $1 \leq p \leq N-1$ where,

1. each variable x_i appears twice on the *two* parallel axes $\bar{X}_i$ and $\bar{X'}_i$,
2. d_i^* is the distance from the y-axis to the $\bar{X}_i^*$ axes,
3. permutations such as $i_1 i_2 \ldots i_q$ consist of unique integers in $[1, 2, \ldots, N]$, and
4. $i_k^* = i_k$ when $j \leq k$, or $= i'_k$ for $j \geq k$.

Note the important special cases:

- when $p = 1$ (i.e. for lines) Eq. (10) yields $N-1$ points having *two* indices,
- when $p = 2$ and $N = 3$ (i.e. plane in 3-D) Eq. (10) yields *two* points having *three* indices, and
- when $p = N-1$ (i.e. hyperplane) Eq. (10) yields $N-1$ points having N indices.

The indexed point representation of p-flats turns out to be robust in the sense that it still works for "approximate" coplanarity. More precisely, when points are nearly coplanar (i.e. perturbing the coefficients of the planar equation to generate a "slab"), the construction in Fig. 7 yields lines $\bar{L}$ intersecting at several but *closely* spaced points and the pattern is unmistakable. So this construction can function as a "slab" detector. In fact, with some modification it can identify or cluster points in several slabs. Gennings et al [5] in an elegant application used this property to detect interacting drugs in chemotherapy.

Recently Chatterjee [2] methodically studied the transformation dualities mentioned earlier and showed that a *projective transformation* (i.e. translation, rotation, scaling and perspective) becomes another projective transformation. So a multidimensional object can still be recognized as the *same* object after it has been acted on by such a transformation. Further, using

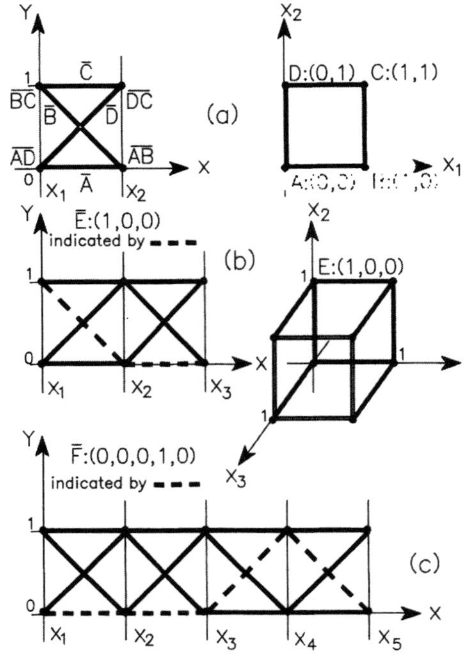

Fig. 8. (a) A square, (b) 3-D cube, and (c) 5-D hypercube – all with unit side.

Eickemeyer's representation, he obtained representations for polytopes in N-D where adjacencies of the faces of various orders can be directly visualized. He extended these to obtain some very useful notions for proximity and tolerances having applications in computer vision and solid modeling. In a remarkable result, he generalized the results on gconics (see section on duality in 2-D) so that, roughly speaking, a body in R^N is convex if the envelope of the points representing its faces of all orders forms an *hstar*. For the first time, convexity in N-D can be seen!

"Law of inopportune timing:"
"If a fortunate event occurs, it tends to happen a bit too soon or a bit too late."

4 Representation Mapping

We provide an outline of the recursive construction but without the technical details which are beyond the scope of this narrative. The program, is to construct a mapping:

$$\mathcal{I} : 2^{P^N} \rightarrow 2^{P^2} \times 2^{[1,2,\ldots,N]} \tag{11}$$

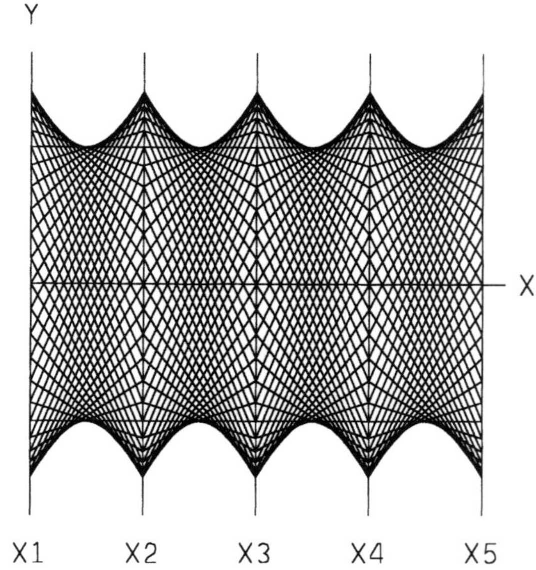

Fig. 9. A sphere in 5-D

where $2^A = [B \subseteq A]$ denotes the power set of a set A, and R^N is embedded in the projective N-space P^N. By the way it can be trivially seen that the cardinality of the domain and range of this mapping are the same. Typically the image $\mathcal{I}(U) = \bar{U}$ of a $U \subset P^N$ consists of several disconnected two-dimensional subsets, each *indexed* by a subset of $[1, 2, ..., N]$; the indexing being an essential part of the representation. The mapping $\mathcal{I}$ provides a unique representation since it is one-to-one with $\bar{U}_1 = \bar{U}_2 - U_1 = U_2$. We have already seen how to construct the representation $\mathcal{I}(\pi_2)$ as the envelope of the polygonal lines connecting the $\mathcal{I}(l)$ where $l \subset \pi_2$ (i.e. the lines contained in π_2). In general, $\mathcal{I}(\pi_p)$ is obtained as the envelope of the polygonal lines connecting (according to the indexing) the points of $\bar{\pi}_{p-1}$ and also consists of indexed points. It is the 3pt collinearity property and its higher dimensional generalization that are crucial in this stagewise construction.

5 Hypersurfaces

The representation of smooth hypersurfaces was originally obtained as the envelope of their surface points represented by polygonal lines. This is a straightforward way which has some technical problems that can be overcome. Such is the case with the 5-D sphere shown in Fig. 9. For this representation a useful *interior point algorithm* was found. In Fig. 10 the representation of a convex hypersurface in 20-D where a polygonal line is constructed repre-

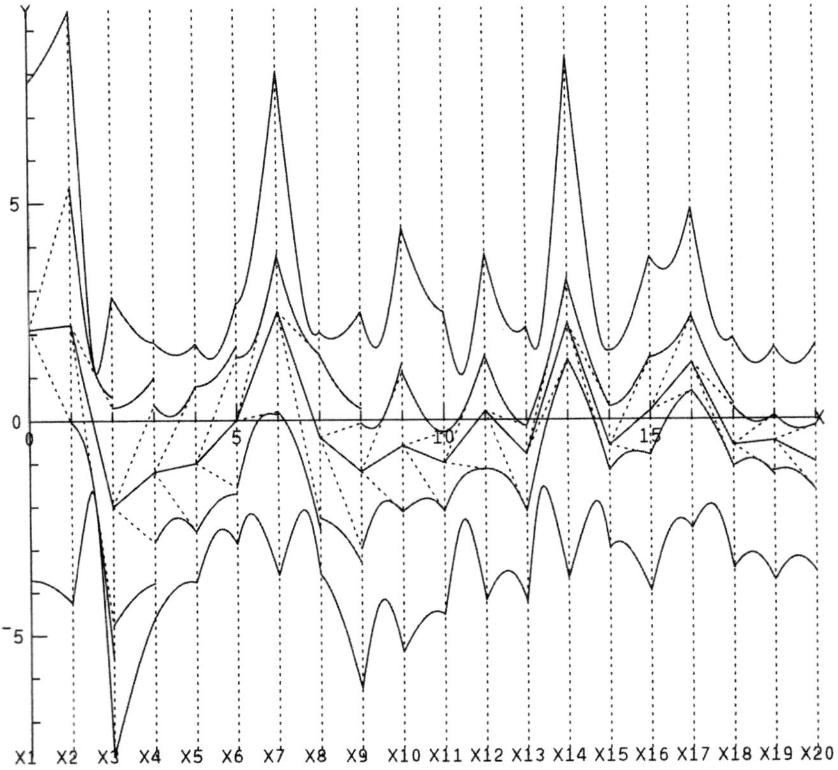

Fig. 10. A convex hypersurface in 20-D and interior point algorithm.

senting an interior point. In fact, also exterior and boundary points can be constructed. The intermediary envelopes shown provide an L_1 neighborhood of the point and thus giving local curvature information. Such a representation is of great interest in intelligent process control as well as nonlinear modeling in general.

Hung [6] recently obtained the representation of smooth surfaces by considering them as the envelope of their tangent hyperplanes. That is in R^N a hyperplane is represented by $N-1$ points with N indices all being adjacent permutations of $[1, 2, ..., N]$. So from each point of a smooth hypersurface σ, $N-1$ points are obtained for the representation of the tangent hyperplane at that point. The representation $\bar{\sigma}$ consists of N point sets, $\bar{\sigma_i}$, where the i stands for one of the permutations. On these, certain grids can be constructed for the reconstruction of the points of σ. It turns out that the $\bar{\sigma_i}$ are *curves* (i.e planar sets with just a boundary and no interior points) $\Longleftrightarrow$ σ is a *developable* surface. Further a grid on the $\bar{\sigma_i}$ can be constructed with one family of grid curves being *straight lines* $\Longleftrightarrow$ the surface is *ruled*. These, as well as the previous results, show how obvious visible features of the representation reveal geometrical structure of the surface being represented. They also provide us with *multidimensional "vision"*. Dimensionality need not be a "curse" after all; to us it is more of a blessing.

References

1. E.W. BASSETT, Ibm's ibm fix. *Industrial Computing*, 14(41):23–25, 1995.
2. A. CHATTERJEE, *Visualizing Multidimensioal Polytopes and Topologies for Tolerances*. Ph.D. Thesis USC, 1995.
3. A. CHATTERJEE, P. P. DAS, AND S. BHATTACHARYA, Visualization in linear programming using parallel coordinates. *Pattern Recognition*, 26-11:1725–36, 1993.
4. J. EICKEMEYER, *Visualizing p-flats in N-space using Parallel Coordinates*. Ph.D. Thesis UCLA, 1992.
5. C. GENNINGS, K. S. DAWSON, W. H. CARTER, AND R. H. MYERS, Interpreting plots of a multidimensional dose-response surface in a parallel coordinate systems. *Biometrics*, 46:719–35, 1990.
6. C.K. HUNG, *A New Representation of Surfaces Using Parallel Coordinates*. Submitted for Publication, 1996.
7. A. INSELBERG, *N-Dimensional Graphics, Part I – Lines and Hyperplanes, in IBM LASC Tech. Rep. G320-2711, 140 pages*. IBM LA Scientific Center, 1981.
8. A. INSELBERG, The plane with parallel coordinates. *Visual Computer*, 1:69–97, 1985.
9. A. INSELBERG, *Parallel Coordinates : A Guide for the Perplexed, in Hot Topics Proc. of IEEE Conf. on Visualization, 35-38*. IEEE Comp. Soc., Los Alamitos, CA, 1996.
10. A. INSELBERG AND B. DIMSDALE, *Parallel Coordinates: A Tool For Visualizing Multidimensional Geometry, in Proc. of IEEE Conf. on Vis. '90, 361-378*. IEEE Comp. Soc., Los Alamitos, CA, 1990.
11. A. INSELBERG AND B. DIMSDALE, Multidimensional lines i: Representation. *SIAM J. of Applied Math.*, 54-2:559–577, 1994.
12. A. INSELBERG AND B. DIMSDALE, Multidimensional lines ii: Proximity and applications. *SIAM J. of Applied Math.*, 54-2:578–596, 1994.
13. A. INSELBERG, M. REIF, AND T. CHOMUT, Convexity algorithms in parallel coordinates. *J. ACM*, 34:765–801, 1987.
14. B. H. MCCORMICK, T. A. DEFANTI, AND M. D. BROWN, *Visualization in Scientific Computing*. Computer Graphics 21-6, ACM SIGGRAPH, New York, 1987.
15. E. R. TUFTE, *The Visual Display of Quantitative Information*. Graphic Press, Connecticut, 1983.
16. E. R. TUFTE, *Envisioning Information*. Graphic Press, Connecticut, 1990.

"Better is the end of a thing than the beginning of it"........ Ecclesiastes 7:8

Hierarchical Techniques for Global Illumination Computations – Recent Trends and Developments

Philipp Slusallek, Marc Stamminger, and Hans-Peter Seidel

Computer Graphics Group, University of Erlangen, Germany

Abstract. Since the beginning of computer graphics, one of the primary goals has been to create convincingly realistic images of three-dimensional environments that would be impossible to distinguish from photographs of the real scene. The goal to create photo-realistic images has lead to the development of completely new software techniques for dealing with the inherent geometric and optical complexity of real world scenes. This paper gives an overview of advanced algorithms for photo-realistic rendering and in particular discusses hierarchical techniques for global illumination computations.

1 Introduction

Photo-realistic rendering has a strong influence on many areas of computer graphics that seems to become even bigger with 3D techniques entering the consumer market. Although the advances in photo-realistic rendering are no longer as spectacular and visible to the public as with early ray-tracing or radiosity images, it is a very active field of computer graphics research with substantial new developments. This paper tries to highlight some of these new developments and focuses on hierarchical techniques for global illumination computations.

Software for photo-realistic rendering can be thought of consisting of three main components: *scene description*, *lighting computation*, and *image generation*. The first component deals with the problem of describing 3D environments with the necessary level of detail and physical accuracy that allows for generating realistic images. This component is also concerned with the interface that a rendering system offers for importing scene descriptions from external modeling programs.

A key component of convincingly realistic presentations is the simulation of the rich details that are due to lighting effects, such as soft shadows and correct interreflections between objects. The proper simulation of these global illumination effects can avoid the "computer" or "artificial" touch of computer generated images. Once the distribution of light in a scene has been computed, we can take a snapshot of the light field with a virtual camera and generate an image from it. This last component of photo-realistic rendering deals with problems like hidden surface removal, shading, motion-blur, and anti-aliasing.

The methods and techniques that have been developed for global illumination computations are not restricted to generating realistic images only. These methods can also be modified to work with other forms of radiation (like heat, radar, radio, etc.). Furthermore, many of the advanced techniques used in realistic lighting computations like efficient hierarchical methods, clustering, and variance reduction techniques can equally be applied to other engineering applications where they may be less widely known.

Lets start with a brief review of the underlying physics and mathematics, before we discuss some of the new techniques for computing realistic images.

2 Fundamentals

The fundamental quantity in physically-based rendering is *radiance*. The quantity radiance $L(y, \boldsymbol{\omega}, \nu)$ is the power radiated from a point y in a certain direction $\boldsymbol{\omega}$ per unit projected area perpendicular to that direction and per unit solid angle. The units of radiance are $[W\ m^{-2}\ sr^{-1}]$. The dependency on the frequency ν of the radiation is usually ignored in lighting computations and the equations are solved for each frequency band (e.g. RGB) separately.

The distribution of light in a scene is completely described by the radiance distribution in space. All other quantities can be derived from it. Radiance is invariant along its direction of propagation, provided there are no absorption, emission, or scattering effects by participating media (like smoke or fog). Additionally, the amount of light received at the retina of the human eye is proportional to the radiance emitted by the viewed surface. These properties make radiance the appropriate quantity for illumination computations in a rendering system.

Light is emitted at light sources and interacts with other surfaces in the environment, being absorbed, reflected, or transmitted. The propagation of light in the absence of participating media is fully described by the radiance equation [6,24]

$$L(y, \boldsymbol{\omega}) = L_e(y, \boldsymbol{\omega}) + \int_{\Omega} f_{rt}(\boldsymbol{\omega_i}, y, \boldsymbol{\omega}) L_i(y, \boldsymbol{\omega_i}) |\cos\theta_y\ d\omega_i|. \tag{1}$$

It states that the radiance reflected at a point y in direction $\boldsymbol{\omega}$ is the sum of the self emitted radiance L_e and the reflected radiance. The latter is computed by integrating over the fraction of incident light from all directions that is scattered into the outgoing direction ω. The scattering characteristics of a surface is described by the *bidirectional reflection distribution function* f_{rt} (BRDF). The incident light L_i itself contains contributions from light reflected off other surfaces, thus coupling it to outgoing radiance L at all other visible points in the environment. It is this global nature of the radiance equation and the embedded problem of visibility together with complex scene arrangements that makes it so hard to compute truly photo-realistic images.

We are often not interested in the full *directional* distribution of light at a point. In this case, we can eliminate the directional parameter and switch to *radiosity* for describing the light distribution. This results in the radiosity equation

$$B(y) = B_e(y) + \frac{\rho(y)}{\pi} \int_{x \in S} B(x) G(x,y) V(x,y) dx, \qquad (2)$$

where we changed the integration domain to all surfaces in the environment. This explicitly adds the visibility function V and the geometric form factor $G(x,y) = \cos\theta_x \cos\theta_y / |x-y|^2$ to the integrand [6,24].

2.1 Finite Element Approach

Computing the illumination in some environment requires the solution of the radiance or radiosity equations (1) and (2). As Fredholm equations of the second kind [1], they cannot be solved analytically, except for trivial cases. Instead, numerical algorithms for computing a solution must be used, all of which are derived from two basic numerical approaches: *Finite Element* and *Monte-Carlo* (or point sampling) techniques. In this paper, we focus on the hierarchical Finite Element approach.

In the finite element approach, the illumination is described as a function representing outgoing radiance or radiosity over the surfaces in the scene. This technique takes advantage of the fact that reflected radiance and radiosity are usually piecewise smooth. Finite element techniques are based on a subdivision of the environment into small surface patches and compute a functional representation of the illumination on each of them.

With this approach Finite-Element algorithms try to exploit coherence in the light distribution in the environment. Finite element techniques are in a sense at the opposite side of the solution spectrum compared to Monte-Carlo algorithms: computation at points versus surfaces, deterministic versus stochastic computations, and explicit versus implicit use of coherence. Unfortunately, approximating the light distribution by smooth functions results in biased solutions — a problem that true Monte-Carlo methods can avoid completely. However, this bias is generally not a problem for most applications.

After representing the distribution of light in a finite-dimensional function space, we can use the standard Galerkin technique to reduce the radiance or radiosity equation to a linear system describing the exchange of light between surface patches. Although many standard techniques are available for solving linear systems, only very few of them can actually be used. Compared to other disciplines the linear system used for lighting computations have some unique features: The computation of the coefficients is very costly and requires knowledge about the scene as a whole. Depending on the complexity of the scene, the full matrix can become extremely large. It is thus mandatory to avoid computing or even storing coefficients if at all possible. This requirement has lead to a number of unique approaches for representing and solving the linear system that will be described in Section 3 in more detail.

2.2 Hybrid Approaches

There are many hybrid techniques that use a combination of Finite-Element techniques and the stochastic Monte-Carlo approach. They usually handle certain aspects of the integral equation (1) as special cases and try to exploit some form of coherence. This often results in a better approximation to the exact solution or in faster computations for certain classes of environments.

A common approach is to separate the indirect illumination from the illumination received directly from light sources. In a typical scene the irradiance received indirectly varies much slower and smoother than the direct illumination, where shadow boundaries are a major source of discontinuities. As a result, the indirect illumination can be approximated by fewer samples or fewer basis functions [29,17,32].

Another approach that is often used, is to separate the BRDF into a diffuse component and a specular or mirror component. The smoother diffuse reflection component can then be approximated using a different technique than the highly localized mirror reflection component [30,23,3].

3 Hierarchical Techniques

After this brief review of the available techniques for computing illumination in a virtual environment, this section concentrates on the use of hierarchical Finite-Element techniques. The use of hierarchies has made Finite-Element techniques practical and applicable to a large variety of virtual environments. It allows for scaling the algorithms with scene complexity, available computation time, and required accuracy.

As described in the previous section, Finite-Element techniques offer a deterministic approach to compute the illumination in a virtual environment. The main issues are the time and memory complexity of the algorithm.

In contrast to pure Monte-Carlo techniques, which require little to no memory for storing intermediate results, the Finite-Element algorithms must store the representation of illumination that is computed for all surfaces. For complex scenes this can become a considerable problem. Furthermore, the matrix of the linear system must be represented in some form during the solution process. Given a number of patches n in the environment, early Finite-Element algorithms compute the light transport between any pair of them, thus resulting in quadratic complexity $O(n^2)$ in the number of patches.

The introduction of hierarchical algorithms have changed this situation dramatically. Hierarchical techniques allow for automatically selecting the level of detail at which light should be transported between patches. Together with clustering techniques, described in the next section, this allows for algorithms that have only *linear* complexity $O(n)$ in the number of input surfaces.

Lets take a closer look on how this can be achieved.

3.1 The Basic Hierarchical Algorithm

Finite element algorithms require the discretization of the environment into patches. For a given functional representation of illumination on a patch, the accuracy of the computation is determined by the discretization of the environment. Many small patches allow for a more detailed and accurate computation. What is required is an adaptive discretization into smaller patches where it significantly improves accuracy, for example along shadow boundaries or in other areas where illumination changes rapidly.

The hierarchical approach allows for exactly this kind of adaptive computation. Instead of discretizing the environment into a fine mesh of patches before starting the computation, the algorithm refines the initial coarse set of surfaces during the course of the computation wherever this becomes necessary. For recursive refinement, patches are usually represented in the form of *quadtrees* on triangles, quadrilaterals, or other parametric surfaces (see Figure 1).

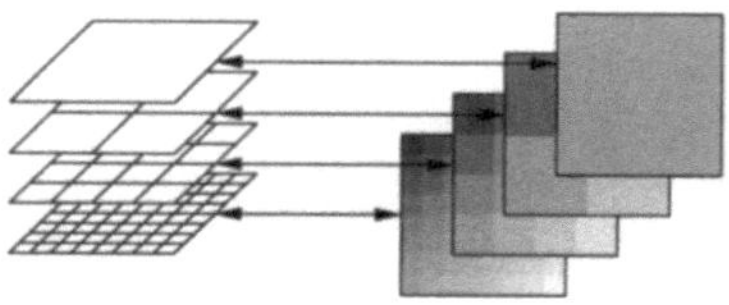

Fig. 1. Representation of illumination at different levels of a hierarchy. Patches in higher levels represent the average illumination of their children.

When considering the transport of light between surfaces, the algorithm chooses the appropriate level of discretization that is required for meeting a user selected error criterion. For two nearby patches that exchange a large amount of radiation, which may result in large illumination gradients on the receiver, a detailed subdivision of either the sender or the receiver or both is required. On the other hand, if the two surfaces are far apart or exchange only very little energy, it might be sufficient to compute the exchange between large patches in upper levels of the quadtrees [11] (see Figure 2).

During the computation the hierarchical algorithms automatically selects the coarsest level at which the accuracy requirements can be met. The same surface may receive and send illumination at various levels of the discretization hierarchy. It is therefore necessary to keep the different levels consistent. If energy is received at some level, this energy must also be distributed or *pushed* down to the children of the receiving patch. On the other hand, the energy must also be propagated or *pulled* to upper levels in the hierarchy. Starting at the leaf nodes in the hierarchy, parent patches then compute their illumination by summing up the energy of their children. The whole process

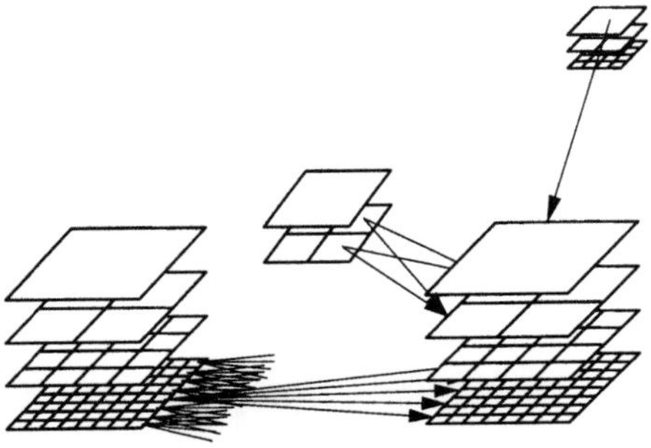

Fig. 2. Light is transported via links between various levels in two hierarchies depending on the configuration of the sending and receiving patch. A refinement procedure selects the coarsest level at which the transport can be computed within a user selected error threshold.

can be described as a single *push/pull* procedure, that keeps the different representations of illumination consistent [11].

Representing and Solving the Linear System A matrix is certainly not an appropriate way to represent the exchange coefficients of the linear system. The complex interactions between different levels in the hierarchy can best be represented by storing the exchange coefficients explicitly as *links* between the two patches. The flexible structure of this representation allows for simple additions and deletion of the links. These dynamic changes would be difficult to implement in a matrix structure.

The solution of the linear system now proceeds by using an iterative solver for linear systems. The most commonly used iterative solver is the Gauss-Seidel algorithm, which considers each surface in turn and *gathers* light via all links that connect to the hierarchy of patches on the receiver. Similarly, the Southwell algorithm can be used, which *shoots* energy via these links to other surfaces. However, due to the fact that the adaptively created links all transport similar amounts of energy, the original advantage of this solver is lost [2,8,11].

Before energy is actually transported via a link, the iterative solver must check the accuracy of the link. Due to the changes in illumination, links may become inaccurate and may need to be refined during the course of the iterations. The refinement of links is guided by a *refiner*, that is discussed in more detail in Section 3.1 and Section 5.

Many more iterative solvers are known for linear systems. Most of these algorithms are difficult to adapt to the link representation of the exchange coefficients. Furthermore, the adaptive creation of links together with the computation of the correct exchange coefficient by far dominates the solution time compared to the actual transport of energy via the links. Thus, somewhat faster iterative solvers would not have much effect and would complicate the hierarchical algorithm compared to the simple Gauss-Seidel and Southwell solvers.

Refinement Criteria The crucial part of any hierarchical algorithm is the *refiner* (often also called oracle). It determines if an exchange coefficient between two patches can adequately be approximated within a user selected error threshold. Otherwise the patches need to be subdivided. The particular design of the refiner determines to a very large extent the accuracy, memory requirements, and running time of a hierarchical algorithm. As a consequence, it is not meaningful to compare hierarchical algorithms without specifying the particular design of the refiner and the chosen parameters.

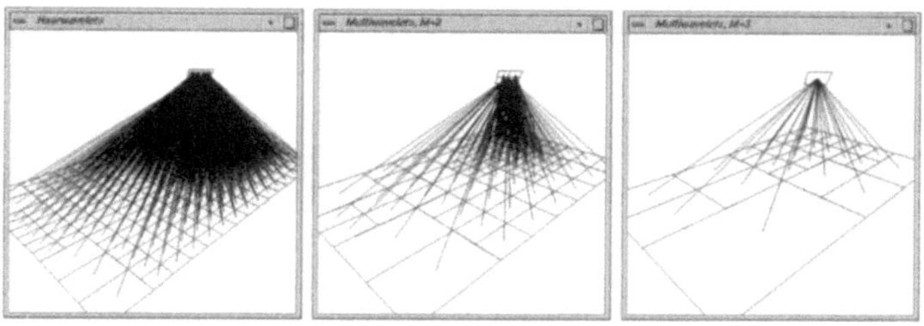

Fig. 3. Links between two patches generated by using Multi-Wavelets of various orders. Constant: 1 coefficient per link (left), Bi-linear: 16 coefficients (center), bi-quadratic: 81 coefficients (right).

For hierarchical radiosity computations several different refiners have been proposed. The most important refiners use the exchange coefficient itself (F-refinement) [11], resulting radiance on the receiver (BF-refinement), or reflected flux (BFA-refinement) for estimating the error that is committed by transporting light across a link. The BFA-refiner seems to work best for all kinds of environments and is used in most implementations.

These refiners are based on the observation that the error of a link can be bounded above by a constant factor times the exchange coefficient of the link. Weighting this error with the transported energy and the area of the receiver results in the last two refiners. The BFA-refiner is well behaved and usually works without a minimum size for patches, which is imperative for the other two refiners. Several modified versions of these refiners have also been proposed (e.g. [12]).

Wavelets and Higher Order Function Spaces Most of the hierarchical algorithms use simple constant basis functions for representing illumination on a patch, requiring only a single coefficient for this representation. The Galerkin construction of Finite-Element algorithms also works for higher order basis functions [6]. The formal mathematical basis for a hierarchical construction is the Wavelet theory that describes hierarchical function spaces [7]. Hierarchical wavelet algorithms have been used in various ways in rendering, e.g. for radiosity computations [10,20], or for full radiance computations [5,17,14].

Higher order functional representations allow for using less but larger patches, because these functions have a higher approximation power. This can result in reduced subdivision and consequently in less patches and links. However, higher order functions are also more costly in terms of computing and storing exchange coefficients. For instance bi-quadratic basis functions require storing 81 exchange coefficients per link (between each pair of the nine coefficients on each patch), instead of storing only a single coefficient per link as with constant basis functions (see Figure 3).

This drastic increase in storage requirements is the main disadvantage of higher order wavelet algorithms, which becomes most apparent at shadow boundaries. Even the larger approximation power of bi-quadratic or bi-cubic functions cannot adequately represent these features and still requires subdivision resulting in additional links that are much more expensive than in the constant case. In these situations the larger approximation power cannot compensate the increased memory requirements for storing links [31]. One possibility to avoid subdivision in these situations is by approximately representing visibility separately using a *shadow mask* [28,32].

4 Clustering

The main problem with hierarchical algorithms as described in the previous section is that the algorithms still have quadratic complexity in the number of input surfaces. The algorithms must still consider each pair of input surfaces even if the energy transfer between them is negligible. For the case of scenes with many small surfaces instead of a few large surfaces, this *initial linking* usually dominates the total computation time. The problem is that the benefits of linear complexity of the hierarchical algorithms are only exploited during adaptive refinement. A simple solution would be to avoid initial linking wherever possible [12].

Another approach extends the hierarchical algorithm. In order to use the benefits of adaptive computation of hierarchical algorithms, we must extend the hierarchy that exists below the level of a surface to a full hierarchy covering the whole scene. This can be done by grouping related surfaces into *clusters*. These clusters can in turn be grouped hierarchically into larger clusters until the whole scene is contained in a single *root cluster* [21,16,9].

Although "hand made" cluster hierarchies often perform slightly better [27], automatic clustering algorithms are preferred, because they are applicable to general and complex scenes. Here, hierarchical bounding volume algorithms seem to give best results compared to octrees or k-d trees [18,27] . The latter two often construct clusters that consist of unrelated patches, which can result in large lighting error.

Fig. 4. The model of a railway station consisting of approximately 32000 patches and 24 light sources. The lighting simulation based on clusters was computed in less than 10 minutes on a MIPS R8K CPU. Only 0.02 percent of the potential links where actually created. The image on the right shows the clustering hierarchy that was used.

The important feature of a cluster is that it can receive and emit energy for all of its contents. This allows to exchange light between a large group of surfaces in a single operation. An important difference to the use of planar surfaces is that a cluster may exchange energy with itself. Consequently, the processing starts by refining the first link in the scene, which is a self link from the root cluster to itself. Since the exchange of light at this level would generally introduce too much error, the refiner would replace this link, open up the cluster, and instead link its child clusters or surfaces before recursing.

This clustering strategy allows for quickly computing coarse approximations in a short amount of time. By choosing a large error threshold only a few large clusters would ever be considered. Although this would result in a coarse approximation, it may already allow a first estimate of the lighting effect in a scene (see Figure 4). By adjusting the threshold the user can continuously change the level of accuracy and speed of the computation. This is a valuable feature for interactive applications.

Even if all surfaces in a cluster are purely diffuse, the cluster will receive and reflect light with a directionally dependent characteristics. However, in the simplest case, the light exchange of a cluster and its environment is modeled as that of an isotropic volume that emits and scatters light in all directions evenly. Although this is an extremely crude approximation, it already works quite well in most cases. In order to avoid some of the resulting artifacts, light energy received by a cluster can immediately be propagated to all surfaces in its hierarchy. This allows to use the direction of the incident light while it is still known and computing its contribution to each surface separately [16]. Note, that this is different and still much more efficient than computing the form factor to each surface separately.

The cluster algorithms can also be extended for using non-isotropic interactions with its environment by storing the directional distribution of emitted energy with each cluster [19].

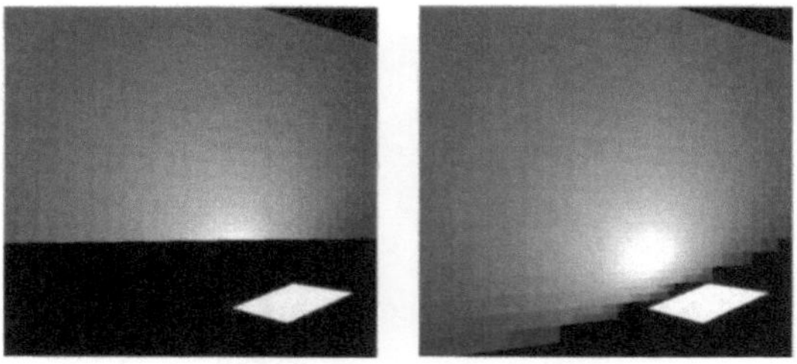

Fig. 5. Sampling problem for a small light source located directly in front large patch causing a small but bright spot: Any point sampling strategy is likely to miss the spot and omit important light contributions during subdivision (left). Correctly bounding the transport quantity allows for locating hot spots on a patch (right).

5 Refiners Based on Bounded Transport

As already noted in Section 3.1, the refiner is the crucial part of any hierarchical algorithm. Optimizing the refiner can have large effects on the quality, speed, and memory requirements of a computation. Early refiners for lighting simulations used simple criteria and point sampling to estimate light transport. By accurately bounding the input parameters of the computation, it is possible to obtain tight but conservative bounds on the amount of light transport over a receiving patch. These bounds offer a new opportunity for guiding the refinement procedure for normal as well as for hierarchical algorithms [26,25].

One problem with the refiners as presented in Section 3.1 is that they only estimate an upper bound of some transport quantity (i.e. the form factor or radiosity). The lower bound is always assumed to be zero. This approach ignores much of the information that is available or can easily be computed for a link [16]. Such information includes geometric quantities like the extent of sender and receiver, their distance, and the variation of normal vectors across each surface. The variation of normal vectors is of particular importance for the ability to treat curved surfaces efficiently.

A second problem of these refiners is the way the transport quantities are computed. These refiners generally use some form of point sampling for estimating them. They often use only a single sample for computing the exchange coefficient at the "center" of a patch, e.g. using the disk to point form factor approximation. The transport quantity may then be further modulated by estimating visibility of the link with sample rays that are sent between the two patches in question.

The sampling problem already becomes apparent for a small light source located directly in front of a large patch (see Figure 5). Although there might be a considerable transport of light to and from a small spot on the large patch, any point sampling strategy across the large patch is likely to miss this hot-spot. Another example for the sampling problem becomes apparent if we consider curved surfaces. In this case, the geometric parameters (locations, distances, and normals), on which the computation of the exchange coefficients is based, vary strongly over the surface, again making any point sampling strategy highly unreliable.

A solution is to compute upper and lower bounds for any of the parameters that are used as input to the computation of the exchange coefficients. If tight conservative bounds are computed for the input parameters, the resulting exchange coefficients can also be bounded by applying interval or affine arithmetic [13,4,22]. The location of points and their normals can efficiently be bound with the cone of normals technique [15] and bounding boxes. Based on this data, bounds on the set of connecting rays, cosines factors, distances, and solid angles can be computed. Bounds can also be computed for visibility between patches. Finally, these terms are combined for obtaining a conservatively bounded estimate of the exchange coefficient [26].

During the computation, triple values <min, estimate, max> are maintained that denote an upper and lower bound as well as an estimate of the correct value. These values can then be used by the new refinement strategy for deciding where to refine links. Links are refined, when the difference between upper and lower bound becomes too large. The new refiner also has enough information for deciding whether to refine the sender or the receiver [25]. The decision is based on their relative influence on the error. This is an important benefit over previous refiners, where no such information is available.

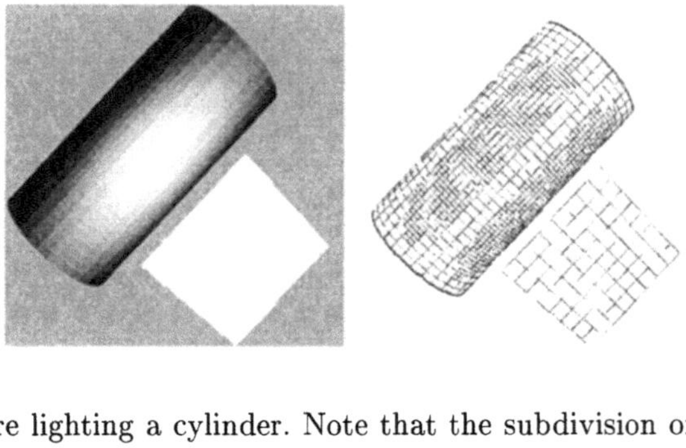

Fig. 6. A square lighting a cylinder. Note that the subdivision on the cylinder at the center coarser the in the neighborhood. This is due to the slower varying cosine term in the center.

Figure 6 shows the effect of this new refinement strategy for the light exchange between a planar patch and a cylinder. The cylinder is refined less in the center, where most of energy is transfered, but the illumination is mostly constant. This is where the older refiners would concentrate their effort. Instead the new refiner shifts the focus to areas with higher illumination gradients as indicated by larger differences between upper and lower bounds. Thus, it refines near the silhouette edges of the cylinder where there is still a large amount of received light and where both the cosine factor at the cylinder and the distance varies considerably.

The interface to geometric objects for obtaining the required geometric information can be kept very simple. This allows a wide variety of objects to implement it [26]. As shown in [25] the same strategy can be extended to work equally well for environment with clusters. Although it is more difficult to compute bounds for clusters, the algorithms and refiners still benefit considerably for having accurate bounds available. The bounds can also be used to estimate the self illumination of non-convex curved surfaces and clusters [26].

6 Conclusions

Photo-realistic rendering has come a long way in reaching the goal of generating images that are hard to distinguish from photographs of the same scene. In the process of this research activity many interesting new techniques have been developed that may also be applied in other engineering applications.

In particular the use of modern mathematical approaches like hierarchical representations and algorithms combined with the clever use of efficient data structures and programming techniques has led to new robust, accurate, and efficient algorithms that allow to simulate the physical process on which photo-realistic image generation is based.

References

1. K. E. Atkinson *A Survey of Numerical Methods for the Solution of Fredholm Integral Equations of the Second Kind.* Society for Industrial and Applied Mathematics, Philadelphia, 1976.
2. M. Cohen, S. E. Chen, J. R. Wallace, and D. P. Greenberg A progressive refinement approach to fast radiosity image generation. *Computer Graphics (SIGGRAPH '88 Proceedings)*, 22(4):75–84, August 1988.
3. S. E. Chen, H. E. Rushmeier, G. Miller, and D. Turner A progressive multi-pass method for global illumination. *Computer Graphics (SIGGRAPH '91 Proceedings)*, 25(4):165–174, July 1991.
4. J. L. D. Comba and J. Stolfi Affine arithmetic and its applications to computer graphics. In *Anais do VII Sibgrapi*, pages 9–18, 1993. Available from http://www.dcc.unicamp.br/~stolfi/EXPORT/papers/affine-arith.

5. P. H. Christensen, E. J. Stollnitz, D. Salesin, and T. D. DeRose Wavelet radiance. In *Fifth Eurographics Workshop on Rendering*, pages 287–301, Darmstadt, June 1994.

6. M. F. Cohen and J. R Wallace *Radiosity and Realistic Image Synthesis.* Academic Press, 1993.

7. I. Daubechies *Ten Lectures on Wavelets.* SIAM Philadelphia, Pennsylvania, 1992.

8. S. Gortler, M. F. Cohen, and P. Slusallek Radiosity and relaxation methods. *IEEE Computer Graphics & Applications*, 14(6):48–58, November 1994.

9. S. Gibson and R. J. Hubbold Efficient hierarchical refinement and clustering for radiosity in complex environements. *Computer Graphics Forum*, 15(5):297–310, December 1996.

10. S. J. Gortler, P. Schröder, M. Cohen, and P. M. Hanrahan Wavelet radiosity. *Computer Graphics (SIGGRAPH '93 Proceedings)*, 27:221–230, August 1993.

11. P. Hanrahan, D. Salzmann, and L. Aupperle A rapid hierarchical radiosity algorithm. *Computer Graphics (SIGGRAPH '91 Proceedings)*, 25(4):197–206, 1991.

12. N. Holzschuch, F. Sillion, and G. Drettakis An efficient progressive refinement strategy for hierarchical radiosity. In *Photorealistic Rendering Techniques (Proceedings Fifth Eurographics Workshop on Rendering)*, pages 343–357, Darmstadt, Germany, June 1994. Springer.

13. R. E. Moore *Interval Analysis.* Prentice-Hall, 1966.

14. S. Pattanaik and K. Bouatouch Haar wavelet: A solution to global illumination with general surface properties. In *Photorealistic Rendering Techniques (Proceedings of Fourth Eurographics Workshop on Rendering)*, pages 281–294. Eurographics, Springer, June 1994.

15. L. A Shirman and S. S. Abi-Ezzi The cone of normals technique for fast processing of curved patches. *Computer Graphics Forum (EUROGRAPHICS '93 Proceedings)*, 12(3):261–272, September 1993.

16. B. Smits, J. Arvo, and D. Greenberg A clustering algorithm for radiosity in complex environments. *Computer Graphics (SIGGRAPH '94 Proceedings)*, pages 435–442, July 1994.

17. P. Schröder *Wavelet Algorithms for Illumination Computations.* PhD thesis, Princeton University, November 1994.

18. F. Sillion and G. Drettakis Feature-based control of visibility error: A multi-resolution clustering algorithm for global illumination. *Computer Graphics (SIGGRAPH '95 Proceedings)*, pages 145–152, August 1995.

19. F. Sillion, G. Drettakis, and C. Soler A clustering algorithm for radiance calculation in general environments. In P.M. Hanrahan and W. Purgathofer, editors, *Rendering Techniques '95 (Proceedings of Sixth Eurographics Workshop on Rendering)*, pages 196–205. Springer, August 1995.

20. P. Schröder, S. Gortler, M. F. Cohen, and P. Hanrahan Wavelet projections for radiosity. *Proceedings on the Fourth Eurographics Workshop on Rendering*, pages 95–104, June 1993.

21. F. Sillion A unified hierarchical algorithm for global illumination with scattering volumes and object clusters. *IEEE Transactions on Visualization and Computer Graphics*, 1(3), September 1995.

22. J. M. Snyder Interval analysis for computer graphics. In *Computer Graphics (SIGGRAPH '92 Proceedings)*, pages 121–130, July 1992.

23. F. X. Sillion and C. Puech A general two-pass method integrating specular and diffuse reflection. *Computer Graphics (SIGGRAPH '89 Proceedings)*, 23(3):335–344, July 1989.

24. F. Sillion and C. Puech *Radiosity & Global Illumination.* Morgan Kaufmann, 1994.

25. M. Stamminger, P. Slusallek, and H.-P. Seidel Bounded clustering – finding good bounds on clustered light transport. Technical Report TR-97-1, Universität Erlangen, IMMD 9, 1997. http://www9.informatik.uni-erlangen.de/Publications/.

26. M. Stamminger, P. Slusallek, and H.-P. Seidel Bounded radiosity – illumination on general surfaces and clusters. *Computer Graphics Forum (EUROGRAPHICS '97 Proceedings)*, 16(3), September 1997. to appear.

27. M. Stamminger, P. Slusallek, and H.-P. Seidel Isotropic clustering for hierarchical radiosity — implementation and experiences. In *Proceedings Fifth International Conference in Central Europe on Computer Graphics and Visualization — WSCG '97*, 1997.

28. P. Slusallek, M. Schröder, M. Stamminger, and H.-P. Seidel Smart links and efficient reconstruction for wavelet radiosity. In *Rendering Techniques '95 (Proceedings Sixth Eurographics Workshop on Rendering)*, pages 240–251, Dublin, June 1995. Springer.

29. G. J. Ward The RADIANCE lighting simulation and rendering system. *Computer Graphics (SIGGRAPH '94 Proceedings)*, pages 459–472, July 1994.

30. J. R. Wallace, M. F. Cohen, and D. P. Greenberg A two-pass solution to the rendering equation: A synthesis of ray tracing and radiosity methods. *Computer Graphics (SIGGRAPH '87 Proceedings)*, 21(4):311–320, July 1987.

31. A. J. Willmott and P. S. Heckbert An empirical comparison of progressive and wavelet radiosity. In J. Dorsey and Ph. Slusallek, editors, *Rendering Techniques '97*, pages 175–186, Saint-Etienne, June 1997. Springer.

32. H. R. Zatz Galerkin radiosity: A higher order solution method for global illumination. *Computer Graphics (SIGGRAPH '93 Proceedings)*, pages 213–220, August 1993.

Two-Dimensional Image Rotation

Ivan Sterling[1] and Thomas Sterling[2]

[1] Department of Mathematics, University of Toledo, Toledo, Ohio, USA
[2] 1715 Chandler, Ann Arbor, Michigan, USA

Abstract. We study the problem of approximating a rotation of the plane, $\alpha : R^2 \longrightarrow R^2$ $\alpha(x,y) = (x\cos\theta + y\sin\theta, y\cos\theta - x\sin\theta)$, by a bijection $\beta : Z^2 \longrightarrow Z^2$. We show by an explicit construction that one may choose β so that $\sup_{z\in Z^2} |\alpha(z) - \beta(z)| \leq \frac{1}{\sqrt{2}}\frac{1+r}{\sqrt{1+r^2}}$, where $r = \tan(\theta/2)$. The scheme is based on those invented and patented by the second author in 1994.

1 Introduction

The need for image rotation arises in many scientific and industrial settings. A typical example of the need for rotation of an image by an arbitrary angle arises in the case of digital storage of an image by digital scanning of a document, such as an engineering drawing. When the drawing is fed into the scanner, it may be slightly skewed. Accordingly, the digitized image as it is initially stored in memory will reflect the skewing of the document relative to the scanner. When the stored image is produced on the computer screen, it will be skewed correspondingly, relative to the screen. It may be desirable to rotate the image as viewed on the screen to a desired alignment with the screen. Another example of the need for rotating an image arises in presenting the image on the computer screen or in a printed document at a rotational position different from that represented by its bit-mapping in memory.

The obvious way to rotate an image is to store the original image in one area of memory and transfer each dot from one image to the other. The problem with this method is that, as the image is rotated, the dots from the original image will fall between the dot locations in the second area of memory. For this reason, interpolation should be applied to each dot. Using that approach, there must be a series of computations performed for each dot, which will result in low speed operation (or high cost). There is a need for a simplified high speed algorithm for rotating images.

Another problem with the naive "nearest-neighbor" type algorithms is that they are neither 1-1 nor onto. This causes difficulty in ordering the storage, (for example, a pixel could be empty, white, black, doubly black, etc.), and it causes difficulty when iterating the algorithm (the image would gradually disappear!).

It is therefore desirable to approximate nearest-neighbor type accuracy using 1-1 and onto algorithms. Such schemes exist in the image processing literature. The best seems to be Paeth's scheme [2]. Which is, however, less accurate than ours (see the end of the next section).

All the schemes for rotating images in this paper are modifications of the schemes invented and patented by the second author in 1994 [4]. These schemes are fast, simple and accurate. The mathematical theory of these schemes was first investigated in [3].

Some old [3] and new results are contained in this paper. The new results are of two kinds. First, we prove that for infinitely many angles our schemes are the best possible ones. Second, we investigate a selection of angles for which we can prove that our schemes are not the best possible ones. Both of these developments were indicated by numerical results presented in [3], but required further work to prove. A more detailed summary of the paper is given at the end of the next section.

2 Mathematical Statement of the Problem

We study the problem of approximating planar rotations (about the origin)

$$\alpha : R^2 \longrightarrow R^2, \quad \alpha(0,0) = (0,0),$$

by *1-1*, *onto* mappings

$$\beta : Z^2 \longrightarrow Z^2,$$

of the integer lattice $Z^2 = \{(x,y) \in R^2 | x,y \in Z\}$. β is said to approximate α with error

$$e(\beta, \alpha) = \sup_{z \in Z^2} \{|\beta(z) - \alpha(z)|\}.$$

A *scheme* S associates to every α a β. That is, if K is the space of rotations and L is the space of 1-1, onto lattice maps, then

$$S : K \longrightarrow L.$$

The error of a scheme S is

$$E_S = \sup_{\alpha \in K} \{e(S(\alpha), \alpha)\}.$$

We seek schemes S whose error E_S is as low as possible.

In Section 5 we define a scheme T and recall that

$$E_T \leq \frac{\sqrt{2+\sqrt{2}}}{2} \approx .924 \ . \tag{1}$$

The estimate in Equation (1) is a special case of the main theorem in [3]. This theorem gives the error bound for rotation by an angle θ, $(0 \leq \theta \leq \frac{\pi}{4})$ for the T-Scheme to be

$$\frac{1}{\sqrt{2}} \frac{1+r}{\sqrt{1+r^2}},$$

where $r = \tan\left(\frac{\theta}{2}\right)$.

In Section 6 we discuss the known lower bounds. In particular, we recall that for every scheme S

$$E_S \geq \frac{2+\sqrt{2}}{4} \approx .853 \ .$$

We also give infinitely many examples where our T-Scheme attains the lower bound and hence, is optimal.

Finally in Section 10 we discuss cases where our schemes are not optimal. Using computer searches we have discovered a variety of subtle ways to improve these cases.

It should be mentioned that many patented schemes, S, invented before those in [4] (in particular those patented in [1]) have $E_S = \infty$.

For every angle, Paeth's [2] scheme is less accurate than our scheme and is never optimal.

$$E_{Paeth} = \frac{\sqrt{5}}{2} \approx 1.12 \ .$$

3 Notation

Let $J_R = [0, \pi/4]$ and $I_R = [0, \sqrt{2}-1]$. The schemes defined in this paper are defined for angles $\theta \in J_R$. If $\tan\frac{\theta}{2} = r$, then $\sin\theta = \frac{2r}{1+r^2}$ and $\cos\theta = \frac{1-r^2}{1+r^2}$. $\tan\frac{\theta}{2}$ maps J_R 1-1, onto I_R and, without loss of generality, it suffices to consider angles in J_R. For $r \in I_R$,

$$\alpha_r := \text{rotation by the angle } 2\tan^{-1} r.$$

Remark 1. We work in pixel coordinates (in particular, y increases as it goes down).

Remark 2. The blocks defined below *do* contain their left and upper edges, but *not* their right or lower edges (blocks do not contain their upper right corners, nor their lower left corners).

Remark 3. $[x] := Floor(x) := x - x \ mod\, 1$ (i.e. the greatest integer less than or equal to x). Note that $x - 1 < [x] \leq x < [x] + 1$.

4 The P Array

A block is a rectangle $B(x, y, w, h)$ in the plane with upper left corner at (x, y), with width w and height h. We tile the plane into an array P of blocks $B(x, y, w, h)$ described below. The array P is transformed, block by block, to a new array P' of blocks $B'(x', y', w', h')$.

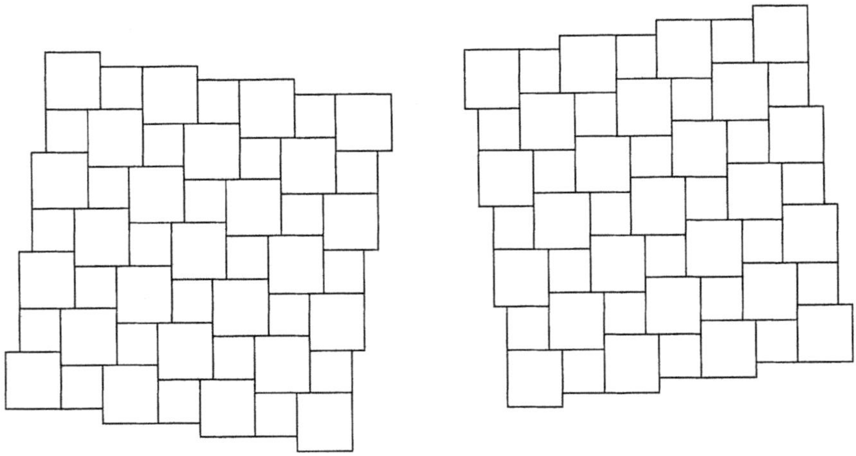

Fig. 1. P and P' Arrays for r = m/n = 1/7

Assume that $\tan\frac{\theta}{2} = r \in I_R$. Let $R = \frac{1}{r}$. The idea is that P will give a tiling which is tilted by about an angle $-\frac{\theta}{2}$ and P' will tilt by roughly $\frac{\theta}{2}$ (see Figures 1 and 2). The blocks in the arrays are labeled by integers a and b. The location of any block in the array P can be determined as follows: the upper left corner of any block $B(a,b)$ (where a+b is even) has an x-coordinate $x_\ell(a,b)$ and a y-coordinate $y_u(a,b)$ given by

$$x_\ell(a,b) = [\frac{aR - b + \zeta}{2}],$$
$$y_u(a,b) = [\frac{a + bR + \eta}{2}].$$

The choice of ζ and η is central to optimizing the scheme (see Section 6). In [4] the choice was $\zeta = 0$ and $\eta = \frac{1}{m}$ (whenever $r = \frac{m}{n}$). For our T-Scheme we choose

$$\zeta = \frac{3-R}{2}, \eta = \frac{3-R}{2}.$$

The size of the block B(a,b) is defined by determining the location of the lower and right sides. The lower and right sides have the same x and y coordinates as the adjoining sides of the contiguous blocks. The x-coordinate of the right side $x_r(a,b)$ is

$$x_r(a,b) = x_\ell(a+1, b-1) = [\frac{aR - b + R + 1 + \zeta}{2}],$$

and the y-coordinate of the lower side $y_\ell(a,b)$ is

$$y_\ell(a,b) = y_u(a+1,b+1) = [\frac{a+bR+1+R+\eta}{2}].$$

The bounds of the blocks having $a+b$ odd are determined from the four blocks that surround it according to the following:

$$x_\ell(a,b) = x_r(a-1,b) = [\frac{aR-b+1+\zeta}{2}],$$

$$y_u(a,b) = y_\ell(a,b-1) = [\frac{a+bR+1+\eta}{2}],$$

$$x_r(a,b) = x_\ell(a+1,b) = [\frac{aR-b+R+\zeta}{2}],$$

$$y_\ell(a,b) = y_u(a,b+1) = [\frac{a+bR+R+\eta}{2}].$$

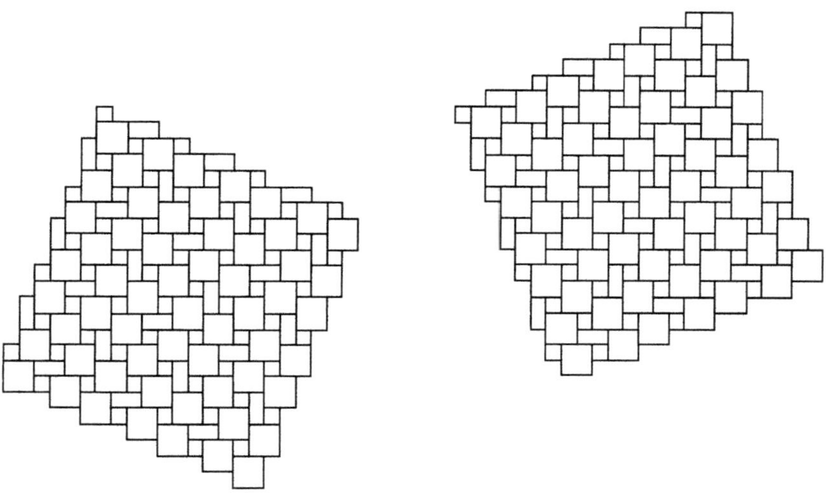

Fig. 2. P and P' Arrays for r = m/n = 2/5

Finally we have

$$w = x_r - x_\ell,$$

$$h = y_\ell - y_u.$$

Lemma 4. *Every point in R^2 is in one and only one block.*

5 The T-Scheme

Having calculated the position and size of the blocks $B(x, y, w, h)$, the next step is to calculate the position to which the blocks are to be moved. The blocks in the new array are specified by $B'(x', y', w', h')$.

$$x' = \begin{cases} [(aR + b + \zeta')/2] & \text{if } a+b \text{ is even} \\ [(aR + b + 1 + \zeta')/2] & \text{if } a+b \text{ is odd,} \end{cases}$$

$$y' = \begin{cases} [(-a + bR + \eta')/2] & \text{if } a+b \text{ is even} \\ [(-a + bR + 1 + \eta')/2] & \text{if } a+b \text{ is odd,} \end{cases}$$

$$w' = w \ , \ h' = h.$$

For our T-Scheme we choose $\zeta' = \zeta$ and $\eta' = \eta$.

Given $r \in I_R$ the transformation of P into P' defines a map $T(\alpha_r)$ as follows: by Lemma 4 we have that every point $z = (x, y) \in Z^2$ is (uniquely) of the form

$$x = x_\ell(a, b) + u,$$
$$y = y_u(a, b) + v,$$

with $0 \le u < w(a, b)$ and $0 \le v < h(a, b)$.

$$T(\alpha_r)(x, y) := (x', y')$$

where

$$x' = x'_\ell(a, b) + u,$$
$$y' = y'_u(a, b) + v.$$

Theorem 5. *For all $r \in I_R$, $T(\alpha_r)$ is 1-1, onto, and $e(T(\alpha_r), \alpha_r) \le \frac{1}{\sqrt{2}} \frac{1+r}{\sqrt{1+r^2}}$.*

6 The Modified T-Scheme

The T-Scheme described above was chosen for ease of exposition and is not optimal. It can happen that the choice

$$\zeta = \zeta' = \eta = \eta' = \frac{3 - R}{2}$$

is not the best one. In particular, sometimes the choice

$$\zeta = \frac{5 - R}{2}, \zeta' = \frac{1 - R}{2}, \eta = \frac{3 - 3R}{2}, \eta' = \frac{3 - 3R}{2}$$

is better (and sometimes it is worse). We define the Modified T-Scheme by choosing the better of these two choices (we suspect there is never a better choice). In Figure 3 we show the error bound given by Theorem 5 and also the actual errors (using the Modified T-Scheme) for a sample of angles.

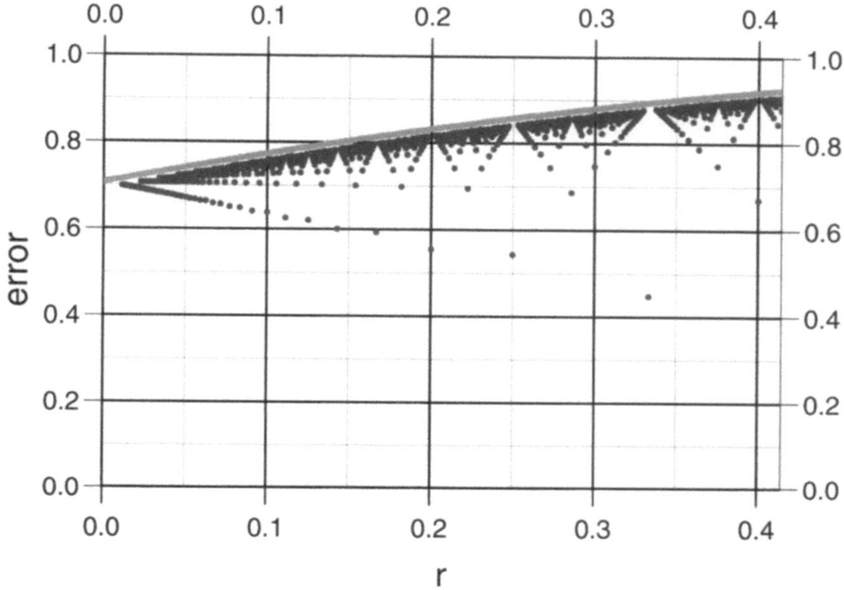

Fig. 3. Modified T-Scheme

7 Periodicity

Note that if r is rational with $r = \frac{m}{n}$ then P is periodic in the following sense:

$$\begin{aligned} x_\ell(a+2m,b) &= x_\ell(a,b)+n, \\ y_u(a+2m,b) &= y_u(a,b)+m, \\ x_\ell(a,b+2m) &= x_\ell(a,b)-m, \\ y_u(a,b+2m) &= y_u(a,b)+n. \end{aligned}$$

The $4m^2$ blocks given, as we vary from a to $a+2m$ and from b to $b+2m$, form a fundamental domain F. There are m^2+n^2 "pixels" in each fundamental domain. When computationally studying the rational case, it suffices to restrict consideration to a fundamental domain.

8 Lower Bounds

8.1 Lower Bounds for "Rational" Angles

Proposition 6. *For all S, $E_S > .787$. In fact, for all β, $e(\beta, \alpha_{``\sqrt{2}-1"}) > .787$.*

Proof. Assume not. Say β approximates rotation by $\pi/4$ with error less than .787. β must send the point $(0,0)$ to $(0,0)$. The point $(1,0)$ gets sent into some square and β must send it to a corner of this square which (i) isn't already in the image of β and (ii) is within a distance .787 of the image point. Now the point $(1,1)$ likewise, etc. This forms a simple tree of possibilities which can be checked until one obtains a contradiction. We did this for some selected angles (see Figure 4). The maximum we found (via a deeper search than shown in Figure 4) was .787 at "$\pi/4$".

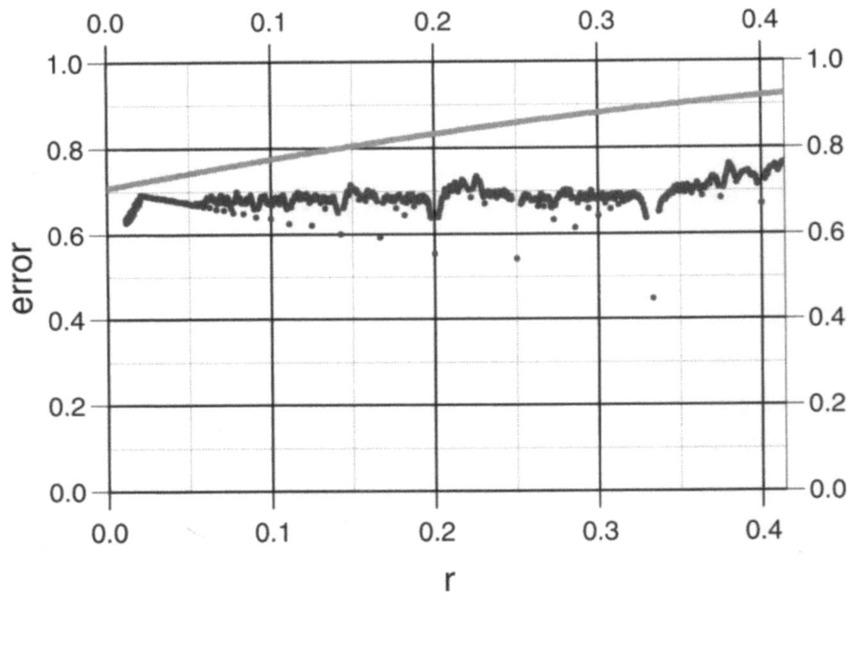

Fig. 4. Lower Bounds for "Rational" Angles

Remark 7. Note that the graphs indicate that the Modified T-Scheme is not optimal. The program actually constructs candidates for better β's; we investigated these and present our results in Section 10. In the simplest case, $\frac{2}{7}$, we have found the computer's candidate and have checked that it is indeed better.

8.2 Lower Bounds for "Irrational" Angles

As indicated by Figure 4 and by experience, the "worst" case seems to occur at $\pi/4$.

Example 8. (Due to Richard Stong.) Suppose we are trying to approximate a rotation $x \longrightarrow a(x)$ by an angle of $\pi/4$ by a bijection. There will be some point $x = (n, 0)$ such that $a(x) = (n/\sqrt{2}, n/\sqrt{2})$ is very close to $(m+0.5, m+0.5)$. Therefore, the images of the four points $(n,-1)$, $(n,1)$, $(n-1,0)$ and $(n+1,0)$ will be approximately at $(m+0.5+\frac{\sqrt{2}}{2}, m+0.5+\frac{\sqrt{2}}{2})$, $(m+0.5-\frac{\sqrt{2}}{2}, m+0.5+\frac{\sqrt{2}}{2})$, $(m+0.5-\frac{\sqrt{2}}{2}, m+0.5-\frac{\sqrt{2}}{2})$, and $(m+0.5+\frac{\sqrt{2}}{2}, m+0.5-\frac{\sqrt{2}}{2})$. The images of these five points will all be reasonably close to the four lattice points which are vertices of the unit square containing $(m+0.5, m+0.5)$. However, the approximating bijection must pair one of these five points to a point not on that square. Up to symmetry it doesn't matter which, and we see that one of the five images must be moved a distance D which approximately satisfies

$$D^2 = (7 - 4\sqrt{2})/2.$$

Hence D is about 0.819.

By generalizing this example we have

Theorem 9. *For all S, $E_S \geq \frac{2+\sqrt{2}}{4} \approx .853..$ In fact, for all β, $e(\beta, \alpha_{\sqrt{2}-1}) \geq \frac{2+\sqrt{2}}{4}$.*

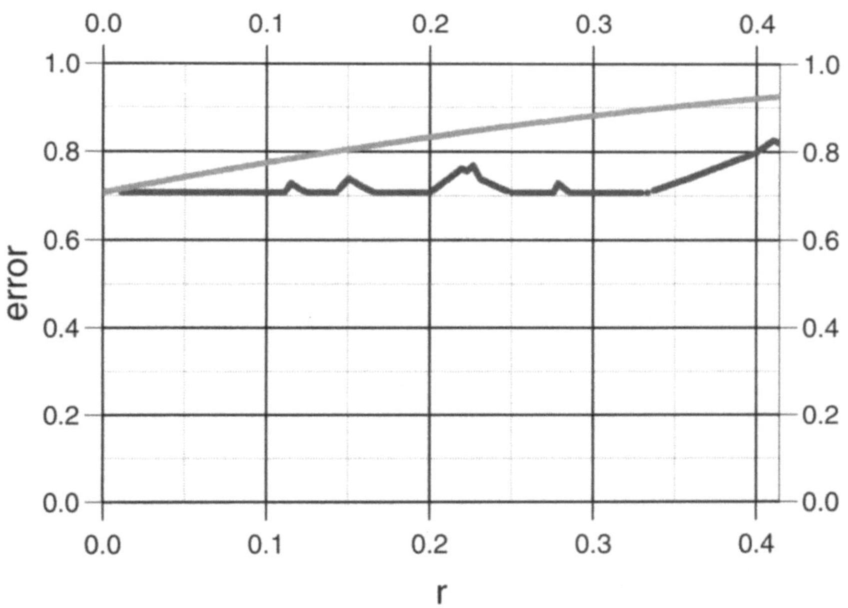

Fig. 5. Lower Bounds for "Irrational" Angles

For "irrational" angles some pixel is sent infinitesimally close to the center of a square. Modifying the computer program of Section 8.1 accordingly, we obtain Figure 5 which shows lower bounds for irrationals infinitesimally close to the rationals calculated by the computer.

9 Optimal Cases

The most important open question is: what is $M := \inf_S\{E_S\}$ where S ranges over all schemes? (We have proven that $.853 < \frac{2+\sqrt{2}}{4} \le M \le \sqrt{\frac{2+\sqrt{2}}{4}} < .924$.) More precisely: Does there exist a scheme (or set of schemes) S^* with $E_{S^*} = M$? Can it (or they) be described explicitly? And more generally: what is the function $M(\alpha) := \inf_{\beta\in L}\{e(\beta,\alpha)\}$? (We have proven that $Lower(\alpha_r) < M(\alpha_r) \le \frac{1}{\sqrt{2}}\frac{1+r}{\sqrt{1+r^2}}$ where *Lower* is the lower graph in Figure 4 [Figure 5 for "irrationals"], which we've only checked for some selected angles.)

As discussed the the next section it seems quite difficult to find $M(\alpha)$ for all α. However we can now find $M(\alpha)$ for at least some α's. In this section we find countably many such α's and a few more are discussed in the next section.

Remark 10. The following theorem was motivated by comparing the lowest "line" of dots in Figure 3 with the lowest line of dots in Figure 4.

Theorem 11. *For all odd n,*

$$M(\alpha_{1/n}) = \frac{\sqrt{2}}{2}\frac{n-1}{\sqrt{n^2+1}}.$$

Proof. We prove that for these angles the modified T-scheme is optimal. In other words we prove that $e(T(\alpha_{1/n}),\alpha_{1/n}) = \frac{\sqrt{2}}{2}\frac{n-1}{\sqrt{n^2+1}}$ and that for any scheme S we have $e(S(\alpha_{1/n}),\alpha_{1/n}) \ge \frac{\sqrt{2}}{2}\frac{n-1}{\sqrt{n^2+1}}$. Given n odd the P array breaks up into squares of sides $(n+1)/2$ and $(n-1)/2$ (see for example Figure 1). The centers of these squares get sent to centers of squares in the P' array. The worst case occurs at the corner of the larger square, and a simple calculation using modular arithmetic shows that these points miss their targets by exactly $\frac{\sqrt{2}}{2}\frac{n-1}{\sqrt{n^2+1}}$. In carrying out this calculation one also sees that the targets are the nearest neighbors - which completes the proof.

Example 12. If $n = 7$ we have an error of exactly 0.6, which precisely agrees with Figures 3 and 4.

Remark 13. When n is even, we believe the modified T-Scheme is still optimal. Note that in this case the nearest neighbor algorithm is not 1-1.

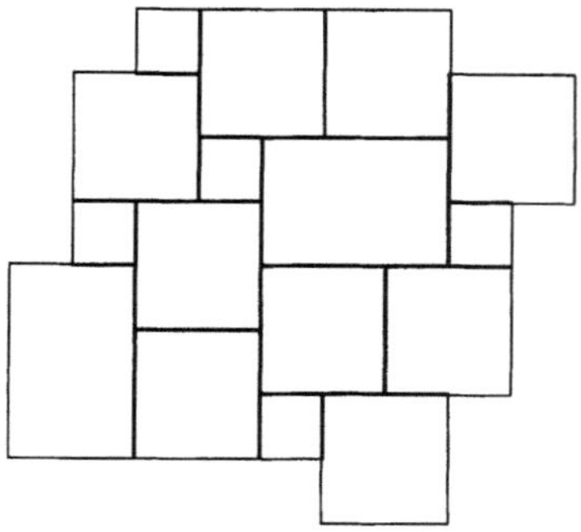

Fig. 6. Fundamental Domain for Optimal Tiling for 2/7

10 Non-optimal Cases

We believe that, except for those cases discussed above, the modified T-scheme is not optimal for the rational case. This belief is based on comparing Figures 3 and 4. On a case by case basis one can attempt to find optimal schemes. For ratios of small integers we have found optimal mappings with the computer. We have found optimal mappings for 2/5, 2/7, 3/8, 2/9, 3/10, 2/11, 2/13, 2/15, 4/19. The tiling showing the blocks which move together for the cases 2/7 and 2/9 are shown in Figures 6 and 7. These mappings are improvements over the modified T-scheme, but share some of the qualities of the T-scheme in that there is a row of tiles tilted at $\frac{1}{2}\theta$ mapped into tiles tilted at $\frac{1}{2}\theta$. Unlike the modified T-scheme the tiles do not always differ by one from their neighbors and the tiles are not always rectangular.

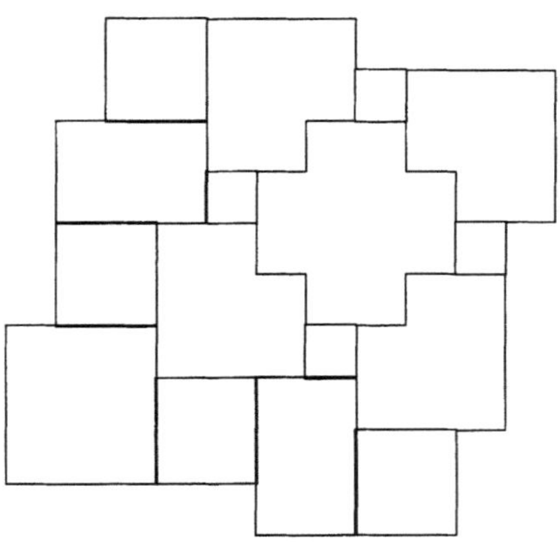

Fig. 7. Fundamental Domain for Optimal Tiling for 2/9

11 Miscellaneous Comments and Questions

- Some irrational algebraic numbers admit constructions not available for transcendental numbers. It may be that the theory is quite different for these two cases.
- What can be said about approximating rotations (about the origin)

$$\alpha : R^3 \longrightarrow R^3, \quad \alpha(0,0) = (0,0),$$

  by *1-1*, *onto* mappings

$$\beta : Z^3 \longrightarrow Z^3,$$

  of the integer lattice $Z^3 = \{(x, y, z) \in R^3 | x, y, z \in Z\}$. Or with 3 replaced by N?
- What happens if one measures the goodness of a scheme by how well it does "on the average" as opposed to "in the worst case" as we have here?
- What happens if the approximations are not required to be 1-1? Onto? If approximations are not required to be 1-1 nor onto, then clearly the "nearest neighbor" scheme is theoretically optimal, although it may be CPU intensive. An interesting related question: When is the nearest neighbor algorithm 1-1 and onto? Although not proved here, this is the case in Theorem 11.
- The circle of ideas in this paper seem to be topological and combinatorial in nature. However, there may be some meaningful (or obvious?) connection with number theory or the theory of symplectic iterators (rotations of the plane are indeed simple examples of area preserving maps, and discretizations of these have received much attention lately).
- Richard Stong observed, using Hall's Marriage Theorem, that one gets a scheme that works unless one can find a configuration as above where the rotates of k points are within r of only a total of $k - 1$ points. This condition certainly seems to simplify the problem, but we cannot yet prove which "clusters" are optimal, nor which angles are "worst".

References

1. KANG ET. AL. *U.S. Pat. No. 4,829,452*, issued May 9, 1989.
2. A. PAETH *A fast algorithm for general raster rotation*, Proceedings, Graphics Interface '86, Canadian Information Processing Society, Vancouver, pp. 77-81.
3. I. STERLING, AND T. STERLING *Approximating Planer Rotation*, to appear in the Journal of Discrete and Computational Geometry.
4. T. STERLING *Image Rotation Using Block Transfers*, U.S. Pat. No. 5,359,706, Oct. 25, 1994.

An Object-Oriented Interactive System for Scientific Simulations: Design and Applications

A.C. Telea and C.W.A.M. van Overveld

Department of Mathematics and Computing Science, Eindhoven University of Technology, The Netherlands

Abstract. Better insight in complex physical processes requires integration of scientific visualization and numerical simulation in a single interactive framework. This paper presents an object-oriented environment which combines the tasks of numerical simulation, visualization, simulation specification, run-time monitoring and steering.

We first review the different existing approaches to the above tasks and outline their relative limitations. Next, we present a model for a framework which attempts to provide a general approach to the tasks of simulation specification and steering in an object-oriented manner. An implementation of the framework is described.

We have built an object-oriented library for finite element computations and integrated it into the simulation system. An example in which our system has been used to solve a practical engineering problem illustrates the combination of object-oriented numerics and interactivity.

1 Introduction

Scientific visualization has been massively used in numerous fields in order to get an insight into various complex physical processes. Interactivity, seen as the ability of the user to examine and modify the universe she observes, has become a critical requirement of simulation and visualization tools, whether they represent numerical simulations of physical processes, computer animations or virtual reality world models. Interaction can come in the form of changing visualization parameters, in which case we have a *visualization system*, or changing parameters of the modelled process, in which case we have a more general *simulation system*.

Modelling capabilities represent another essential requirement for simulation systems which should provide an easy, natural way to specify the simulated universe in terms of high-level, modular entities which closely parallel the concepts of the real problem to be described.

Although many simulation and visualization systems exist, few of them provide a generic framework combining the abstractions required for modelling complex physically-based or virtual universes together with full interaction freedom with all the simulation parameters.

We have attempted to answer to the above requirements by designing a general purpose system for scientific simulations. The proposed system addresses the tasks of simulation specification and interactivity in a uniform manner, via a high-level object-oriented user interface.

The organization of this article is as follows. Section 2 presents an overview of the existing types of simulation frameworks and outlines their relative limitations. Section 3 presents the conceptual model and the design of the proposed system. For this we firstly introduce the concepts of dependency graphs and relationships in object-oriented programming. Then, we show how we combine object-oriented specifications with constraint management into a homogeneous, interactive environment. Although the presented simulation system is general purpose, we have concentrated on support for simulations using the finite element (FEM) method. Section 4 introduces a FEM simulation library that we have built and integrated in our system. Sections 6 and 7 illustrate the functionality of the FEM library by a couple of simulation examples. The last section discusses the directions of our ongoing research.

2 Previous Work

The simplest simulation frameworks come as libraries dedicated to a limited range of operations, such as geometrical modelling [1], linear algebra [2] or visualization [3]. Object-oriented libraries such as Diffpack [5] or LAPACK++ [6] provide a more abstract application programming interface (API) by which the user can represent simulation concepts as *objects*.

Specification of complex simulations can be however only partially done by such libraries. Besides modelling the simulation's entities by objects, the programmer should represent the *relationships* between these entities. For example, a numerical simulation can have many parameters depending on each other in complex ways. Such dependencies impose *constraints* on the time evolution of the parameters they involve. Since it is complicated and error-prone for the user to 'steer' such a simulation by explicitly changing all its parameters and maintaining the constraints, a *constraint management* mechanism is provided to specify and automatically enforce constraint relationships. A good example of a simulation library offering constraint management is OpenInventor [4].

Adding *interactivity* to simulation systems takes yet another step in modelling reality. While some systems allow only the monitoring of time dependent data, interactive steering systems practically integrate numerical computations and visualization in one tool, which can monitor but also interactively steer a running simulation. Haber and McNabb [7] and Marshall et al. [8] give a good survey of interactive simulation systems.

Dataflow systems like AVS [9], Khoros, Iris Explorer or apE provide some of the above features. The simulation is interactively specified by means of a graphical user interface (GUI) which allows connecting various computational modules in a directed graph, called a flow network. While the simulation runs, data flows from its source through modules which perform various operations on it up to the modules which perform the effective visualization.

Although powerful, most dataflow systems have a series of limitations. Explicit (by value) data transfer between modules is mainly used, which is time and memory consuming in case of large data sets. Secondly, most dataflow systems use purely procedural (state-less) modules, which often don't offer the abstraction level required for modelling complex processes (e.g. simulations described by coupled partial differential equations). Moreover, the dataflow model can express constraints only by pipelining modules in a flow network. We would like a more abstract, possibly object-oriented way of specifying relationships between the entities involved in a simulation. Finally, the integration of many existing object-oriented libraries in a dataflow system raises serious problems. These are partly caused by the system's interface inability to deal with object-oriented entities and relationships, partly by the dynamic and interactive nature of a simulation which may conflict with the library's design philosophy.

The simulation framework we propose uses an object-oriented approach to simulation specification, user interaction and visualization. Constraint specification can be done either in the manner provided by dataflow systems or in a more powerful, objectual way, via an object-oriented GUI of a special design. The next section presents these issues.

3 Conceptual Model and Design of the Simulation System

We shall firstly present a conceptual model of a simulation, which is used by the software system as a simulation basic representation form. Atop of this basic model, the system uses a more sophisticated specification paradigm, which we designed by combining the object-oriented and constraint specification policies. We present the advantages of the combined specification approach and show how it is mapped on the basic representation.

3.1 Conceptual Model

The conceptual model is based on the notion of *state*, defined as the set of parameters that fully characterize the system at a given time instant. These parameters (also called state variables) can model physical quantities of the system (e.g. simulation time, velocity of a body in an animation, pressure and vorticity of a fluid in a flow simulation) but can also be parameters of the visualization system monitoring a simulation or the convergence rate of a linear algebra solver.

A second concept is the *dependency* or law. If a state parameter b depends on a state parameter a, then whenever a changes, b must change as well in order to maintain a constitutive law of the process. For example, if the position x of a body depends on time t, this can be expressed by a law $x = x(t)$. More complex laws can express the dependency of the temperature in each point of a body on the temperature on its boundary by means of a diffusion partial differential equation (PDE).

The set of all state variables and dependencies of a simulation constitutes the system's dependency graph.

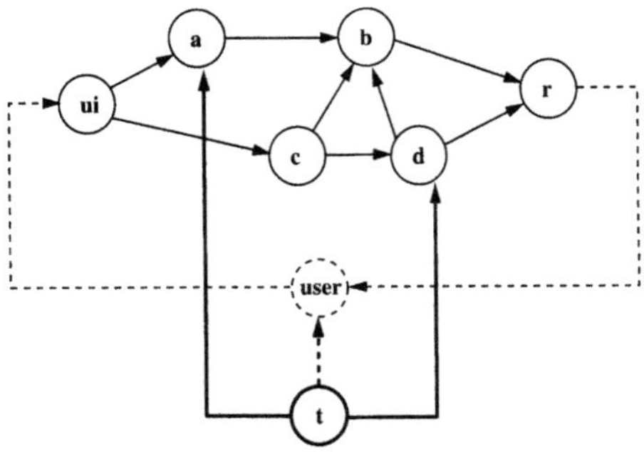

Fig. 1. Dependency graph illustrating state variables and laws. A directed arc (e.g. from node a to b) represents a dependency of the form $b = b(a)$

Figure 1 depicts such a dependency graph. A dependency $b = b(a)$ can be implemented as a a functional or procedural module which evaluates the state parameter b given the parameter a. The system's parameters directly controlled by the user (i.e. the user interface) are modelled by the state parameter ui, while the state parameter r represents all parameters directly observable by the user (e.g. graphics or numerical output). An interesting feature of the model is that the human user can be represented by a parameter determining the ui parameter and depending on the system's r parameter via a complex feedback law. Time t is the only independent parameter since all other parameters are, directly or indirectly, seen as functions of time. The human user's implicit dependency on time is represented by a dotted arrow.

A dependency graph is a complete specification of an arbitrarily complex simulation. Indeed, all entities in the simulation are completely characterized by their state parameters, the system's behaviour in time is described by the laws or dependencies between these parameters and user interaction with the system is described in terms of the ui and r state parameters. When a state parameter changes, the system traverses the dependency graph from the corresponding node and uses the existing laws to evaluate all the parameters encountered during the traversal.

Although the above model is general and very convenient for expressing and enforcing dependencies, the description of a simulation in terms of state variables is a too low level modelling paradigm. A different approach views a simulation in terms of *objects*, which practically group subsets of state variables and allow the user to treat them as a whole by means of specialized methods. Although objects elegantly model simulation concepts, they are unable to *directly* express and enforce dependences between their parameters.

We have combined the two paradigms into one system. Firstly, we use *objects* (implemented in the C++ programming language) to model the entities of our simulation. Then, we express *dependencies* between these objects in order to describe the laws of our simulation. Finally, we provide a mechanism which automatically enforces these dependencies.

Interactivity introduces a third degree of freedom in the system's design. The user should be able to create, destroy, modify and examine objects while the simulation is running. We designed an object-oriented graphical user interface (GUI) which associates an objectual 'widget' to each class in the system. The GUI also offers an object-oriented means to managing dependencies between objects.

We have kept the three design issues involved in the system (object-oriented programming, constraint programming and interactivity) orthogonal. That is, simulation objects can be designed independently on the constraint specification mechanism (see [12] about a discussion on the problems which arise) *and* on the interaction paradigm the system uses. Firstly, this allows us to use class libraries which have been designed independently on our simulation system. Constraint management can be transparently added to such classes without having to reprogram them. The same is true for the object-oriented GUI widgets associated to the classes. In conclusion, we can change or upgrade any component of the system without changing the other two.

3.2 Design of an Object-Oriented Simulation System

The simulation system consists of three main parts, implementing the three main functionalities previously outlined: the object manager, the dependency manager and the interaction manager (see Fig. 2 for an overview). All these parts communicate together by sharing the dependency graph simulation description.

The *object manager* is the interface between the system and the simulation-specific class libraries. It keeps a registry of all the classes known by the simulation system and allows the user to dynamically create, destroy and copy instances of any of these classes at run-time via a GUI. Instances are referenced by names, exactly like objects in the C++ language. Since the object manager does not use any information on simulation classes besides their names, any application-specific class library can be easily integrated into the system.

The *dependency manager* offers a way to interactively express dependencies between simulation objects. Dependencies are represented by objects which are similar to the computational modules encountered in dataflow systems. Besides such dependencies, the system introduces the capability

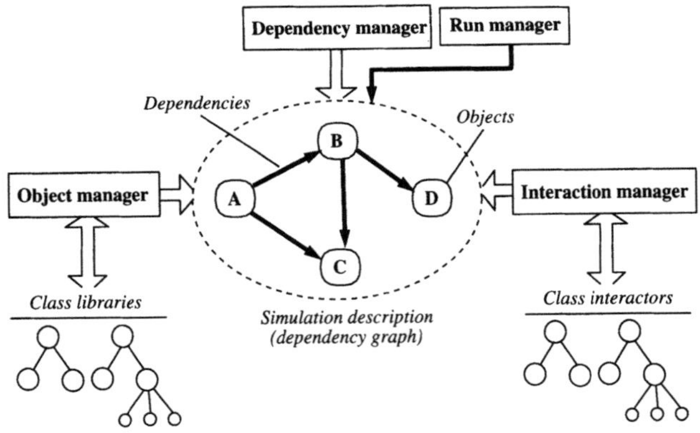

Fig. 2. Overview of the object-oriented simulation system.

to interactively build 'has-a' and 'uses-a' relationships which are specific to object-oriented programming.

A class `A` *has-a* class `B` if `A` has `B` as a member. Similarly, class `A` *uses-a* class `B` if `A` has a pointer to `B` as a member (Fig. 3 b). Together with inheritance (the *is-a* relationship), the *has-a* and *uses-a* relationships are the fundamental tools for expressing data dependencies in object-oriented class libraries. While *is-a* is a 'static' relationship (class hierarchies are constructed at compile-time), the *has-a* and *uses-a* relationships are established at run-time.

The dependency manager allows the user to create or destroy *has-a* and *uses-a* relationships between classes in an interactive way. The objectual *has-a* and *uses-a* relationships are automatically translated into a low level dependency graph similar to the one presented in the previous section. The nodes of this graph are the simulation objects and the arcs are the relationships between objects. A node has a set of typed *ports* which represent the publicly accessible data members of its object (i.e. its state). An arc is a connection between two ports of compatible types, meaning that there is a dependency between the data members corresponding to the ports (Fig. 3 c). Arcs can express both *has-a* (by-value) data dependencies (data is copied from one port to another to enforce the equality constraint) or *uses-a* (by-reference) dependencies (the two ports share the same physical data member). The user's attempts to create invalid relationships are prevented by an object-oriented run-time type checking component. If the user wishes however to have total control over the existing constraints, she can directly interact with the dependency graph and add or remove connections between ports in a way similar to the network management of the Oorange system [11].

A special component of the dependency manager, called the *run manager*, uses the dependency graph to propagate changes when an object is modified, thus ensuring the constraint satisfaction completely transparent to the user.

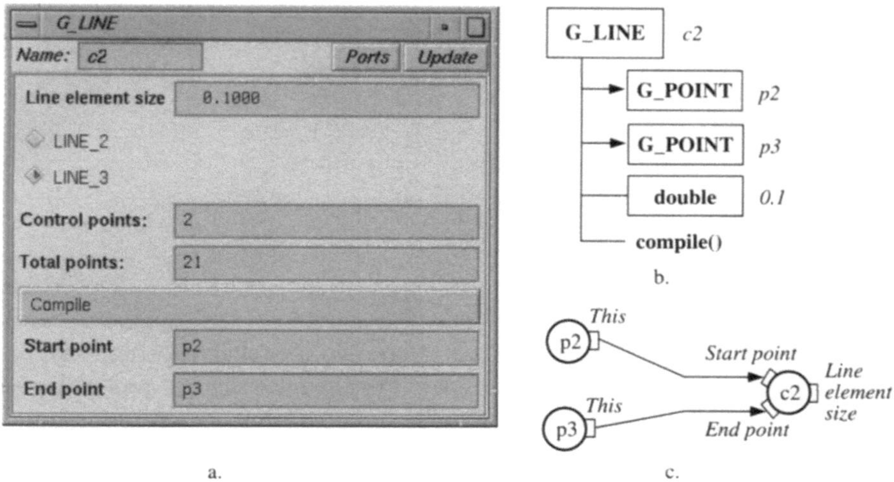

Fig. 3. Objects, constraints and interactors. **a)** Interactor for a class *G_LINE*. **b)** Structure of class *G_LINE* (arrows are *uses* relationships, while lines are *has* relationships). **c)** Dependency graph for a *G_LINE* object (port names are written in italics).

The *interaction manager* offers a GUI to all objects in the system. The user can visually examine all the data members of any existing object, change their values (and see the effects the changes have on other objects if the changed items are involved in dependencies) or call the object's methods. Each class in the system has an *interactor*, which is a GUI widget displaying all public methods and data members of that class (Fig. 3 a). As sketched in Fig. 2, there is a one to one correspondence between simulation class hierarchies and interactor class hierarchies. This allows the programmer of a new simulation class which inherits from an existing class to rapidly derive an interactor from the one of the original class. Creating GUI interactors for existing C++ classes is facilitated by a set of predefined basic interactors for the fundamental C++ types (integers, floats, booleans, enumerations, typed pointers, arrays, etc).

To illustrate the relationship between classes, interactors and constraints, we shall use a very simple example of a line class (Figure 3. A class *G_LINE* represents a line as two references to two *G_POINT* objects being the line's end points, a double being the size of the element obtained when the line is meshed, a method `compile()` and some other less relevant data (Fig. 3 b). Three objects have been created: the line *c2* and the two end points *p1* and *p2*. We say that *c2 uses p1* and *p2 has* a double member (unnamed). Figure 3 a shows the interactor for class *G_LINE*, which allows visual control over all the line's members, e.g. change the start and end points or the line element size, call the `compile()` method or change the line object's name.

Figure 3 c) pictures the dependency graph, showing how *c1* depends on both *p1* and *p2*. The dependency graph is automatically modified as the user employs the GUI interactor to change, for example, the values of the `Start point` or `End point` widgets.

4 A Finite Elements Object-Oriented Library

Most of the existing finite element (FE) applications come in form of packages whose input is given as batch files and output is visualized in a postprocessing (post simulation analysis) phase. This separation of modelling, computations and result visualization limits the user's freedom to change and examine parameters of the FEM simulation to the preprocessing and postprocessing stages.

We have addressed the above limitation by designing an object-oriented library for finite element methods (FEM) and integrating it in the general-purpose simulation system previously presented. The library can be used also standalone, similarly to other object-oriented FE libraries such as Diffpack [5].

The combination of the OO FEM library with the simulation system creates a "virtual simulation and mathematical research" laboratory in which the problem specification, computation and result visualization tasks are fully interactive. End users of a simulation such as engineers can *steer* the ongoing process and monitor its evolution without quitting the simulation environment in order to redefine input files or recompile. Researchers can run FEM problems interactively and experiment with different numerical techniques or monitor error or convergence rates.

5 Structure of a Generic FE Simulation

Although functionally different, most FE simulations exhibit a similar generic structure. This structure and the FE library classes which occur in it are presented in the following.

A generic FE simulation has a dependency graph which consists of three main parts (Fig. 4), which correspond roughly to the modelling, computation and result visualization stages presented in the previous section. The main difference between such a dependency graph and the 'classical' 3-stages pipeline is that objects belonging to different simulation stages can be connected. This less clear separation between stages corresponds to an interleaving of modelling, computations and visualization and results in an increased interactivity.

A top-down presentation of the FE library classes found in each stage follows:

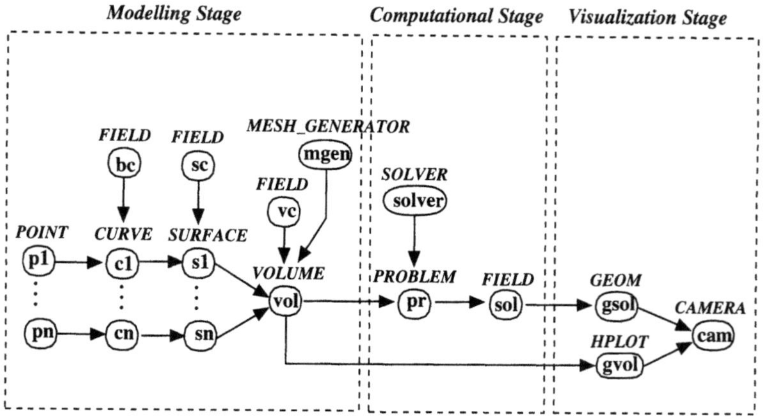

Fig. 4. Dependency graph for a generic FE simulation illustrating the three simulation stages with their respective `objects` and their *classes*.

5.1 Modelling Stage

Modelling comprises the specification of a geometric domain, boundary conditions and a PDE to be solved. These tasks are implemented by specific classes as follows:

Geometrical modelling: Geometries are specified by *POINT, CURVE, SURFACE* and *VOLUME* classes (Fig. 4, objects `p1..pn`, `c1..cn`, `s1..sn` and `vol`). Several *MESH_GENERATOR* classes (`mgen`) are available for discretization of the geometrical domain.

Boundary conditions: Boundary conditions of several types (e.g. Neumann, Dirichlet) are specified on the curves and/or surfaces of the geometrical domain (`bc`, `sc`). Besides its type, a condition uses also a *FIELD* object which represents an analytical or discrete function of position and gives the values prescribed for that condition.

PDE: The PDE is modelled by a *PROBLEM* class which contains specialized methods for that PDE type (e.g. building the stiffness matrix).

5.2 Computation Stage

The key class of this stage is the problem created during modelling (actually *PROBLEM* belongs to both the modelling and computation stages). In this stage, the problem is solved and its solution is written to a *FIELD* object `sol`. Other computation classes include several matrix types (sparse, diagonal, etc), iterative solvers (bi-conjugate gradient, generalized minimum residual, etc) and preconditioners, which overall form an OO linear algebra library similar to SparseLib++ or IML++ [14]. A number of 'low-level' FEM operations (building a stiffness matrix, renumbering schemes, etc) are implemented by private classes thus shielding the user from such technical details.

5.3 Visualization

This final stage comprises a set of classes which permit interactive visualization of the various data objects produced. Two types of classes are involved in this stage. The first type is used to *map* various data entities into objects which can be graphically represented (e.g. geometries *GEOM*, scalar and vector plots, Gouraud-shaded height plot classes *HPLOT*, etc). The second class type represents the *CAMERAs*, i.e. objects that graphically display the output of mappers and are interactively controlled by the user. Visualization classes are functionally similar to AVS's mapper and data output modules or to vtk's [10] rendering classes.

6 Example of Modelling a PDE: The Wave Equation

In this section we shall illustrate the FE library by a simple simulation based on a wave equation $\Delta u + c^2 \frac{\partial^2 u}{\partial t^2} = 0$ solved on a square domain with essential boundary conditions equal to zero. The wave is initiated by an exponential excitation function excitation $=$ height $*$ $\exp^{-\frac{x^2+y^2}{width}}$ centered in some point on the geometrical domain. The simulation will generate the time dependent PDE solutions and present them to the user as 3D elevation plots. During the simulation, the user can change the position, amplitude and width parameters of the excitation and then superimpose the new excitation over the current PDE solution. The simulation will resemble dropping water droplets in a square bucket. The dependency graph for this simulation (Fig. 5) is presented in the following.

Geometry: The square's geometry is defined by four *POINT* objects `p1..p4`, four *LINE* specializations of *CURVE* `l1..l4` and a *SURFACE* `square` (the square itself). Note how the line objects share the point objects.

Boundary conditions: All boundaries have zero essential boundary conditions, specified by a specialization *CONST_FIELD* of the *FIELD* class (object `bc`) which models a constant field with zero value. This object is shared by all the curves.

Excitation: The excitation `excit` is described by a specialization *EXP_FIELD* of the *FIELD* class which models an exponential function. Note that `excit` depends on the floating-point objects `x, y, height, width` which are directly user controlled.

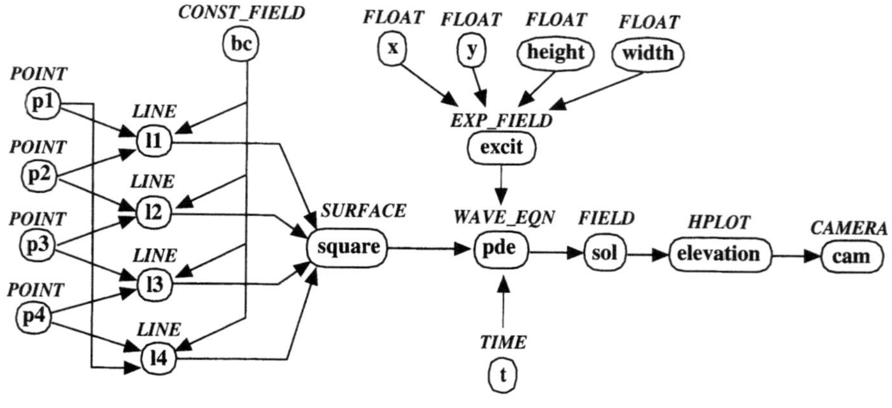

Fig. 5. Dependency graph for the wave equation simulation.

PDE: The PDE is described by an instance `pde` of the class *WAVE_EQN* which contains all specific methods for the wave PDE. This object will get automatically triggered when any of its dependencies is changed (e.g. time, excitation, geometrical domain, etc) and write a new solution to the `sol` *FIELD* object.

Time dependency: Time dependency of the PDE is modelled by a *TIME* class object `t` on which the *WAVE_EQN* object `pde` depends and which plays the role of the time node depicted in Fig. 1. The *TIME* object runs as a coroutine which advances time with a certain increment and triggers computation of a new solution.

Visualization: There is a *CAMERA* `cam` which views a *HPLOT* height plot object `elevation` of the PDE's solution.

Interactivity The user can modify the excitation directly by changing the `x, y, height, width` parameters (i.e. 'drop' a new droplet in the bucket), interact with the camera's controls, change the time step to control the simulation speed, deform the domain by moving one of the points `p1..p4`, start or stop the time `t` to pause or unpause the animation or experiment with the solver's or mesh generator parameters. Briefly, all parameters of all objects are fully accessible for modification during the simulation.

Simulating a diffusion or Navier-Stokes PDE would involve only minor modifications in the above setup (replacement of the *WAVE_EQN* object by a *DIFFUSION* or *NAVIER_STOKES* object and adjustment of boundary conditions and excitation to physically relevant values).

7 Use of the Simulation System in Engineering Problems

We used the FEM class library and the simulation system to implement a more complex numerical simulation of an electrochemical drilling (ECD) process (an electrolytic process in which the anode acts like a drill that advances into a metallic plate acting as a cathode). The anode speed and the voltage applied between the drill and the plate must be varied in time in a well controlled way in order to produce holes of a complex geometry (see color plate 44, p. 389). The problem is to find the correct voltage and speed variations in time that produce holes with the desired geometry. The simulation that we have constructed enables an engineer to vary the process parameters in real time in order to simulate the drilling of holes.

The user can control the variation of all process parameters by means of different GUI widgets like sliders and to monitor the process evolution in real time by selectively zooming in the areas of interest and/or choosing different visualization metaphors (color plate 45 e, p. 390). As parameters are changed, the system automatically performs new FE computations, domain re-meshing and presents the new solution to the user (color plate 44, p. 389). The numerical analyst can experiment by interactively changing the mesh generator or the solver objects used in the FEM simulation with different ones and monitor the convergence rate of the FE solver or change its tolerance if desired. A simple extension would be to construct an object which automatically adapts the solver's tolerance or the mesh refinement to the solution's gradient.

The ECD process is a good test case for the FEM simulation system since it essentially relies on real time user control and evolution monitoring.

8 Conclusion

Better control of complex numerical simulations of physical processes demands a general purpose simulation system which integrates modelling, computing and visualization into a single environment. This paper has presented the conceptual model and structure of a simulation system which combines the data dependency paradigm of dataflow applications with an object-oriented modelling philosophy. The result is an integrated environment in which simulations can be interactively built, steered and monitored. The system offers the power of imperative programming via an object-oriented GUI allowing visual manipulation of objects, methods and data members. Secondly, constraint programming is provided by means of an object-oriented visual specification of data dependencies.

We have designed an object-oriented library for finite elements and integrated it into the presented simulation system. The combination between object-oriented numerics and interactive simulation specification yields an environment where both high level, intuitive simulation steering and fine control over numerical aspects are available. We illustrate this combination by a set of examples including a practical engineering problem.

Current research goals include both the enhancement of the FEM object-oriented library with new solvers, preconditioners and support for other PDE types and investigation of better ways to interactively describe constraints in scientific simulations. A possible development considers constraint specification in terms of implicit or explicit equations or laws and an automatic conversion of these to the dependency graph representation. Another research issue regards the implementation of a full-fledged run-time C++ interpreter which should add the possibility to understand and execute complex procedural descriptions of simulations and possibly even the definition of new classes at run-time.

References

1. J. Barry, *GEOMPACK - A Software Package for the Generation of Meshes using Geometric Algorithms*, Adv. Eng. Software *13*, pp. 325–331.
2. S. Carney, M. A. Heroux, G. Li, and K. Wu, *A Revised Proposal for a Sparse BLAS Toolkit*, Army High Performance Computing Research Center Technical Report 94-034, June 1994.
3. J. Neider, T. Davis, M. Woo, *The OpenGL Programming Guide*, Addison-Wesley, 1993.
4. J. Wernecke, *The Inventor Mentor: Programming Object-Oriented 3D Graphics with Open Inventor*, Addison-Wesley, 1993.
5. A. M. Bruaset, H. P. Langtangen, *A Comprehensive Set of Tools for Solving Partial Differential Equations: Diffpack*, Numerical Methods and Software Tools in Industrial Mathematics, (M. Daehlen and A.-Tveito, eds.), 1996.
6. J. J. Dongarra, R. Pozo, D. Walker, *LAPACK++: A Design Overview of Object-Oriented Extensions for High Performance Linear Algebra*, Proceedings of Supercomputing '93, IEEE Press, 1993, 162–171.
7. R. B. Haber, D. McNabb, *Visualization idioms: a conceptual method for visualization systems*, In *Scientific Visualization: Advances and Challenges*, Academic Press, 1994.
8. R. Marshall, J. Kempf, S. Dyer, and C. C. Yen, *Visualization methods and simulation steering for a 3D turbulence model of Lake Erie*, Computer Graphics *24*, 1990.
9. C. Upson, T. Faulhaber, D. Kamins, D. Laidlaw, D. Schlegel, J. Vroom, R. Gurwitz, and A. van Dam, *The Application Visualization System: A Computational Environment for Scientific Visualization.*, IEEE Computer Graphics and Applications, July 1989, 30–42.
10. W. Schroeder, K. Martin, B. Lorensen, *The Visualization Toolkit: An Object-Oriented Approach to 3D Graphics*, Prentice Hall, 1990

11. C. Gunn, A. Ortmann, U. Pinkall, K. Polthier, U. Schwarz, *Oorange: A Virtual Laboratory for Experimental Mathematics*, Sonderforschungsbereich 288, Technical University Berlin. URL http://www-sfb288.math.tu-berlin.de/oorange/OorangeDoc.html

12. B. N. Freeman-Benson, A. Borning, *Integrating Constraints with an Object-Oriented Language*, Proceedings ECOOP'92 – European Conference on Object-Oriented Programming, (O. Lehrmann Madsen, ed.), Utrecht, 1992.

13. D. H. H. Ingalls, *A Simple Technique for Handling Multiple Polymorphism*, In *Proceedings of OOPSLA '86, Object-Oriented Programming Systems, Languages and Applications*, pp. 347–349, November 1986.

14. R. Pozo, K. A. Remington, A. Lumsdaine, *SparseLib++: A Sparse Matrix Class Library, Reference Guide*, World Wide Web document `http://math.nist.gov/iml++/`, April 1996.

Part IV

Geometric Visualization Techniques

Auditory Morse Analysis of Triangulated Manifolds

Ulrike Axen and Herbert Edelsbrunner

Department of Computer Science,
University of Illinois, Urbana, USA

Abstract. Visualization of high-dimensional or large geometric data sets is inherently difficult, so we experiment with the use of audio to display the shape and connectivity of these data sets. Sonification is used as both an addition to and a substitution for the visual display. We describe a new algorithm called *wave traversal* that provides a necessary intermediate step to sonification of the data; it produces an ordered sequence of subsets, called *waves*, that allows us to map the data to time. In this paper we focus in detail on the mathematics of wave traversal, in particular, how wave traversal can be used as a discrete Morse function.

1 Audio as an Experimental and Analytic Tool

Visualization has become an important tool for mathematicians, allowing them to see complicated spaces and to further their insight into the nature of these spaces. No amount of imagination can equal a visual ride through hyperbolic space, for example, as evidenced by the video *Not Knot* [6] or the CAVE^{TM} application *postEuclidean Walkabout* [5]. Visualization is limited by our ability to see in only three dimensions, however, and by our ability to see for only short distances, often due to occlusion. The latter problem becomes particularly acute when we use a two dimensional display to visualize a three dimensional world.

These observations have led us to investigate *sonification*. Sonification is "the use of data to control a sound generator for the purpose of monitoring and analysis of the data..." [8, p. 187]. Kramer points out in [7] that the ability of sound to display multiple variables simultaneously can be used to increase the display dimensionality of a visual system, or it can be used by itself to display a high-dimensional system. We are investigating sonification of simplicial complexes of dimensions three and higher, with and without a visual display.

One problem associated with using geometric data to control an audio signal is the static nature of the data. Sound is perceived through time; it has an intrinsic time dimension. We describe in the next section an algorithm for making the data dynamic. This algorithm, called *wave traversal*, outputs an ordered sequence of subsets of the data. We then show how to make wave traversal on triangulated manifolds into a discrete Morse function. Thus wave traversal is both an intermediate step to sonification and an analytic tool. In this paper we focus on the mathematics of wave traversal, but we describe now briefly its application to sonification.

We apply maps, called *transfer functions*, from the output of the wave traversal into the audio domain. For example, the relative size of each subset in the sequence might control the carrier frequency (in Hertz) of a tone, and the number of components of each subset might be mapped to a parameter that modulates the frequency. The order of the sequence of subsets gives us a natural map to the time dimension. Critical points are mapped to specific signals so that they can be easily distinguished from sound that is controlled by other data.

Wave traversal does not solve the specific problem of how to map geometric data to parameters of a sound synthesis algorithm, however. So, we still need to supply arbitrary but meaningful maps from data to sound. These maps have been developed with collaborators from the National Center for Supercomputing Applications and refined with experimentation. Some details of the sonification can be found in [1] and [2]. User training is an important part of this process; the audio signals must first be interpreted for a new user before they become meaningful.

2 Wave Traversal

For definitions of terms from piecewise linear topology in this section, see [10], for example. Let K be a finite simplicial complex and assume K is connected. Let the distance between two vertices u and v of K be defined as the number of edges in the shortest path from u to v. We will denote this distance by $d(u,v)$ and refer to u as the start vertex. For a simplex $\sigma \in K$ with dimension greater than 0, define the distance $d(u,\sigma)$ to be equal to the minimum $d(u,v)$ over all vertices v of σ.

Definition 1. Let $W_{K,u}(i), i \geq 0$, be the set of $\sigma \in K$ such that

1. $d(u,v) = i$ for all vertices $v \in \sigma$, and
2. σ is face of some simplex $\tau \in K$ with $d(u,\tau) = i-1$.

For example, $W_{K,u}(0) = \{u\}$. Condition 1 implies that waves are progressively further from u as the index i increases. Condition 2 guarantees that each wave has dimension strictly less than the dimension of K. We will assume that the complex K and the start vertex u are fixed, and simplify our notation to $W(i) = W_{K,u}(i)$. Let D be the maximum $d(u,v)$ for all $v \in K$ and define $W = \{W(i) \mid 0 \leq i \leq D\}$. We call the algorithm that produces W *wave traversal* because of the analogy to a wave moving through a medium, see Fig. 1. We will refer to $W(i)$ as a *wave* and W as the *wave subcomplex*.

Let $S = S_K$ be the set of simplices of K that do not belong to W. Def. 1 implies that all vertices of K are in W, so S contains only simplices of dimension 1 or higher. Let $S(i)$ be the set of simplices in S such that $d(u,\sigma) = i$.

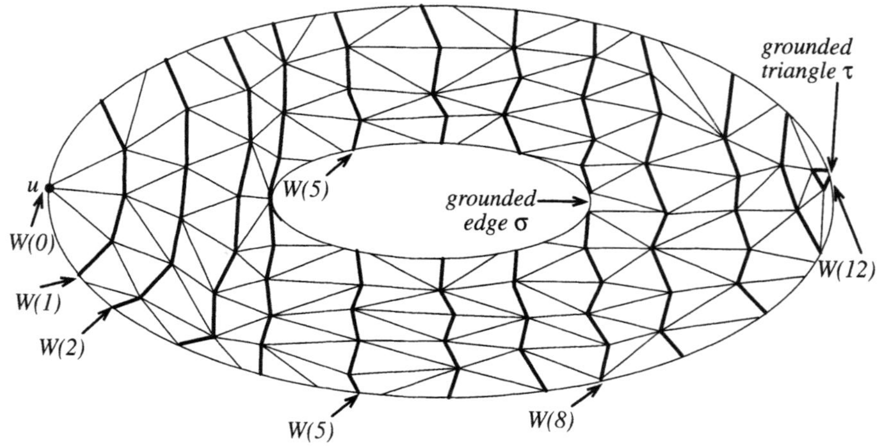

Fig. 1. Waves on a 2-dimensional torus. $W(8)$ includes both vertices of edge σ but not σ itself. Therefore, σ is a grounded edge in $S(8)$. $W(12)$ includes all the edges of triangle τ but not τ itself, so τ is a grounded triangle in $S(12)$.

We would like to extend $d(u, v)$ linearly to a continuous function $d(u, x)$ over all points of K, so that $d(u, x) = i$ for all points of $\sigma \in W(i)$, and $d(u, x)$ varies from i to $i+1$ for all points of $\sigma \in S(i)$. For most $\sigma \in S(i)$ this presents no problem, since some $v \in \sigma$ belong to $W(i)$ and some belong to $W(i+1)$. However, it is possible to have a simplex in $S(i)$ which has all vertices in $W(i)$. We call such a simplex a *grounded simplex.* This can happen, for example, when $W(D)$ is the boundary of a simplex of maximal dimension, or when $W(i)$ consists of two components which divide all the vertices of a simplex in $S(i)$ between them, see Fig. 1. Note that in these cases, grounded simplices occur at critical places where the connectivity of wave $W(i+1)$ is different from wave $W(i)$. We will investigate this relationship further in Sec. 4.

Let $\sigma_i \prec \sigma_j$ denote σ_i is a proper face of σ_j. We consider the *barycentric subdivision* sd$K$ of K, whose vertices are the barycenters b_σ of the simplices $\sigma \in K$. sd$K$ is the collection of all simplices of the form $b_{\sigma_0} b_{\sigma_1} \dots b_{\sigma_k}$, where $\sigma_0 \prec \sigma_1 \prec \dots \prec \sigma_k$. Let $W_{\mathrm{sd}K}$ be the wave subcomplex of sd$K$, and let $S_{\mathrm{sd}K}$ be defined accordingly. The start vertex u is defined to be the same for both W_K and $W_{\mathrm{sd}K}$. Somewhat surprisingly, we can eliminate grounded simplices from the wave traversal using barycentric subdivision, and we state this as a theorem and prove this in the rest of this section.

Theorem 1. All vertices of a simplex $\sigma \in \mathrm{sd}K$ have distance i if and only if $\sigma \in W_{\mathrm{sd}K}(i)$. That is, sd$K$ contains no grounded simplices.

We first prove a lemma which says that the distances in K are doubled in sd$K$.

Lemma 2. *For vertices u and v in K, $d(b_u, b_v) = 2d(u, v)$.*

Proof. We will show that there is a shortest path from b_u to b_v in sd$K$ that is restricted to sd$(K^{(1)})$, where $K^{(1)}$ is the 1-skeleton consisting of all edges and vertices of K. The lemma then follows. Let $\mathrm{fd}(b_\sigma)$ (for *former dimension* of b_σ) be equal to $\dim(\sigma)$. We want to show that there is a shortest path from b_u to b_v in sd$K$ whose vertices b_σ have only $\mathrm{fd}(b_\sigma) = 0$ or 1. Note that $\mathrm{fd}(b_\sigma) \neq \mathrm{fd}(b_\tau)$ if b_σ is adjacent to b_τ. Also, no 3 contiguous vertices in the shortest path, b_ρ, b_σ, b_τ, can have $\mathrm{fd}(b_\rho) < \mathrm{fd}(b_\sigma) < \mathrm{fd}(b_\tau)$, since this would imply that $\rho \prec \sigma \prec \tau$, and so there would be an edge from b_ρ to b_τ. Similarly we cannot have $\mathrm{fd}(b_\rho) > \mathrm{fd}(b_\sigma) > \mathrm{fd}(b_\tau)$. Thus, any shortest path from b_u to b_v must have a sequence of former dimensions of the form

$$0 < \mathrm{fd}(b_1) > \mathrm{fd}(b_2) < \mathrm{fd}(b_3) > \cdots < \mathrm{fd}(b_n) > 0 \ .$$

$\mathrm{fd}(b_2), \mathrm{fd}(b_4), \ldots, \mathrm{fd}(b_{n-1})$ are all locally minimum in this sequence, and so we can replace the vertices $b_2, b_4, \ldots, b_{n-1}$ with original vertices of K to get a path of the same length with

$$0 < \mathrm{fd}(b_1) > 0 < \mathrm{fd}(b_3) > \cdots < \mathrm{fd}(b_n) > 0$$

(see Fig. 2). Now vertices $b_1, b_3, \ldots, b_n$ are all between two original vertices, and so can be replaced by the barycenters of the edges connecting those vertices in K. □

Lemma 2 suggests that every other wave in sd$K$ is also a wave in K, or rather the barycentric subdivision of one. We present a technical lemma before proving that this is indeed the case. Call a simplex $\Sigma \in K$ that has $d(u,v) = 2i$ in sd$K$ for all vertices $v \in \Sigma$ a *dip*, a *flat*, or a *bump* if $d(u, b_\Sigma) = 2i-1$, $2i$, or $2i+1$, respectively.

Lemma 3. *1. K contains no dips.*
2. If $\Sigma \in K$ is a flat then $d(u,w)$ is the same for every vertex $w \in$ sdΣ.

Proof. 1. Suppose K contains a dip Σ. Then $d(u,b_\Sigma) = 2i-1$ and $d(u,x) = 2i-2$ for the predecessor of b_Σ along some shortest path from u to b_Σ. If x is a vertex of sdΣ then it is connected by an edge to at least one vertex $v \in \Sigma$, and if $x \notin$ sdΣ then it is the barycenter of a coface [1] of Σ and therefore connected by an edge to every vertex $v \in \Sigma$. In either case we have a contradiction because $d(u,x) = 2i-2$ and $d(u,v) = 2i$.

2. We have $d(u,b_\Sigma) = 2i$ if Σ is a flat in K. By claim 1, no vertex $w \in$ sdΣ has $d(u,w) = 2i-1$, because then $w = b_T$ for a face T of Σ and T would be a dip. So, $d(u,w) \geq 2i$. Let x again be the predecessor of b_Σ along a shortest path from u. $d(u,x) = 2i-1$ and $x \notin$ sdΣ. Therefore x is the barycenter of a coface of Σ and thus connected by an edge to every vertex $w \in$ sdΣ. So $d(u,w) \leq 2i$. Therefore $d(u,w) = 2i$ as claimed. □

[1] A *coface* of Σ is a simplex which has Σ as a face.

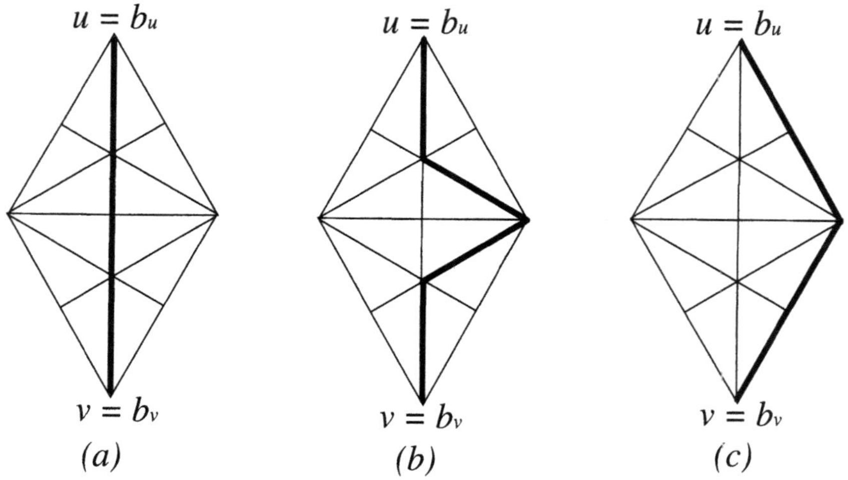

Fig. 2. Shortest paths from b_u to b_v. In (a), the path goes through the barycenters of the two triangles and their common edge. The fd sequence is $0 < 2 > 1 < 2 > 0$. Since the barycenter of the edge has former dimension which is a local minimum, we can replace it with an original vertex to get (b), with fd sequence of $0 < 2 > 0 < 2 > 0$. Finally, we can replace the barycenters of the two triangles, now between original vertices, with barycenters of edges to get (c), a shortest path contained entirely in $\mathrm{sd}(K^{(1)})$.

By Lemma 3, a simplex in K whose vertices all have the same distance from u is either a bump, or a flat whose faces are also flats. Lemma 3 is used in the proof that every other wave in $\mathrm{sd}K$ is the barycentric subdivision of a wave in K.

Lemma 4. $\mathrm{sd}W_K(i) = W_{\mathrm{sd}K}(2i)$.

Proof. First we show $W_{\mathrm{sd}K}(2i) \subseteq \mathrm{sd}W_K(i)$. Let σ be a k-simplex in $W_{\mathrm{sd}K}(2i)$ and let $\Sigma \in K$ be the simplex of lowest dimension that contains σ. By definition, $d(u,v) = 2i$ for all $v \in \sigma$; in particular, $d(u, b_\Sigma) = 2i$. All original vertices $w \in \Sigma$ are adjacent to b_Σ, and so must have distance $2i-1, 2i,$ or $2i+1$. Lemma 2 implies that $d(u,w)$ must be $2i$ in $\mathrm{sd}K$, and so $d(u,w) = i$ in K. In other words, Σ is a flat.

Now, σ is face of a simplex $\tau \in \mathrm{sd}K$ with $d(u,\tau) = 2i-1$. We may assume that τ has dimension $k+1$. Let T be the simplex of lowest dimension in K that contains τ. Since Σ is a flat it cannot contain τ and therefore $\Sigma \prec T$. Also note that the vertex of τ at distance $2i-1$ is the barycenter b_T. The predecessor x of b_T along a shortest path from u to b_T has distance $2i-2$. Assume that no vertex of T has distance $2i-2$. Then x is not one of the vertices of T and so it must be the barycenter of a coface of T or of a face of T. If x is the barycenter of a coface of T, then x is adjacent to a vertex v of Σ, implying that $d(u,v) < 2i$, a contradiction. If x is the barycenter of a face of T, then at least one vertex w of this face has distance $2i-2$, or else that face is a dip, which would contradict Lemma 3(1). But $d(u,w)=2i-2$ is also a contradiction. Thus x must be a vertex of T and $d(u,T)=i-1$ in K.

Now we will show that sd $W_K(i) \subseteq W_{\mathrm{sd}K}(2i)$. Let Σ be a k-simplex in $W_K(i)$. Σ is face of a $(k+1)$-simplex T with $d(u,T)=i-1$. The vertices of Σ have distance $2i$ in sd$K$ and the other vertex of T has distance $2i-2$. Since b_T is adjacent to all of these, $d(u,b_T)=2i-1$. All $(k+1)$-simplices $\tau \in$ sd T with a k-face in sd Σ thus have distance $2i-1$. By Lemma 3(1), the barycenter of Σ cannot have distance $2i-1$ and it has distance at most $2i$ because it is adjacent to b_T. So, by Lemma 3(2), all vertices of sdΣ have distance $2i$ from u. Hence all simplices σ in sd$K$ contained in Σ satisfy both conditions for belonging to wave $2i$ of sd$K$. □

It follows from Lemma 4 that the barycenters of all simplices in S_K belong to odd-numbered waves of sd$K$. We can show a stronger result.

Lemma 5. *The barycenters of all simplices in $S_K(i)$ belong to $W_{\mathrm{sd}K}(2i+1)$.*

Proof. If $\Sigma \in S_K(i)$ has vertices at distance i and $i+1$ then $d(u,b_\Sigma)=2i+1$ in sd$K$ and b_Σ belongs to $W_{\mathrm{sd}K}(2i+1)$ as claimed. The only other case is that Σ is grounded, that is, all its vertices have distance $2i$ in sd$K$. By Lemma 3(1) we have $d(u,b_\Sigma) \neq 2i-1$, and by Lemma 4 we have $d(u,b_\Sigma) \neq 2i$. The only remaining possibility is $d(u,b_\Sigma)=2i+1$. □

Now we have the tools to prove Theorem 1, which says that if we first take the barycentric subdivision of a complex, we can eliminate the second condition of Def. 1; in sd$K$ equidistance of vertices is the sole determining condition for inclusion of a simplex into a wave.

Proof (of Theorem 1). We show that the assumption that $\sigma \in$ sd$K$ is grounded leads to a contradiction. Let Σ be the simplex of lowest dimension in K that contains σ; b_Σ is a vertex of σ. Σ cannot belong to a wave of K, otherwise Lemma 4 implies that σ belongs to a wave of sd$K$ and can therefore not be grounded.

We may therefore assume that $\Sigma \in S_K(j)$. By Lemma 5, $d(u,b_\Sigma)=2j+1$. For σ to be grounded it must have all vertices at distance $2j+1$. Besides σ, sdΣ also contains simplices τ that contain σ and one original vertex v of Σ. If $d(u,v)=2j$ for any such τ then σ is in a wave and cannot be grounded, a contradiction. If $d(u,v)=2j+2$ for all choices of τ then we have a face of Σ that is a dip; its vertices are the vertices v of the simplices τ, and its barycenter is a vertex of σ. This contradicts Lemma 3(1).

Since $S_{\mathrm{sd}K}$ contains no grounded simplices, we may conclude that if all vertices of $\sigma \in$ sd$K$ have distance i, then $\sigma \in W_{\mathrm{sd}K}(i)$. In other words, we do not need the second condition of Def. 1. □

Figures 3 and 4 illustrate the cases where $\Sigma \in K$ is a triangle not contained in $W(i)$ for any i. In Fig. 3 Σ is a grounded simplex, and in Fig. 4 it is not.

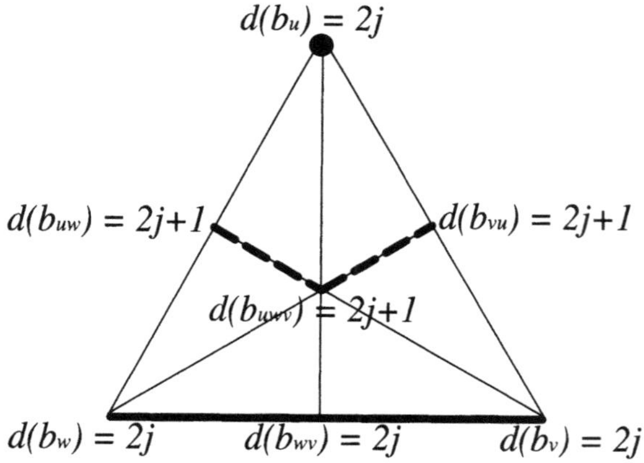

Fig. 3. Before subdivision, vertex u and the segment wv were in $W(j)$, while triangle $\Sigma = uwv$ was in $S(j)$, a grounded simplex. After barycentric subdivision, all simplices with equidistant vertices belong to W.

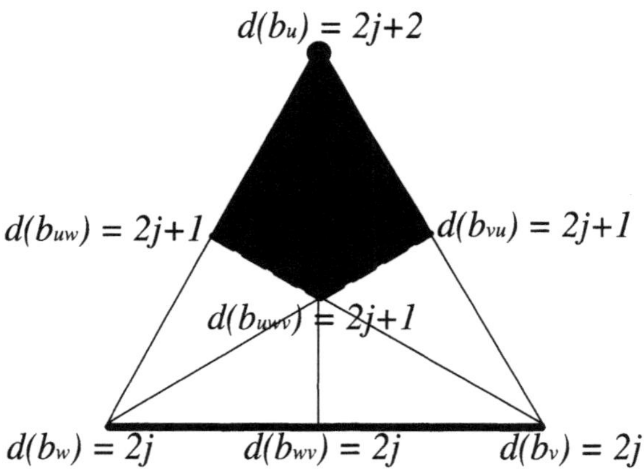

Fig. 4. Before subdivision, vertex u was in $W(j+1)$ and the segment wv was in $W(j)$, while triangle $\Sigma = uwv$ was in $S(j)$, *not* a grounded simplex. After barycentric subdivision, the white triangles belong to $S(2j)$ and the black ones belong to $S(2j+1)$.

3 Wave Traversal as a Morse Function

Morse theory relates the critical points of a smooth function on a smooth manifold to the connectivity of the manifold; details of Morse theory can be found in [9]. A smooth function that is often used for Morse analysis is a height function $h : M \to \mathbb{R}$ that maps a point of M to its distance from a hyperplane, see Fig. 5. In this section, we will construct a piecewise linear function f analogous to the smooth height function h, and show how to isolate the critical points of f so that we obtain a discrete Morse function.

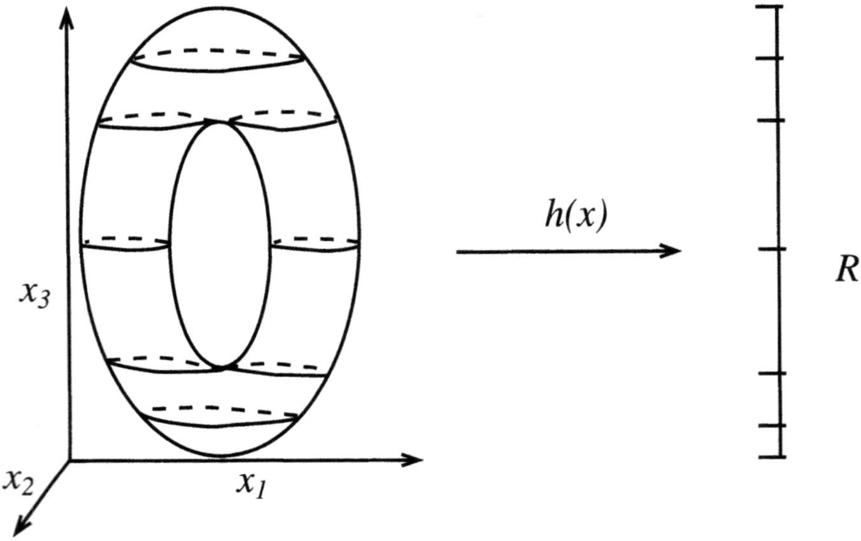

Fig. 5. $h : M \to \mathbb{R}$ takes points of the 2-torus to real numbers, $(x_1, x_2, x_3) \to x_3$.

In classical Morse theory, critical points of h are found by examining the gradient of h, and the index of each critical point is found by examining the matrix of second derivatives of h evaluated at the critical point p. If this matrix has full rank, the critical point p is non-degenerate. For our purposes, though, we are interested in a geometric interpretation of critical points and their indices, and we will present these as Banchoff does in [3].

Assume M is a smooth 2-manifold in $\mathbb{R}^3$. If $h : M \to \mathbb{R}$ is a Morse function, then critical points of h are isolated and have only three types: minima, saddle points, and maxima, of index 0, 1, and 2, respectively. If p is a critical point for h then the tangent plane to M at p is horizontal. Consider a small circle about p on M. If p is a minimum or maximum, the plane through p does not intersect the circle. If p is a saddle point, it intersects the circle in 4 points. Another way to describe this is that the tangent plane at a saddle point p divides a small disk neighborhood of p on M into "four separate pieces." The horizontal plane through a *regular* or non-critical point q is not the tangent plane, and therefore meets a small circle about q in two points, and divides a small disk neighborhood of q on M into two pieces.

We will use these observations to find critical points of our discrete function. In the following we assume that K is a triangulated oriented closed surface; i.e., K is homeomorphic to an oriented compact 2-manifold. We also assume that K is embedded in $\mathbb{R}^3$, although what follows is applicable to abstract manifolds as well. Finally, we assume now that K is the first barycentric subdivision of another simplicial complex, so that we can apply the results of the previous section, in particular Theorem 1.

We extend the distance function for the vertices linearly to all points of K. Since K contains no grounded simplices, we do not lose the correspondence between distance and waves. Let $x \in \sigma$ and let $t_v = t_v(x)$ be its *barycentric coordinates*, where $t_v = 0$ if $v \notin \sigma$,

$$1 = \sum_v t_v \ , \text{ and } x = \sum_v t_v \cdot v \ .$$

Definition 6. The distance from u to x is

$$d(u, x) = \sum_v t_v \cdot d(u, v) \ .$$

Observe that $d(u, x)$ is defined over the entire set of points of K and is continuous. If we embed K in $\mathbb{R}^4$ so that

$$x = (x_1, x_2, x_3) \rightarrow (x_1, x_2, x_3, d(u, x)) \ ,$$

then $f(x) = d(u, x)$ becomes a height function.

Lemma 7. *All critical points of f are in the wave subcomplex W.*

Proof. We will show that all points $x \in |S|$, where $|S| = \bigcup_{\sigma \in S} \operatorname{int} \sigma$, are regular. Consider a point $x \in \operatorname{int} \sigma$, with $\sigma \in S$. Since grounded simplices have been eliminated, σ has two vertices at different heights and is therefore not contained in the horizontal hyperplane passing through x. It follows that this hyperplane cuts a suitable neighborhood of x into precisely two pieces, compare [3, p. 478]. □

Observe that each wave $W(i)$ is in a horizontal hyperplane in $\mathbb{R}^4$. A vertex $v \in W(i)$ is a regular point of f if $W(i)$ cuts the star of v, $\text{St}(v)$, into exactly two pieces, one a subset of $S(i-1)$ and the other a subset of $S(i)$ (Fig. 6). Similarly, all points x of an edge $e \in W(i)$ are regular if $W(i)$ divides $\text{St}(e)$ into two pieces, where $\text{St}(e)$ contains e and the two triangles that share e.

f has only one minimum and it is non-degenerate; it is the start vertex u. All simplices in $\text{St}(u)$, except u, belong to $S(0)$. Similarly, for an isolated local maximum $v \in W(i)$, $\text{St}(v)$ consists only of v and simplices in $S(i-1)$.

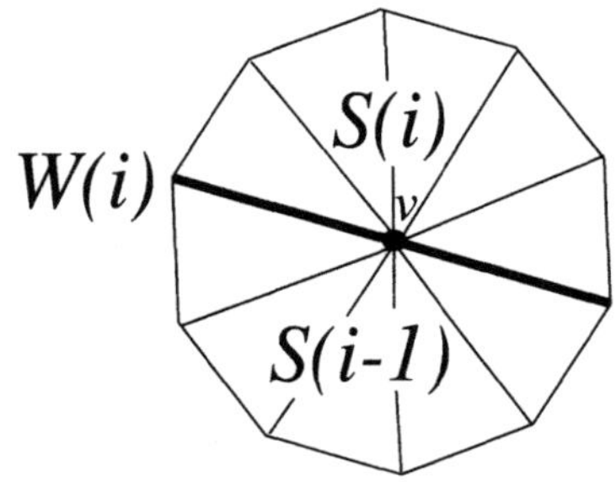

Fig. 6. The star of a regular point $v \in W(i)$. $W(i)$ divides the star into two pieces, one a subset of $S(i-1)$ and the other a subset of $S(i)$.

If $v \in W(i)$ is an isolated saddle point, then $W(i)$ divides $\mathrm{St}(v)$ into 4 pieces, alternating between $S(i-1)$ and $S(i)$ (Fig. 7(*a*)). It is also possible that several saddle points come together at a single location. The star of an m-fold saddle point is cut by $W(i)$ into $2m+2$ pieces. A 2-fold saddle point is also know as a "monkey" saddle.

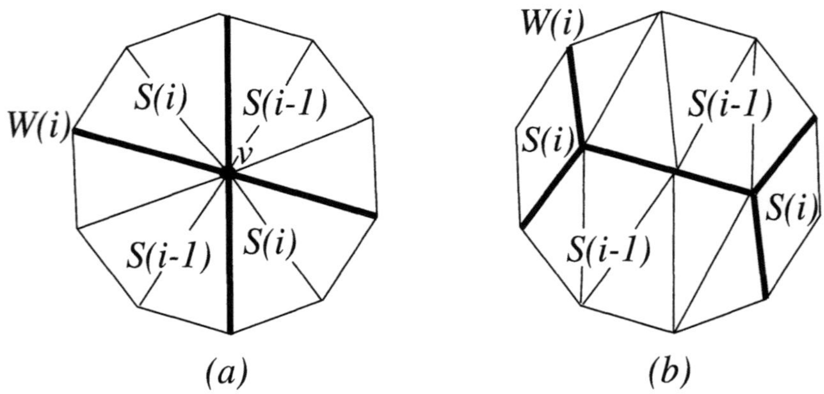

Fig. 7. (*a*) illustrates the star of an isolated saddle point v. (*b*) illustrates a component of a critical subcomplex.

Critical points of f are not isolated in general, so we need a way to identify and resolve these degeneracies. A degeneracy occurs when one or more adjacent edges $e \in W(i)$ all have $\mathrm{St}(e) \subseteq W(i) \cup S(i-1)$, see for example Fig. 7(*b*). More formally, we define a *(degenerate) critical subcomplex* C of $W(i)$ as the closure of all edges e such that $\mathrm{St}(e) \subseteq W(i) \cup S(i-1)$. Note that it is possible for some vertices in C to have stars which also contain simplices in $S(i)$.

Since C is itself a simplicial complex, wave traversal is defined for its components and we use it to isolate critical points. We choose an arbitrary vertex w in a component C_p of C and compute $d(w, x)$ on C_p. [2] Then we replace f on C_p with

$$f(x) = d(u, w) - \varepsilon \cdot d(w, x) \ , \tag{1}$$

for small $\varepsilon > 0$. After redefining f on critical subcomplexes, we need to recalculate f on the rest of K.

Theorem 2. Let x and y be two vertices of C_p connected by an edge e. Then $f(x) \neq f(y)$, i.e., (1) isolates critical points of f in C.

Proof. Suppose $d(w, x) = d(w, y) = j$. Then edge e is in wave j of the wave traversal on C_p, since otherwise it would be grounded. But C_p is a 1-dimensional complex, so wave j must be 0-dimensional and cannot therefore contain e. Therefore $d(w, x) \neq d(w, y)$ and as a consequence $f(x) \neq f(y)$. □

We now illustrate with some examples. Suppose C_p comprises an entire connected component of wave $W(i)$. Then $\mathrm{St}(v) \subseteq W(i) \cup S(i-1)$ for all $v \in C_p$ because K is a manifold. The new start vertex w becomes a local maximum for f. If C_p contains no closed curve, then it contains no further critical point. If C_p contains a closed curve, then (1) isolates both a saddle point and a maximum. This happens, for example, when $W(D)$ is a (homologically) non-trivial closed curve on the torus, see Fig. 8. We prove this in the following claim.

Claim 8. *Let C_p comprise an entire connected component of $W(i)$ and suppose C_p contains a closed curve P. (1) isolates both a maximum and a saddle point on C_p.*

Proof. We recalculate f on C_p with (1) and as in the general case, w is an isolated maximum. Since there are no grounded simplices in the wave traversal on C_p, there is a vertex x on P which has a minimum value for f on P. This vertex x becomes an isolated saddle point, because f increases from x in two directions on C_p, and f decreases in two directions in $S(i-1)$. □

In fact, it is clear from the proof that (1) will isolate one saddle point for *each* closed curve in C_p. To give an intuitive understanding of why this must be, we prove the following claim.

Claim 9. *Let C_p be as in Claim 8. The closed curve P in C_p must be non-trivial.*

[2] It is a fact that there are no grounded simplices in this wave traversal, but we will not prove it here.

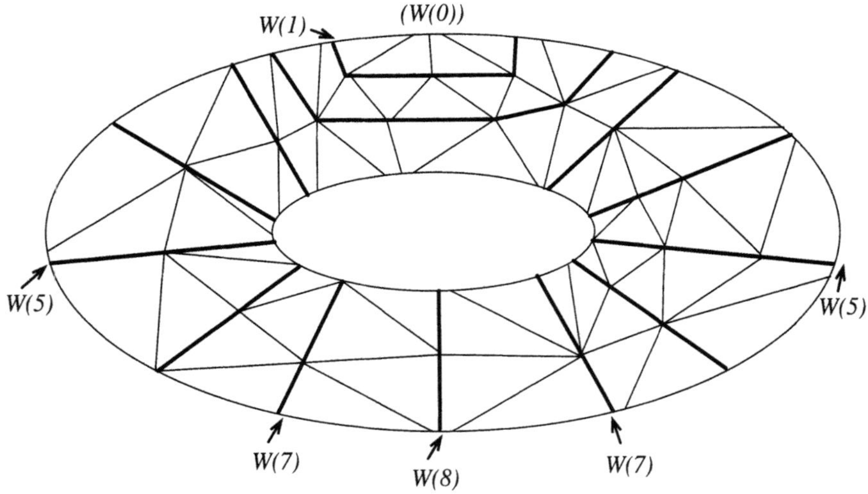

Fig. 8. $W(8)$ has a maximum and a saddle point.

Proof. Assume that P is a trivial closed curve on the surface K. P divides K into two pieces, only one of which contains the start vertex u. Any path from u to a vertex x in the other piece, and in particular the shortest path, must go through P, and so $d(u,x) = i+1$. Therefore $\mathrm{St}(v)$ for some $v \in C_p$ must have simplices in $S(i)$, a contradiction. If the piece that does not contain u does not contain any vertices at all, then it must contain a grounded simplex, which is also impossible. □

Now suppose C_p is a subcomplex of a larger connected component of $W(i)$, see Fig. 7(b). In this figure, C_p consists of two edges and their three endpoints. The two outer endpoints include triangles from $S(i)$ in their stars. Call these vertices x and y and suppose w is different from both. After recalculating f on C_p, w again becomes an isolated maximum and both x and y become isolated saddle points. If $w = x$, then y becomes the only isolated critical point on C_p; it is a saddle point. Since a maximum and a saddle point effectively cancel each other out, the net result in both cases is a saddle point.

When only one vertex $v \in C_p$ has simplices in $S(i)$, then a similar case analysis for w shows that the critical points isolated by (1) all cancel; this is analogous to the "shoe" saddle on a continuous manifold.

4 Computation of Waves, Critical Points and Sound

Computing the waves on sd$K$ is straightforward. We find $f(v) = d(u,v)$ for each vertex using *breadth-first search* [4]. We alter this algorithm slightly because we do not compute sd$K$ explicitly first; we perform the breadth-first search on an implicit subdivision of K. We add a simplex to $W(i) = W_{\mathrm{sd}K}(i)$ if all its vertices have $f(v) = i$.

We locate critical subcomplexes in odd-numbered waves by locating grounded simplices of the original complex K. It is a fact that the barycenter b_Σ of a grounded simplex Σ is either an isolated critical point of f, or it is in a degenerate critical subcomplex. For example, in Fig. 3 the barycenter of the triangle is in a degenerate critical subcomplex. If Σ is not grounded, then b_Σ is a regular point of f, see Fig. 4. In even-numbered waves, we analyze the star of each simplex to locate critical subcomplexes.

We find the connected components of critical subcomplexes and identify them using the cases illustrated in Sec. 3. For example, if a critical subcomplex C_p contains no closed curves and $\mathrm{St}(v) \subseteq W(i) \cup S(i-1)$ for all $v \in C_p$, then C_p contains a local maximum for f and no other critical points.

To compute the sound, we map properties of waves and critical points dynamically to parameters of a sound synthesis algorithm, so that we listen to the process (the wave traversal) as well as the analysis (the result of the Morse analysis). The sonification of the wave traversal itself conveys to a trained user some of the features of each wave, for example, its relative size and number of components. Critical points are mapped to specific sound signals, so that they can be distinguished from the underlying bed of sound produced by the wave traversal. These signals indicate the type and number of critical points found at a particular wave step. The user then gains knowledge of the shape and connectivity of the data by listening to the entire composition. The ideas in this paper are adapted to sonification of general 2- and 3-complexes.

The algorithms are implemented in C++ on the SGI platform, with a visual display for complexes of dimension 3 or less. A program called vss (for vanilla sound server) provides the sound synthesis algorithms and an interface for controlling these algorithms with our data (see http://www.ncsa.uiuc.edu/-VEG/audio).

References

1. U. Axen and I. Choi, *Using additive sound synthesis to analyze simplicial complexes*, Proceedings of the Second International Conference on Auditory Display, ICAD '94 (G. Kramer and S. Smith, eds.), Santa Fe Institute, 1995, pp. 31–43.
2. ———, *Investigating geometric data with sound*, web site http://www.santafe.edu/ icad/ICAD96/proc96/axen.htm, 1997.
3. T. F. Banchoff, *Critical points and curvature for embedded polyhedral surfaces*, American Mathematical Monthly **77** (1970), 475–485.
4. T. H. Cormen, C. E. Leiserson, and R. L. Rivest, *Introduction to algorithms*, MIT Press, Cambridge, Massachusetts, 1990.
5. G. Francis, C. Hartmann, J. Mason, U. Axen, A. Arias, P. McCreary, and G. Chappell, *posteuclidean walkabout*, VROOM at SIGGRAPH 94, 1994.

6. C. Gunn, D. Maxwell, and S. Levy, *Not knot*, Video produced by the Geometry Center, University of Minnesota, A. K. Peters, Wellesley, MA.

7. G. Kramer, *An introduction to auditory display*, Auditory Display: Sonification, Audification, and Auditory Interfaces (G. Kramer, ed.), Addison-Wesley Publishing Company, 1994, 1–77.

8. ———, *Some organizing principles for representing data with sound*, Auditory Display: Sonification, Audification, and Auditory Interfaces (G. Kramer, ed.), Addison-Wesley Publishing Company, 1994, 185–221.

9. J. Milnor, *Morse theory*, Princeton University Press, Princeton, New Jersey, 1963.

10. J. R. Munkres, *Algebraic topology*, Addison-Wesley, Reading, Massachusetts, 1984.

Computing Sphere Eversions

George Francis, John M. Sullivan, and Chris Hartman

Mathematics Department, University of Illinois, Urbana, USA

Abstract. We consider several tools for computing and visualizing sphere eversions. First, we discuss a family of rotationally symmetric eversions driven computationally by minimizing the Willmore bending energy. Next, we describe programs to compute and display the double locus of an immersed surface and to track this along a homotopy. Finally, we consider ways to implement computationally the various eversions originally drawn by hand; this requires interpolation of splined curves in time and space.

1 Introduction

In an earlier paper [14], we described a minimax sphere eversion, driven automatically by minimization of the Willmore bending energy for surfaces. Here, we consider three extensions of that work, which are of interest in the areas of optimal geometry, splined surfaces and regular homotopy theory. These fields are some of the many areas of geometry and computation which converge so nicely on the problem of finding new ways to evert the sphere. We consider this paper a continuation of [14], to which we refer the reader for background; further information on sphere eversions can be found in [20,10,24].

Symmetry has always been a great aid to understanding complicated geometrical structures, including sphere eversions. Twenty-five years ago, Bernard Morin proposed the family of tobacco-pouch eversions, with increasing rotational symmetry. In 1983, Rob Kusner observed that certain Willmore surfaces could serve as good halfway models for these eversions. Two years ago, we succeeded in implementing the first minimax eversion, driven by minimizing the Willmore energy in Ken Brakke's *Evolver*. This eversion, with two-fold symmetry, was topologically the simplest of the tobacco-pouch eversions. However, that computation did not make use of the symmetry. If instead we enforce the symmetry, and work with only one fundamental domain, we gain efficiency and the ability to compute the higher-order eversions, as we describe in Section 2.

An immersed surface can be profitably investigated by examining its self-intersections or *double locus*. In a regular homotopy, like a sphere eversion, we have a continuous family of double loci X_t at different times. In their seminal paper [1], Morin and Apéry demonstrated the significance of the *double locus surface* $X \subset \mathbb{R}^4$ obtained by stacking the X_t on top of each other. Events in the regular homotopy correspond to critical points for the height function t on X. For the eversion we presented in [14] (equivalent to one in [1]), counting these critical points shows that X has Euler characteristic -1 and thus is a

Dyck's surface (a sphere with three cross-caps) immersed in $\mathbb{R}^4$, which we want to visualize. This involves computing the double locus at each stage of the eversion, splicing these level curves together into a surface, and then viewing the resulting immersion in $\mathbb{R}^4$ by projection or slicing. We describe the software we used for these three steps in Section 3.

Once we have the software for the middle step (knitting a polyhedral surface together from a discrete sequence of plane curves representing its levels), we can extend it to also allow us to arrange a temporal succession of such layered surfaces into a regular homotopy. This is precisely the visualization exercise expected of the reader of certain classical sphere eversions like the one by Tony Phillips [24]. We want to design our program to work, in such a way that sculpting and rearranging the homotopy in time and space requires the manipulation of only a few control points, with real-time graphics feedback. Our solution uses Catmull-Rom splines in both space and time, as is described in Section 4.

Our first test case is a sphere eversion suggested by Bryce DeWitt [8]. He presented an array of level curves at different times; the conjecture that these could be interpolated by a regular homotopy of nonsingular surfaces has stood unchallenged for twenty years. We hope to resolve this question empirically with the computational tools described in this paper.

The authors wish to express their gratitude to all of the many students and colleagues who have contributed to the projects described herein, but special thanks go to Alex Bourd, Ken Brakke, Ulises Cervantes-Pimentel, Glenn Chappell, Rob Kusner, Stuart Levy, Dennis Roseman, Nick Schmitt, and Matthew Stiak.

2 Symmetric Eversions Driven by Willmore Energy

Earlier we described a minimax sphere eversion [14] computed numerically by minimizing a surface bending energy. In fact, that eversion seems to be only the first in an infinite family of similar eversions with higher-order symmetry. We now have methods to compute these further eversions, and have performed the computations for several examples.

Tobacco-pouch eversions. One way to describe a sphere eversion is to give an immersed sphere, called a *halfway model*, which can be turned inside-out by a rigid motion. In other words, the halfway model has a symmetry which interchanges its inside and outside. More precisely, for an appropriate choice of parameterization, antipodal points on the abstract domain sphere are mapped by the immersion to points related by this symmetry. Given such a halfway model, any regular homotopy which simplifies it down to the round sphere can be extended by symmetry to a sphere eversion.

The halfway models that have been used in this way can be divided into two classes. The first class includes double-covered immersions of a projective plane like Boy's surface. Here the orientation-reversing symmetry is simply the identity map in space: antipodal points of the sphere cover the same point in $\mathbb{RP}^2$ and thus are at the same place in the halfway model. The early sphere eversions of Shapiro [12], Phillips [24] and Kuiper [17] used Boy's surface as a halfway model. In the second class, the halfway model has $2p$-fold rotational symmetry reversing orientation (and thus p-fold symmetry preserving orientation). The original Morin-Froissart halfway model was of this type, with $p = 2$.

In the early seventies, Morin proposed a family of sphere eversions for integers $p > 1$, later called the tobacco-pouch eversions.[1] In 1977, after Nelson Max kindly loaned one of us (Francis) his newly issued film [22], the students in a freshman honors topology seminar helped design an accurate, but only combinatorial, description of the tobacco-pouch eversions [9]. For p even, the halfway model used in these eversions is of the second class, with $2p$-fold rotational symmetry (reversing orientation). For p odd, it is of the first class, a projective plane with p-fold rotational symmetry. In both cases, the entire eversion can proceed maintaining p-fold rotational symmetry.

Morin soon found analytic expressions for the essential steps of these regular homotopies [23] (see [10, p. 116f]), which were further developed by Apéry [1]. Although these formulas are analytically elegant, they do not lead to nice pictures, so we are led to look for nicer (or even optimal) geometric forms for these same eversions, which can be computed automatically.

Willmore-critical spheres. An elastic bending energy for surfaces should be quadratic in the principal curvature, and if symmetric can be reduced by the Gauss-Bonnet theorem to the integral of mean curvature squared, $W = \int H^2\, dA$, known as the Willmore energy [26]. (See [15] for more about the history of this energy, and some early computer experiments minimizing it.)

In the 1980's, Bryant [6] showed that all critical points for this energy (which is Möbius-invariant) among immersed spheres arise as Möbius transformations of minimal surfaces in R^3 with k flat ends, and thus they can be described explicitly by the Weierstrass representation with data on the Riemann sphere. The Willmore energy of such a critical point is $W = 4\pi k$; aside from the round sphere (a global minimum at $W = 4\pi$) the lowest energy examples occur with $k = 4$.

Shortly thereafter, Kusner [18,19] found particular examples of such critical spheres with rotational symmetry, which he proposed (inspired by a draft of [9]) as particularly nice geometric realizations of the halfway models for the tobacco-pouch eversions. He described a minimal surface S_p as the image

[1] Because of their resemblance to the way the French *blague de tabac automatique* closes up.

of the (punctured) Riemann sphere under the map

$$S_p(w) = \mathrm{Re}\left(\frac{\left(i(w^{2p-1} - w),\ w^{2p-1} + w,\ i\frac{p-1}{p}(w^{2p} + 1)\right)}{w^{2p} + \frac{2\sqrt{2p-1}}{p-1}w^p - 1}\right).$$

From this formula, it follows that the orientation-reversing $2p$-fold symmetry $w \mapsto e^{\pi i/p}/\bar{w}$ of the Riemann sphere becomes a rotational symmetry of the surface S_p by angle $\pi - \pi/p$ around the z-axis. Looking at the pth power of this symmetry highlights the distinction between p odd (when $S_p(-1/\bar{w}) = S_p(w)$ so we have a double-covered (punctured) projective plane), and p even (when $S_p(-w)$ is $S_p(w)$ rotated a half-turn around the z-axis).

To get a halfway model with the same rotational symmetry, we apply a Möbius transformation to S_p by inverting in a sphere centered at some point $(0, 0, s)$ along the z-axis. Because S_p passes through the origin (but no other point of the z-axis) we must choose $s \neq 0$ to get a compact image M_0^p. We see no reasons other than esthetic ones to pick any particular value for s; for low p, we have found $s \approx \frac{1}{3}$ gives an appealing halfway model. (Note that the sculpture at Oberwolfach described in [16] is the Boy's surface obtained from these formulas with $p = 3$ and $s = \frac{1}{2}$.)

Minimax symmetric eversions. Each of the immersed spheres M_0^p described above is a critical point for the bending energy W, with an orientation-reversing symmetry of order $2p$. In general, we expect the (Morse) index of a critical point to decrease as more symmetry is imposed. Here the index is not known theoretically, but the numerical experiments we have performed support a reasonable conjecture: if this $2p$-fold symmetry is enforced, M_0^p is stable (a local minimum for the energy), while if only the p-fold orientation-preserving subgroup is enforced, M_0^p is unstable with index one. (With no symmetry imposed, the index would be higher, depending on p.) That is, in the space of immersed spheres with p-fold rotational symmetry, the round sphere minimizes W, while M_0^p is the lowest saddle point (with $W = 8\pi p$).

Therefore, we propose to generate a minimax eversion with this p-fold symmetry, by flowing along a heteroclinic orbit for the gradient flow of W, starting at M_0^p and ending at the round sphere. (Of course this gives only the second half of the eversion; the first half is the same homotopy, reversed in time and orientation, and rotated by $\pi - \pi/p$.)

The gradient flow for W is a fourth-order parabolic flow, which is not well understood. Probably in some cases the flow could start with a smooth surface and pinch off a handle or otherwise produce a singularity. We have only numerical evidence for the fact that this does not happen in our case: computer experiments with Brakke's *Evolver* give an approximation to the flow we want, with no singularities.

It is interesting to note that sphere eversions have often produced with simplification of the halfway model as a guiding principle. For his original eversion, Morin progressively simplified the double locus. Earlier tobacco-pouch eversions were guided by simplification of the apparent contours. It is remarkable that we get essentially the same eversion by reducing the bending energy W to simplify the surface.

Symmetric halfway models in the *Evolver*. Brakke's *Evolver*[2] is a general tool for geometric optimization, like minimization of surface area or bending energies [4]. There are several discretizations of the Willmore energy available in the *Evolver* [15]. In fact, we made use of these before to implement the minimax eversion with $p = 2$, which is equivalent to Morin's original eversion [14]. However, there we did not make use of the two-fold symmetry. This means that the full surface was computed at each stage, wasting effort, and that, as small numerical errors accumulate, the symmetry we start with is not perfectly preserved during the evolution.

However, the *Evolver* can also work with just one fundamental domain of a symmetric surface. For instance, with mirror symmetry, we simply constrain boundary vertices to lie in the bounding mirror planes (and insure that the energy computations deal with them appropriately). There is also a general-purpose symmetry-group mechanism, where edges from one fundamental domain to the next are marked with a "wrap" value in the group. However, with rotational symmetry, there are certain extra difficulties when the surface spans across the axis of rotation. Recently, with Brakke's help, we have implemented the special features needed to overcome these [5].

To implement one of the eversions, we first get a good triangulation of the halfway model, with symmetries. Our explicit formula for the surface is as a map from the Riemann sphere (which we identify with the unit sphere in space), and it is easier to picture the symmetries on this domain. Thus we start by implementing a sphere with $2p$-fold symmetry generated by a flip-rotation. This can be described with just two vertices (one at the pole and one on the equator), three edges connecting them and their images under the group, and two triangles (one in each hemisphere); the complete input file is shown in Fig. 1.

We can evolve this crude sphere to a nicely triangulated round sphere, for instance by minimizing W as we refine the triangulation. We then make use of an auxiliary program which applies the Weierstrass map to each vertex position; this was implemented in C++ so that the formula could be written in complex notation, almost exactly as it appeared above. (The subroutine to map a single point is shown in Fig. 2.)

We then only need to change one line in the file header to tell the *Evolver* that the generating symmetry is now rotation by $\pi - \pi/p$ instead of flip-

[2] Available at URL `http://www.geom.umn.edu/locate/evolver/`.

```
parameter rotation_order = 2*5
parameter cr = cos(2*pi/rotation_order)
parameter sr = sin(2*pi/rotation_order)
view_transform_generators 1
cr sr 0 0   -sr cr 0 0   0 0 -1 0   0 0 0 1
constraint 1 formula x
constraint 2 formula y
constraint 3 formula x^2+y^2+z^2-1
symmetry_group "flip_rotate"
quantity willmore energy modulus rotation_order/4/pi \
                 global_method star_eff_area_sq_mean_curvature

vertices
1 0 0 1 axial_point constraints 1 2
2 1 0 0

edges
1 1 2
2 2 2 wrap 1
3 1 2 wrap 1

faces
1 1 2 -3
2 3 2 -1

read
transform_expr sprintf "%.0fa",rotation_order-1;
set facet density 0;
raw_cells;
mob_offset := .35;
do_dump := {
        set vertex constraint 3; recalc; unset vertex constraint 3;
        dump sprintf "s%.0f.fe", rotation_order/2;
        system sprintf "weier %.0f %f 0 s%.0f.fe h%.0f.fe",
            rotation_order/2, mob_offset, rotation_order/2, rotation_order/2;
}
```

Fig. 1. The input file **sphere.fe** for a sphere with flip-rotation symmetry in the *Evolver*; only the first line (current set for $p = 5$) needs to be changed for different p, since the constraints, symmetries, and viewing transformations are all expressed in terms of this parameter.

```
void calc(int p, double s, double theta, double x[3], double y[3])
{
    double q,r;
    complex z,denom,i,zp,eit;
    i = complex(0,1);
    if (x[2]==1)
        {y[0]=y[1]=0; y[2]=1/t; return;}
    z = complex(x[0],x[1])/(1-x[2]);
    q = 2*sqrt(2*p-1)/(p-1);
    zp = pow(z,p);
    denom = -i*(zp*zp+q*zp-1);
    eit = exp(i*theta/180*M_PI);
    y[0] = real(eit*(zp*zp/z-z)/denom);
    y[1] = real(eit*-i*(zp*zp/z+z)/denom);
    y[2] = real(eit*(p-1)*(zp*zp+1)/denom/p) + s;
    r = y[0]*y[0]+y[1]*y[1]+y[2]*y[2];
    y[0] /= r; y[1] /= r; y[2] /= r;
}
```

Fig. 2. This C++ subroutine maps a point x on a round sphere to the corresponding point on the halfway model of order p. (If $\theta \neq 0$ we get an inversion of an associate minimal surface.) It is part of a standalone program that transforms *Evolver* input files.

rotation by π/p.[3] This gives an approximation of the halfway model with all its symmetry enforced. Using the standard interactive features of the *Evolver*, we can now even out the triangulation while keeping near a minimum for the discrete W. Once we are happy with the model, we can run a special *Evolver* script to duplicate this fundamental domain and thus enforce only half the symmetry (the p-fold rotation that will remain throughout the eversion).

Evolution scripts. We have described how to create an initial data file for the halfway model M_0^p, respecting the p-fold symmetry. The *Evolver* can compute the matrix of second derivatives for the energy W, and as we expected, we find it has a single negative eigenvalue. This eigendirection is the perturbation we should make to push off from the saddle point and start evolving downhill towards the round sphere. In fact, the `saddle` command in the *Evolver* makes exactly this kind of motion in the lowest eigenvalue direction.

The evolution now proceeds much as in the $p = 2$ case [14]. Theoretically, once we have pushed off the initial critical point, we could continue downhill simply with the conjugate gradient method. However, we discovered for $p = 2$ that this was computationally very slow both at the beginning and again near the end when $W \approx 8\pi$, and the surface has the shape of a gastrula. We found that using the `saddle` command repeatedly (between every 50 iterations of conjugate gradient) sped up the computation at these stages. For the eversions with higher p-fold symmetry, we have decided to use these regular `saddle` commands throughout the eversion, since it is not clear *a priori* when they would be needed.

Thus, our script tells the *Evolver* to repeat (as many times as needed) a saddle command (of a certain magnitude) followed by, say, 50 iterations of conjugate gradient and a retriangulation step. This last step, a subroutine `trian`, involves several *Evolver* primitives, refining long or sharply bent edges, weeding out small triangles, *etc.*, as described in [14].

We mentioned that we observed these evolutions to proceed without any singularities forming. However, we must be careful: if there are not enough triangles to model the necks and other highly curved regions of the surface, problems will develop. The `trian` step will create new triangles in regions where they are needed (as well as removing them from other regions which are no longer so delicate). But we need to carefully choose the constants in that routine. For instance, to recompute an eversion with a finer triangulation, we not only must start with more triangles in the halfway model, but also must ensure that `trian` is adjusted to try to keep the triangles smaller. Probably the most important constant to choose is the bound on how bent an edge can be: the maximum allowed dihedral angle between adjacent faces.

[3] Note that for p odd, we now have a double-covered projective plane. The generating symmetry is a rotation which seems to have order p, but we treat it as still having order $2p$ in the symmetry group, to get both sheets of the surface.

On an a suitable computer,[4] these *Evolver* computations take on the order of ten minutes. To study the resulting eversions with interactive graphics, we save a sequence of *topes*, to be fed into our viewer `illiVert`. Generally we save a tope each time through the loop described above. If we wanted to get finer resolution in time (for creating a high quality video) we should use a smaller saddle step and fewer iterations of conjugate gradient each time through the loop.

The resulting eversions. Roughly, these tobacco-pouch eversions all start by pushing the north pole of the sphere down (forming a gastrula) and then through the south pole, creating the first self-intersection, or double curve. Then, around the neck, we see p fingers push up, and intersect each other near the symmetry axis. The middle stages, approaching the halfway model, are harder to understand without powerful visual assistance.

For pictures of the minimax eversion with $p = 2$, we refer the reader to [14]. Several stages in the $p = 3$ eversion are shown in Color Plate 4 on page 368.

It has often been thought that a sphere eversion must include a "twisting" phase (like the one seen clearly in Thurston's eversion [21] where the poles rotate a full turn in opposite directions). In our minimax eversions, there is no obvious twisting. It would be interesting to examine them more carefully, to see how they avoid this step that been used in nearly all earlier eversions.

In [14] we described extensively the double locus of the $p = 2$ minimax eversion. The halfway model is a Morin surface with one quadruple point; at nearby times we see the generic creation of such a point, with six double lines intersecting in four triple points forming the edges and vertices of a tetrahedron whose faces are four sheets of the surface. The sheets move inwards, and the tetrahedron shrinks down to the quadruple point; as the sheets continue to move, it then expands again, inverted.

For $p = 3$ we can give a similar description; near the halfway stage the double locus is like a four-fold cover of the double locus of Boy's surface. As the sheets come together, at the center we see a small cube shrinking, instantaneously creating a single sextuple point before expanding again, inverted. Fig. 3 shows the double locus at a stage near $t = 0$; note that there is one extra loop because the two sheets cross each other an additional time.

For general p, the halfway model has a $2p$-tuple point. At nearby times, the $2p$ sheets of surface are arranged like the faces of the polyhedron dual to a p-gonal antiprism; note that this polyhedron has parallel opposite faces if and only if p is odd. As these faces move inwards and through each other, the polyhedron shrinks to a point and then expands again. Again, it is helpful to focus on the double locus to gain understanding of the eversion.

[4] For example an SGI Onyx with ten R10000 processors running in parallel.

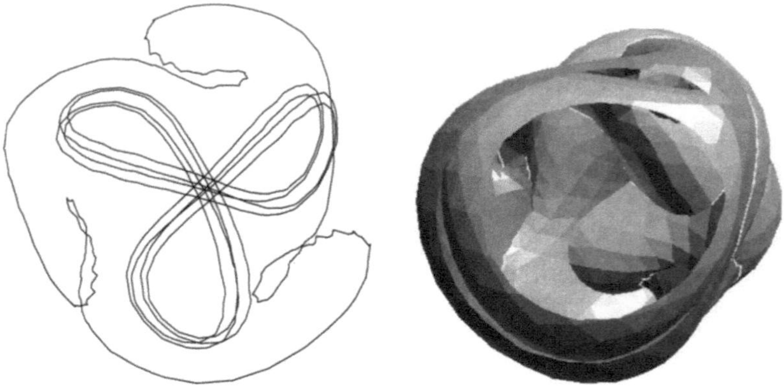

Fig. 3. Near $t = 0$ in the $p = 3$ minimax eversion, we see the double locus (left) and a cutaway view of the whole surface (right). At this stage, the double locus includes a four-fold cover of the three loops in the double locus of Boy's surface, and one extra curve. Compare Color Plate 4 on page 368.

3 Visualizing the Double Locus Surface of an Eversion

The double locus surface of a sphere eversion is a two-dimensional surface immersed in $\mathbb{R}^4$ whose level sets show the self-intersections of different stages of the eversion. We have developed software to compute these level sets, to splice them together into a surface, and to view the resulting surface in projection.

The double locus surface. Suppose $h : \mathbb{S}^2 \times [-1,1] \to \mathbb{R}^3$ is a regular homotopy which parameterizes an eversion of the sphere. The *double locus* of the eversion at time $t \in [-1,1]$ is the subset of $\mathbb{R}^3$ where the immersed surface $M_t = h(\mathbb{S}^2 \times \{t\})$ self-intersects:

$$X_t = \{h(x,t) : h(x,t) = h(y,t) \text{ for some } y \neq x\}.$$

Generically, X_t consists of a collection of curves, which meet at isolated multiple points. However, for those eversions whose halfway model is a doubly covered projective plane, at $t = 0$ we have $X_0 = M_0$, meaning that every point is in the double locus.

We shall assume that for all but a finite set of critical times, the immersed surface M_t intersects itself *transversally*, which means, for all practical purposes, that the surface is in as general position as possible given that it is part of a homotopy. Each X_t consists of the double curves of M_t meeting at triple points, except at the critical values of t, where X_t changes connectivity. In a favorable case, the succession of curves X_t may be regarded as the level curves of a surface $X = \bigcup_t (X_t \times \{t\}) \subset \mathbb{R}^3 \times \mathbb{R} = \mathbb{R}^4$ which is itself immersed in 4-space, and which is called the *double locus surface* of the eversion.

Fig. 4. The double loci of three adjacent topes near $t = 0$ in the $p = 2$ minimax eversion, drawn in cross-eyed stereo. The quadruple point of the halfway model is in the back center, and there are four saddles symmetrically around it, with a fifth in the front upper left. Compare Color Plate 5 on page 369.

Morin and Apéry [1] introduced the double locus surface X because of the connection between events in the homotopy and critical points of the height function t on X in $\mathbb{R}^4$. For example, in the Morin-type eversion described earlier, the double locus of the halfway model M_0 is a bouquet of six loops emanating from the quadruple point, which cross pairwise at the five points where M_0 has double tangency [14, Fig. 4.5]. In a neighborhood of time $t = 0$ and in a neighborhood of each of these points, the situation can be modeled as follows. At the quadruple point, the four planes of a shrinking tetrahedron pass through each other at $t = 0$ and then regrow the tetrahedron; this is not a critical point of the height function t. At each of the other five places, the model is that of a saddle sinking below sea level (an *isthmus event*); these are saddle critical points for t on X. The double loci at three times near $t = 0$ are shown in Fig. 4.

Following through the second half of the eversion, there is a time when two pairs of triple points are destroyed. Here, X_t changes from being two immersed loops to being two embedded loops, but this is not a critical time. However, when these two loops disappear we get two critical times, which are the two local maxima of t on X. Symmetrically, in the first half of the eversion we find just the two local minima for t. On the whole surface X, therefore, t has these four extremal points plus the five saddle points at $t = 0$, so X has Euler characteristic -1. Thus it is the connected sum of a torus with a cross cap, or, by Dyck's theorem [10, p.101f], a sphere with 3 cross caps, which we call *Dyck's surface* [11].

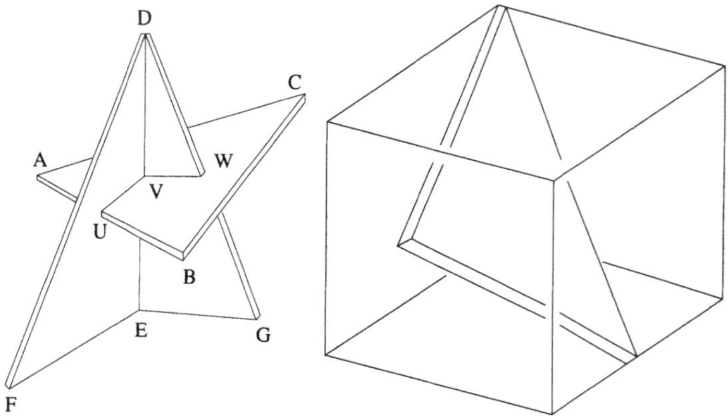

Fig. 5. At the left, the horizontal triangle ABC is pierced by an edge, DE, belonging to two triangles; these meet ABC (along the double curve UVW) in the two different possible ways. At the right, any triangle has at least one vertex at a corner, at least one on an edge, and at most one on a face, of its smallest (coordinate-aligned) bounding box.

Computing a double locus. We want to visualize the double locus surface of an eversion, like the particular immersion of Dyck's surface in $\mathbb{R}^4$ just described. The first step is to compute the double locus at any given time.

A numerical simulation of an eversion is more properly described by a discrete sequence (in time) of simplicial approximations to the immersed spheres $M_t = h(\mathbb{S}^2, t)$. We have coined the word *tope* for such a discrete stage in a homotopy to distinguish it more readily from a so-called *frame* in a computer animation. These topes are not parameterized spheres; in other words, we do not calculate numerical approximations of the function h itself, whose principal role is to simplify the topological description. For economy, we will henceforth use the same notation for smooth objects and their discrete polyhedral approximations.

For each tope M_t, the *Evolver* emits a list of vertices and facets (triangular faces), computed according to the script. We need to compute which facets intersect others. Generically, there are two ways that two triangles in space can meet along a segment: either the two boundaries link, with one edge of each cutting the interior of the other, or a corner of one triangle pierces the other (see Fig. 5). All degenerate cases are defaulted to one of these by using non-strict inequalities. Clearly, the primitive test which we must implement checks whether an edge DE intersects a triangle ABC, and if so, returns the intersection point V. In terms of linear algebra, the putative intersection point V has expansions as convex combinations of ABC and of DE:

$$A + u(B - A) + v(C - A) = V = D + t(E - D),$$

where t, u, v and $u+v$ are all between 0 and 1. We can express this as a matrix equation, to be solved within the constraints:

$$[t\ u\ v\,] \begin{bmatrix} D-E \\ B-A \\ C-A \end{bmatrix} = [D-A]$$

This 3×3 matrix is (nearly) singular, $\delta = (D-E) \cdot (B-A) \times (C-A) \approx 0$, either if the segment FE is (nearly) parallel to the triangle ABC, or if the latter is (nearly) degenerate. For our visualization purposes it was sufficient to ignore these cases. We found it most useful to solve this system by Cramer's Rule. First, we compare the determinant $\tau = (D-A) \cdot (B-A) \times (C-A)$ with δ. If τ and δ have opposite signs, or if $|\tau| > |\delta|$, then D and E are on the same side of the plane ABC, so we abort the search. Otherwise, the segment does intersect the plane at $t = \tau/\delta$, so we compute the determinants in the numerators for u and v, and check the inequalities that guarantee that the intersection point V is inside the triangle.

For larger numbers of facets, it would be best to implement a divide-and-conquer strategy with oct-trees to decide which facets might intersect. For simplicity (and ease of parallelization) we simply used a double loop over all pairs of distinct facets F, F', checking if F' intersects any edge E of F. The only optimization we used was to immediately discard the pair E, F' if their (coördinate-aligned) bounding boxes (see Fig. 5) failed to intersect.

If we do discover an intersection point V of some E and F', we then proceed to find the other endpoint U of the segment UV along which F and F' intersect. U is the intersection either of some other edge of F with F', or of some edge of F' with F.

Of course, if F and F' intersect along UV, then some neighboring triangles intersect in adjoining segments. But we did not make use of this information in our search, and the output of this first stage was simply an unordered collection of segments. The next step was to arrange these into continuous curves, by identifying common vertices. For the $p=2$ minimax eversion, the example we computed, at each stage the double locus X_t can be arranged into exactly two loops.

Splicing curves together into a surface. To mesh these polygons X_t in time, we first automatically resampled the curves so that each of them had the same number of vertices, approximately equally spaced around the curve. At this point, we needed to do two things by hand: to choose aligned base points (from tope to tope) on each loop, and to reorder the vertices correctly near the critical halfway time (where the height function on the double locus surface has saddle points).

Now that the we had a pair of 100-vertex cycles for each of 126 values of t, we could in principle look at a projection of the double locus surface from $\mathbb{R}^4$ to $\mathbb{R}^3$. The splining viewer, `illiLevel`, which was designed for viewing layered surfaces with far fewer control points, and which we will discuss in Section 4, revealed an intolerably "over-sampled" surface which was very slow to render, and also corrugated to an ugly degree. Since we planned to examine the surface more thoroughly with the 4-D viewer `illiSlice`, we felt that little would be gained by resampling to let the splines do more of the work.

Instead, the meshed dataset was next translated into a list of triangular facets suitable for resubmission to the *Evolver*, which has subroutines for culling small (hence superfluous) facets, and for swapping diagonals to give more equiangular triangles. Dennis Roseman helped us reduce in this way a representation of X in terms of 86,912 facets into an eminently manageable model with just 3665 triangles.

Viewing the surface in four-dimensional space. Now we have data for a surface in $\mathbb{R}^4$, given as a collection of triangular facets. We can use Glenn Chappell's `illiSlice` (originally designed to investigate Roseman's knotted surfaces in 4-space [13]) to view it, by means of sections and projections. On an SGI Indigo Extreme II with High Impact graphics, a version written in C (with the `gl` library under Irix 5.3) can render 2000 triangles at the practical limit of 10 frames per second.[5]

Given the surface X in $\mathbb{R}^4 = \{(x, y, z, t)\}$ we can first of all rotate it about the xt, yt, and zt axes to give a new surface X^ρ in $\mathbb{R}^4$. (The other principal rotations are in the 3-D space of the viewer.) Secondly, given two values $t_0 < t_1$, `illiSlice` retriangulates X^ρ into three separate surfaces, the belt $X^\rho_0 = X^\rho[t_0, t_1]$, and the distal parts $X^\rho_\pm$ beyond the belt. These are (separately) orthogonally projected to xyz-space, where they can be rotated and zoomed in the usual `illiView` style. The belt is colored in a range from blue to yellow to indicate which part is nearer to t_0 and which is nearer to t_1. The distal parts are less gaudily colored, and provided mainly as reference, especially for when we are in "slice mode" with $t_1 \approx t_0$. When the belt is wide enough to include all of the surface, we are in "projection mode" and perceive the fourth coordinate on the color scale.

Recall that, unrotated, X is already sliced by the X_t. As t_0 and t_1 are kept close, but moved together in time from the earliest to the last place on X, we see the homotopy of the double locus of the eversion. Rotated 90° in $\mathbb{R}^4$, the slicing and projecting becomes more interesting, and reveals a disturbingly complex immersion of Dyck's surface. The reason for this is not hard to discern. Recall that the *Evolver* script that reduces the Willmore bending energy (from 16π for the halfway model $X_0 = M_0^2$ to 4π for the round sphere) knows nothing about the double locus. In its evolution in time, the

[5] This is the same order of magnitude we expect from the CAVE wholly immersive virtual environment, powered by an Onyx with eight R10000 processors driving two Infinite Reality Engines which each render four images (two stereo walls) per frame.

eversion pulsates, perhaps moving the double locus about unnecessarily. Such excess motion is stacked into spatial complexity by passing from the $X_t \subset \mathbb{R}^3$ to $X \subset \mathbb{R}^4$. It is a wonder that the double locus turned out to be even as simple as it is.

4 Level Curve Methods for Everting Spheres

Many methods used for visualizing sphere eversions can properly be traced all the way back to the late nineteenth century. Then many geometers were preoccupied with the problem of demonstrating the existence (or its impossibility) of surfaces (with prescribed combinatorial or algebraic properties) in real or complex spaces. To be sure, the fact that a sphere could be turned inside out was the generally surprising consequence of a much later theorem by Smale [25] on classifying the space of immersions of surfaces up to regular homotopy. Yet the first concrete visualization of such a homotopy by Arnold Shapiro [12] was achieved with topological methods that had reached their maturity before the turn of the century. Even more to the point, Tony Phillips used precisely the same graphical methods for his celebrated sphere eversion (in the Scientific American cover story [24]) as Werner Boy [3] had used to describe his famous immersion of the real projective plane in $\mathbb{R}^3$ at the turn of the century.

Boy's surface via level sets. The problem Boy solved was to present a nonsingular (we now say 'immersed') surface in 3-space which, like the Klein bottle, was one-sided (nonorientable). However, unlike the Klein bottle, Boy's surface was a projective plane, with the smallest possible connectivity number for a closed nonorientable surface, namely Euler characteristic $\chi = 1$. Singular projective planes such as Steiner's Roman surface were well known, but the Klein bottle had been the simplest immersed nonorientable surface for decades. Ever since Möbius [16], we have known that χ for a nonsingular surface equals the number of maxima and minima minus the number of saddles, for any height function in general position (which we now call a Morse function). Thus the best Boy could hope for was to display a surface with precisely one of each kind of critical point.

The easiest way to remember Boy's levels, shown at the left in Fig. 6, is to begin with the saddle level, B^0, consisting of a figure-8 crossing a circle at two points, with the double point of the figure-8 the outside the circle. The circle is oriented so that it has tangent winding number (Gauss's *amplitudo*) $\tau = +1$. Either orientation of the figure-8 has $\tau = 0$. Following Boy, we take one of the two points where the figure-8 crosses the circle to be the saddle point, while the other is just another double point. We split the saddle point of B^0 in the two possible ways to give immersed circles, $B^{\pm}1$, above and below the saddle. With the orientation inherited from the circle in B^0, each has $\tau = 1$. Let B^1 have two positively oriented loops outside a negative

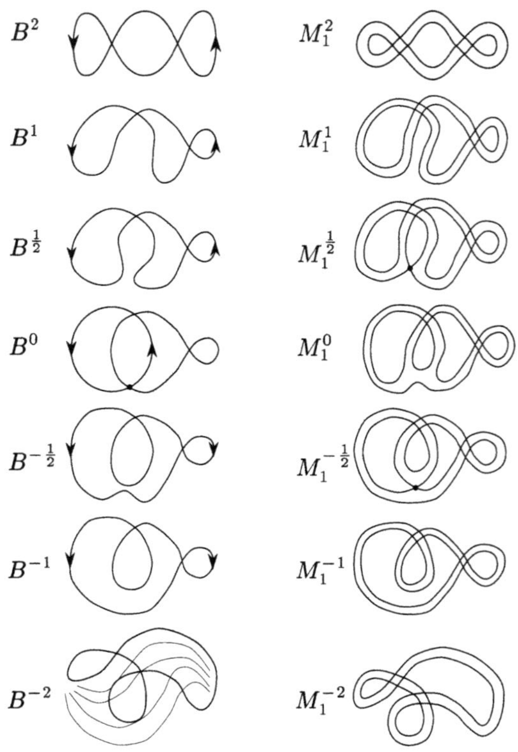

Fig. 6. Level sets of Boy's surface (left) and of a doubling M_1 (right); see text.

circuit, while B^{-1} has a positive loop inside, and a negative loop outside, a positively oriented circuit, as in Fig. 6. It is easy to see that the belt in between, $B[-1,1]$, is an immersed Möbius band with a hole. Hence, all that is left to do to complete Boy's surface is to attach one immersed disc to the top of B^1, and another to the bottom of B^{-1}.

Once B^1 has been straightened out a little (by an isotopy) into B^2 it is easy to see how to move it (by a regular homotopy) into a circle B^3. The belt $B[1,3]$ is thus an immersed cylinder with one double curve. It is harder to move B^{-1} into a circle. Indeed, there is no obvious way to do it. One way is to take hold of an arc opposite the inside loop, and move it across the loop, thereby forming a new positive loop while erasing the negative loop. The result is a curve B^{-2} which is isotopic to B^2, and we proceed as before to deform it to the circle B^{-3}. Whichever way the cylinder $B[-3,-1]$ is immersed, it must have one triple point by a theorem of Banchoff [2]. The double locus of $B[-3,3]$ is a bouquet of three loops emanating from this triple point.

Creating a sphere eversion from Boy's surface. We note that this Boy surface is a combinatorially optimal immersion $\beta : \mathbb{RP}^2 \to \mathbb{R}^3$ of the real projective plane, from the viewpoint of building the surface as a set of level curves of a Morse function. If we compose the double covering $\mathbb{S}^2 \to \mathbb{RP}^2$ (identifying antipodes) with β we obtain an immersion M_0 of the sphere, which covers every point of B twice. Imagine injecting air between the two sheets of M_0 to produce an immersed sphere M_1 in more general position. We shall have an explicit eversion of the sphere if we do two things. The easy part, which everyone started with until Morin's eversion, is to turn M_1 inside out, by moving it (along the normal lines of B) back through M_0 and out the other side to M_{-1}. The hard part is to construct a regular homotopy which moves the complicated immersed sphere M_1 into a round sphere $M_n = \mathbb{S}^2$, in a reasonable number of stages, so that each transition from M_i to M_{i+1} is immediately clear.

Note that the doubling M_1 of Boy's surface, using the same height function, now has two maxima, two minima, and two saddles. If we ask that no further critical levels develop and disappear along the homotopy M_t, this constraint makes the eversion harder to devise, but also easier to follow. This means that the eversion is to be *level-preserving* in the sense that it consists a two-parameter family of plane curves, $M_t^u, 1 \leq t \leq n, -m \leq u \leq m$, where M_1^u are horizontal slices through the inflated double of Boy's surface that we labeled M_1 above.

These slices of M_1 are shown at the right in Fig. 6. M_1^0 is the single immersed circle with $\tau = 1$ running along either side of the saddle level B^0 of Boy's surface. Immediately above and below are two saddles $M_1^{\pm\frac{1}{2}}$. The saddles break apart into two curves at $M_1^{\pm 1}$, which run parallel to $B^{\pm 1}$. Color Plate 6 on page 369 shows these levels spliced together into $M_1[-1, +1]$.

Implementing DeWitt's suggested eversion. At the same Battelle Conference in 1967 where Morin began to develop Marcel Froissart's suggestion for a sphere eversion, the physicist Bryce DeWitt [8] suggested a different one. Sphere eversions were very much the topic of discussion at this conference, and Tony Phillips' drawings [24] were much debated.

In the approach to an eversion outlined above, we have to deform M_1 into some embedded sphere. There are precisely two axially symmetric embeddings of the sphere with two maxima, two minima and two saddles. One is shaped like a dumbbell, the other like a gastrula, both with axis horizontal. Phillips used the former and filled in all the intermediate steps. DeWitt proposed using the latter because is it actually easier to find regular homotopies $M_t^{\pm\frac{1}{2}}$ connecting the saddle levels on M_1 to the corresponding saddles on the gastrula than on the dumbbell. Any such homotopy extends to regular neighborhoods $M_t[\frac{1}{3}, \frac{2}{3}]$ and $M_t[-\frac{2}{3}, -\frac{1}{3}]$ of the saddles. The real difficulty lies in visualizing what to do in between, $M_t[-\frac{1}{3}, \frac{1}{3}]$, to meld these two surface homotopies into each other.

DeWitt designed a two-dimensional array of plane curves. At one side are the horizontal slices of Boy's surface, and on the other side is an immersed sphere which moves to the gastrula embedding in an obvious way. He connected certain critical levels by regular homotopies of the plane curves. The conjecture that these key elements could be interpolated by a continuous succession of surfaces which did not develop singularities during the homotopy has stood unchallenged for 20 years.

Our plan is to check this using suitable computer graphics tools. Furthermore, we plan to use the same general tools to realize Phillips' eversion and many other visualizations based on drawing and deforming complicated plane curves.

Splining curves in time and space. The abstract problem looks like this. Given a 2-parameter family of plane curves, M_{tu}, connect them into a succession of surfaces, M_t, and animate these with smoothly changing shapes. The approach we have taken for `illiLevel` is to use the same number of control points p_{ijk} for each level M_{tu}, and an efficient sampling of key levels $u = k$ and key topes $t = j$. We then obtain model surface homotopies that are easy to edit using piecewise cubic interpolating splines. Although these Catmull-Rom [7] splines do not enjoy great popularity among computer graphicists, we find that they serve our purpose well.

We begin with four points, p_0, p_1, p_2, p_3, in $\mathbb{R}^n$ (or in any vector space, even infinite dimensional), and consider the cubic polynomial $q(t) = m_0 + m_1 t + m_2 t^2 + m_3 t^3$, with coefficients chosen so that

$$q(0) = p_1, \quad q(1) = p_2, \qquad q'(0) = s_1(p_2 - p_0), \quad q'(1) = s_2(p_3 - p_1),$$

where the s_i are pseudo-speeds, described below.

Thus, to compute the coefficients of q we solve the linear system:

$$\begin{bmatrix} 1&0&0&0 \\ 0&1&0&0 \\ 1&1&1&1 \\ 0&1&2&3 \end{bmatrix} \begin{bmatrix} m_0 \\ m_1 \\ m_2 \\ m_3 \end{bmatrix} = \begin{bmatrix} 0 & 1 & 0 & 0 \\ -s_1 & 0 & s_1 & 0 \\ 0 & 0 & 1 & 0 \\ 0 & -s_2 & 0 & s_1 \end{bmatrix} \begin{bmatrix} p_0 \\ p_1 \\ p_2 \\ p_3 \end{bmatrix},$$

or equivalently,

$$\begin{bmatrix} m_0 \\ m_1 \\ m_2 \\ m_3 \end{bmatrix} = \begin{bmatrix} 0 & 1 & 0 & 0 \\ -s_1 & 0 & s_1 & 0 \\ 2s_1 & (s_2 - 3) & (3 - s_1) & -s_2 \\ -s_1 & (2 - s_2) & (s_1 - 2) & s_2 \end{bmatrix} \begin{bmatrix} p_0 \\ p_1 \\ p_2 \\ p_3 \end{bmatrix}.$$

From this expression we see how to spline an arbitrarily long stream p_i of control points. We are free to choose a pseudo-speed s_i at each control point (except the first and the last). The interlacing of the tangents at the control points guarantees that the velocity is continuous at the corners. In practice, we choose all pseudo-speeds to be equal, $s_i = s_*$, but vary s_* interactively. There are two distinguished values for the pseudo-speed, $s_* = 0$ and $s_* = \frac{1}{2}$. With $s_* = 0$, we find that $q(t) = p_1 + (3t^2 - 2t^3)(p_2 - p_1)$. Therefore, for this pseudo-speed the spline runs along the polygon connecting the control points, but with more interpolating points near the corners than in the middle.

In order to distribute the interpolating points more evenly (for evenly chosen values of $t \in [0, 1]$), we choose $s_* = \frac{1}{2}$. In this case, when the $p_i = p_0 + iu$ are uniformly distributed along a straight line, the cubic interpolation degenerates to a truly linear one, $q(t) = p_1 + tu$. In practice, values of s_* closer to 1 makes for better shaped curves and surfaces, though interpolating splines are never likely to form beautiful surfaces.

More interesting is the *tensorial* nature of these splines. For simplicity, consider first a four-by-four array of control points, p_{ij}. If we first spline in one direction, we obtain four cubic polynomials, $q_j(t) = t^a Q_{ai} p_{ij}$, where Q is the matrix determined above and we use the Einstein convention. If now we spline the quadruples produced by the interpolated values of t, using a different time parameter, s, we obtain cubic polynomials $r(s,t) = s^c K_{cj} q_j(t) = s^c t^a Q_{ai} K_{cj} p_{ij}$. We would get exactly the same result if we splined first in the other dimension. Note that we could choose pseudo-speeds independently for each dimension at each control point.

A graphical user interface (written by Matthew Stiak) lets us easily draw and move the level curves M_{tu} (as in Color Plate 6 on page 369). It promises to make `illiLevel` into a versatile visualization tool that handles time and space in a uniform and interchangeable manner.

References

1. F. Apéry, *An algebraic halfway model for the eversion of the sphere*, Tohoku Math. J. *44* (1992), 103–150, with an appendix by B. Morin.
2. T. F. Banchoff, *Triple points and surgery of immersed surfaces*, Proc. Amer. Math. Soc. *46* (1974), 407–413.
3. W. Boy, *Über die Curvatura integra und die Topologie geschlossener Flächen*, Math. Ann. *57* (1903), 151–184.
4. K. A. Brakke, *The Surface Evolver*, Experimental Math. *1*:2 (1992), 141–165.
5. K. A. Brakke and J. M. Sullivan, *Using symmetry features of the surface evolver to study foams*, Visualization and Mathematics (H.-C. Hege and K. Polthier, eds.), Springer Verlag, Heidelberg, 1997, pp. 95–117.
6. R. Bryant, *A duality theorem for Willmore surfaces*, J. Differential Geometry *20* (1984), 23–53.
7. E. Catmull and R. Rom, *A class of interpolating splines*, Computer Aided Geometric Design (R. Barnhill and R. Riesenfeld, eds.), Academic Press, 1974, pp. 317–326.
8. B. DeWitt, *Eversion of the 2-sphere*, Battelle Rencontres, 1967, Benjamin, New York, 1967, pp. 546–557.

9. G. FRANCIS, *Drawing surfaces and their deformations: The Tobacco pouch eversions of the sphere*, Math. Modelling *1* (1980), 273–281.

10. ———, *A topological picturebook*, Springer-Verlag, New York, 1987.

11. ———, *On knot-spanning surfaces: An illustrated essay on topological art*, The Visual Mind (M. EMMER, ed.), Leonardo Books: MIT Press, 1993.

12. G. FRANCIS AND B. MORIN, *Arnold Shapiro's eversion of the sphere*, Math. Intelligencer *2* (1979), 200–203.

13. G. FRANCIS, J. M. SULLIVAN, K. BRAKKE, R. KUSNER, D. ROSEMAN, A. BOURD, C. HARTMAN, G. CHAPPELL, AND J. RUBENSTEIN, *LATERNA matheMAGICA*, Virtual Environments and Distributed Computing at SC'95: GII Testbed and HPC Challenge Applications on the I-WAY (H. KORAB AND M. D. BROWN, eds.), ACM/IEEE Supercomputing'95, 1995.

14. G. FRANCIS, J. M. SULLIVAN, R. B. KUSNER, K. A. BRAKKE, C. HARTMAN, AND G. CHAPPELL, *The minimax sphere eversion*, Visualization and Mathematics (H.-C. HEGE AND K. POLTHIER, eds.), Springer Verlag, Heidelberg, 1997, pp. 3–20.

15. L. HSU, R. KUSNER, AND J. M. SULLIVAN, *Minimizing the squared mean curvature integral for surfaces in space forms*, Experimental Mathematics *1*:3 (1992), 191–207.

16. H. KARCHER AND U. PINKALL, *Die Boysche Fläche in Oberwolfach*, Mitteilungen der DMV *1997*:1, 45–47.

17. N. KUIPER, *Convex immersions of closed surfaces in E^3*, Comm. Helv. *35* (1961), 85–92.

18. R. KUSNER, *Conformal geometry and complete minimal surfaces*, Bull. Amer. Math. Soc. *17* (1987), 291–295.

19. ———, *Comparison surfaces for the Willmore problem*, Pacific J. Math. *138* (1989), 317–345.

20. S. LEVY, *Making waves: A guide to the ideas behind Outside In*, AK Peters, Wellesley, MA, 1995.

21. S. LEVY, D. MAXWELL, AND T. MUNZNER, *Outside In*, AK Peters, Wellesley, MA, 1994, narrated videotape (21 min) produced by the Geometry Center, University of Minnesota.

22. N. L. MAX, *Turning a sphere inside out*, International Film Bureau, Chicago, 1977, narrated videotape (21 min).

23. B. MORIN, *Équations du retournement de la sphére*, Comptes Rendus Acad. Sci. Paris *287* (1978), 879–882.

24. A. PHILLIPS, *Turning a sphere inside out*, Sci. Amer. *214* (1966), 112–120.

25. S. SMALE, *A classification of immersions of the two-sphere*, Trans. Amer. Math. Soc. *90* (1959), 281–290.

26. T. J. WILLMORE, *Note on embedded surfaces*, An. Stiint. Univ "Al. I. Cuza" Iasi Sect. I, a Mat. *11* (1965), 493–496.

Morse Theory for Implicit Surface Modeling

John C. Hart

School of EECS, Washington State University, Pullman, WA 99164-2752

Abstract. Morse theory describes the relationship between a function's critical points and the homotopy type of the function's domain. The theorems of Morse theory were developed specifically for functions on a manifold. This work adapts these theorems for use with parameterized families of implicit surfaces in computer graphics. The result is a theoretical basis for the determination of the global topology of an implicit surface, and supports the interactive modeling of implicit surfaces by direct manipulation of a topologically-correct triangulated representation.

1 Introduction

Implicit surfaces provide a powerful and versatile shape model in computer graphics by representing geometry as the zero-set of a function over three-space, although displaying such surfaces requires a search through space. The display of an implicit surface is hastened by maintaining a triangulation that can be quickly rendered on modern graphics workstations. However, when the implicit surface changes topological type, the triangulation needs to be updated in the neighborhood of the topology change. A recent technique uses the critical points of the function to detect changes in topology and reconfigures the triangulation to correctly reflect the topology of the new surface [10,11].

The fundamental detail missing from these publications is the connection between a function's critical points and the topology of its implicit surface. This connection can be found in Morse theory, but the theorems of Morse theory do not directly apply to the implicit surfaces used in computer graphics. This paper formalizes this connection with obvious but not entirely trivial extensions of theorems from Morse theory to implicit surface topology.

Section 2 summarizes the implicit surface geometric representation and techniques for modeling with implicit surfaces. Section 3 reviews Morse theory, focusing on the connection between critical points and homotopy type. Section 4 applies the results of Morse theory to implicit surfaces. Section 5 concludes with remarks on further applications of Morse theory in computer graphics.

2 The Problem of Modeling with Implicit Surfaces

An implicit surface is defined as the zero-set of a function $f : \mathbb{R}^3 \to \mathbb{R}$. The implicit surface is often a compact manifold, though not always smooth [7], compact (e.g. the cylinder $f(x, y, z) = x^2 + y^2 - 1$), nor even a manifold [3].

Natural geometric primitives, such as the plane, sphere, cylinder, cone and torus, can be described implicitly as the solutions to linear, quadratic and quartic polynomials. These primitives are commonly treated as solids (3-manifolds-with-boundary) by considering the points where the function is negative (or positive) to be in the interior the set. These solid primitives are combined with binary set operations (union, intersection and difference) to form more complex shapes in a procedure known in computer graphics as *constructive solid geometry* (CSG). Implicit surfaces also facilitate the joining of surfaces with a process called *blending* which smoothes the results of a CSG operation.

Perhaps the most popular blending technique in computer graphics is the *blobby* model [1]. The blobby model represents shapes with implicit surfaces defined by functions of the form

$$f(\mathbf{x}) = T - \sum_{i=1}^{N} e^{-k_i F_i(\mathbf{x})} \tag{1}$$

where the functions $F_i : \mathbb{R}^3 \to \mathbb{R}$ implicitly define primitive shapes, the k_i are parameters controlling the strength of the primitives and T is a threshold value. The primitive shapes are often quadric spheres

$$F_i(\mathbf{x}) = (\mathbf{x} - \mathbf{x}_i) \cdot (\mathbf{x} - \mathbf{x}_i) \tag{2}$$

centered about so-called *key points* $\mathbf{x}_i$. The implicit surface is the boundary of a solid, and the function f is negative in this solid. As might be clear from (1) and (2), the blobby model originated as a method for visualizing electron densities in molecules with nuclei at $\mathbf{x}_i$, but has matured into a geometric representation capable of synthesizing a variety of natural and man-made forms [2]. Moreover, in addition to points, other primitives such as lines, polygons, curves and patches [4] can be collected together to form a *skeleton.* The primitives composing this skeleton may be thickened (using a suitable function F_i) into implicit surfaces which may then be blended (using a suitable function f) into a single smooth implicit surface.

For example in Figure 1, the shape on the left is composed of the CSG union of eight spheres whereas the shape on the right is composed of the same eight spheres joined with (1).

While implicit surfaces serve as a powerful shape representation in computer graphics, they are not well suited for interactive modeling. The main impediment is rendering. Whereas other shape descriptions such as the parametric surface yield a surface as the range of a function, an implicit surface must be found in a given region of space. The increased computation required to find the implicit surface makes displaying them at interactive rates difficult.

A rendering method called *ray tracing* displays shapes by following each ray of light backwards from the eye, through each pixel and into the scene.

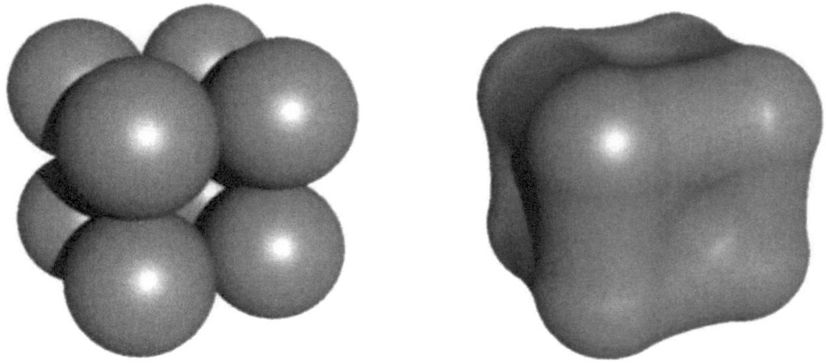

Fig. 1. A ray-traced implicit surface composed of the union of eight spheres (left) and the blended union of eight spheres (right).

Implicit surfaces are well suited for ray tracing. Let $l : \mathbb{R}^+ \to \mathbb{R}^3$ be the parametric definition of a ray. Then the intersection of the ray with the implicit surface of $f(\mathbf{x})$ is determined by finding the zeros of the real function of one variable $f \circ l$. The images in Figure 1 were rendered using such a ray tracing algorithm developed specifically for mathematical visualization [6].

In order to design a blobby model interactively, the implicit surface needs to be displayed in real time. Even with the power of modern graphics workstations, ray tracing remains too costly for interactive applications. Instead, recent techniques visualize the implicit surface in real time by maintaining a simplified approximation. For example, an implicit surface can be interactively manipulated using an efficient visual representation consisting of a system of mutually-repelling particles constrained to the surface, displayed as a collection of disks tangent to the surface [12]. As the surface changes shape due to user interaction, the disks maintain their position on the surface. Figure 2 (left) demonstrates this method of display.

Connecting these particles triangulates the implicit surface, as shown in Figure 2 (right). As the implicit surface changes, the vertices remain on the surface and the triangulation remains intact. However, when the implicit surface changes topological type, the triangulation is no longer a valid representation of the implicit surface. Whereas the particles require only the local tangent information to indicate the surface, the triangulated representation must be aware of any portions of the surface that are newly joined or separated. Morse theory provides the tools necessary to make such a determination.

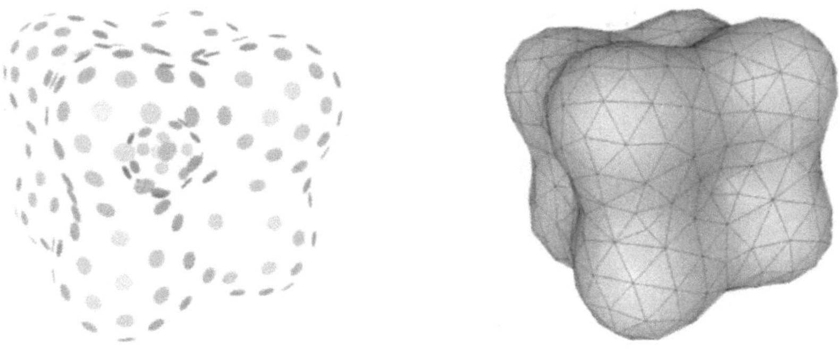

Fig. 2. The blobby cube displayed using a particle system (left) and triangulated (right). Note that the blobby cube is hollow and the particle system rendering reveals the air bubble.

3 Morse Theory

This section reviews elementary Morse theory [8], specifically the classification of critical points of a function on a manifold, and the effect of these critical points on the homotopy type of the manifold. The function is commonly smooth, but Morse theory can be applied to functions of varying smoothness, even piecewise linear. The following development of definitions and theorems require only C^2 (second-derivative) continuity which broadens the variety of implicit surfaces accessible by the theorems. The section relies on some prior knowledge of homotopy theory [9].

Definition 1. Let f be a C^2 real map on a manifold M. A point $p \in M$ is a *critical point* iff its derivatives with respect to a local coordinate system on M vanish.

More specifically, since M is an n-manifold, then there exists a C^2 one-to-one correspondence g between a neighborhood about any point $p \in M$ and an open neighborhood of the origin in $\mathbb{R}^n$ such that $g(p) = \mathbf{x} = (x_1, x_2, \ldots, x_n)$. Then the point $p \in M$ is a critical point with respect to f if the gradient

$$\nabla f = \left(\frac{\partial f \circ g^{-1}(\mathbf{x})}{\partial x_1}, \frac{\partial f \circ g^{-1}(\mathbf{x})}{\partial x_2}, \ldots \frac{\partial f \circ g^{-1}(\mathbf{x})}{\partial x_n} \right) = \mathbf{0}. \qquad (3)$$

Morse theory focuses only on non-degenerate critical points. Such points, also called Morse points, are critical points where the *Hessian*

$$V(f) = \begin{bmatrix} \frac{\partial^2 f}{\partial x_1^2} & \frac{\partial^2 f}{\partial x_1 \partial x_2} & \cdots & \frac{\partial^2 f}{\partial x_1 \partial x_n} \\ \frac{\partial^2 f}{\partial x_2 \partial x_1} & \frac{\partial^2 f}{\partial x_2^2} & \cdots & \frac{\partial^2 f}{\partial x_2 \partial x_n} \\ \vdots & \vdots & \ddots & \vdots \\ \frac{\partial^2 f}{\partial x_n \partial x_1} & \frac{\partial^2 f}{\partial x_n \partial x_2} & \cdots & \frac{\partial^2 f}{\partial x_n^2} \end{bmatrix} \tag{4}$$

has non-zero determinant. Since $\partial^2 f / \partial x_i \partial x_j = \partial^2 f / \partial x_j \partial x_i$, the matrix $V(f)$ is symmetric with real eigenvalues. Let $\lambda_1 \leq \lambda_2 \leq \ldots \leq \lambda_n$ be the eigenvalues of $V(f)$. If any of the eigenvalues is zero, then the critical point is *degenerate.* Otherwise it is called *non-degenerate.* The *index* of the critical point is the number of negative eigenvalues of $V(f)$.

The Morse Lemma states that the neighborhood about a non-degenerate critical point can be deformed into the neighborhood of the non-degenerate critical point of a quadratic function.

Lemma 2. (Morse Lemma) *Let p be a non-degenerate critical point of f with index λ, and let $c = f(p)$. Then there exists a local coordinate system $\mathbf{y} = (y_1, y_2, \ldots, y_n)$ in a neighborhood U of p with p as its origin and*

$$f(\mathbf{y}) = c - y_1^2 - y_2^2 - \cdots - y_\lambda^2 + y_{\lambda+1}^2 + \cdots + y_n^2. \tag{5}$$

Morse theory focuses on determining the homotopy type of a shape based on its critical points. A classic example [5] demonstrates the effects of critical points on homotopy type by observing the portion of a torus below a clipping plane, as the clipping plane moves through the torus. One can observe these same changes by dunking a doughnut into a cup of coffee, as shown in color plage 12, p. 373.

For this example, let M denote the surface of a vertically-oriented torus and let $f(p)$ return the height of point $p \in M$. Assume the bottom of the torus is at height zero and the top is of height one. In general the notation M^a indicates the points $p \in M$ such that $f(p) \leq a$, in this case the portion of the torus up to a height of a.

As the clipping plane traverses up the torus, Figure 3 shows that the changes in the topology of the torus can be described by attaching the appropriate k-cell to the truncated surface. Notice that the dimension of the attached cell equals the index of the critical point passed by the clipping plane.

The following theorem shows that M^a is topologically similar to $M^b \supset M^a$ if there is no critical point in M^a that is not also in M^b.

Theorem 3. [8] *Let $f : M \to \mathbb{R}$ be C^2, let $a < b$ and suppose that the set $f^{-1}[a, b]$ is compact and contains no critical points of f. Then M^a is homeomorphic to M^b.*

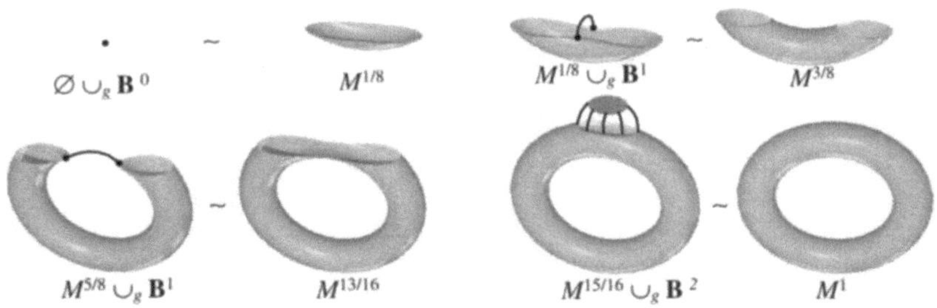

Fig. 3. Homotopy classes of the clipped torus.

Proof. Let the family of continuous maps $\varphi_t : M \to M$ be defined as the solution of the ordinary differential equation

$$\dot{\varphi}_t(p) = \frac{\nabla f(\varphi_t(p))}{||\nabla f(\varphi_t(p))||^2} \tag{6}$$

(where $\dot{\varphi} = d\varphi/dt$) with the initial value $\varphi_0(p) = p$ on $f^{-1}[a,b]$, and let φ continuously go to the identity ($\dot{\varphi}_t(p) = \mathbf{0}$) outside a compact neighborhood of $f^{-1}[a,b]$ not containing any critical points, such that each map φ_t is bijective and continuous with continuous inverse.

The function value of f on the curve $\varphi_t(p)$ generates on M fixing p and varying t changes at the same rate as t changes, since the directional derivative

$$\frac{df(\varphi_t(p))}{dt} = \dot{\varphi}_t(p) \cdot \nabla f(\varphi_t(p)) = 1. \tag{7}$$

Hence the homeomorphism φ_{b-a} carries M^a onto M^b. □

Theorem 4. [8] *Let $f : M \to \mathbb{R}$ be C^2, and let $p \in M$ be a non-degenerate critical point with index λ. Setting $f(p) = c$, suppose that $f^{-1}[c-\varepsilon, c+\varepsilon]$ is compact, and contains no critical point of f other than p, for some $\varepsilon > 0$. Then, for all sufficiently small ε, the set $M^{c+\varepsilon}$ has the homotopy type of $M^{c-\varepsilon}$ with a λ-cell attached.*

Elements of the proof of this theorem will be needed to prove a later proposition. The following is a brief summary of a classic proof [8], which should be consulted for details.

Proof. Using Morse's Lemma, choose a coordinate system $u_1, \ldots, u_n$ in a neighborhood U of p such that

$$f(p) = c - u_1(p)^2 - \ldots - u_\lambda(p)^2 + u_{\lambda+1}(p)^2 + \ldots + u_n(p)^2. \tag{8}$$

We abbreviate $\zeta = u_1^2 + \ldots + u_\lambda^2$ and $\eta = u_{\lambda+1}^2 + \ldots + u_n^2$ such that $f(p) = c - \zeta + \eta$.

Let $\varepsilon > 0$ be sufficiently small such that $f^{-1}[c-\varepsilon, c+\varepsilon]$ is compact and contains no other critical points other than p, and U contains a ball of radius 2ε.

The function $\mu : \mathbb{R} \to \mathbb{R}$ (any differentiable function such that $\mu(0) > \varepsilon, \mu(r) = 0$ for $r \geq 2\varepsilon$, and $-1 < \mu'(r) \leq 0$) locally warps the effects of the function on the manifold as $F(p) = f(p) - \mu(\zeta(p) + 2\eta(p))$.

Given the definition of the new function F, the following four assertions follow and suffice to prove the theorem. Proof of each of the assertions can be found in [8].

Assertion 1. $F^{-1}(-\infty, c+\varepsilon] = M^{c+\varepsilon}$.

Assertion 2. F shares the same critical points as f.

Assertion 3. $F^{-1}(-\infty, c-\varepsilon] \cong M^{c+\varepsilon}$.

Let $e^\lambda \subset M$ be the λ-cell

$$e^\lambda = \{(u_1, \ldots, u_n) : u_1^2 + \cdots + u_\lambda^2 \leq \varepsilon, u_{\lambda+1}^2 + \cdots + u_n^2 = 0\}. \tag{9}$$

Denote $\mathcal{H} = \text{closure}(F^{-1}(-\infty, c-\varepsilon] - M^{c-\varepsilon})$. Note that $e^\lambda \subset \mathcal{H}$.

Assertion 4. $M^{c-\varepsilon} \cup e^\lambda$ is a deformation retract of $M^{c-\varepsilon} \cup \mathcal{H}$. □

4 Application to Implicit Surfaces

The following proposition is a first step at applying the theorems from the previous section to implicit surfaces. It essentially states that two isosurfaces of the same function are topologically similar if there is no critical point in any isosurface between them.

Proposition 5. *Let $f : M \to \mathbb{R}$ be C^2, and such that $f^{-1}[a,b]$ is compact and contains no critical points. Then $f^{-1}(a) \cong f^{-1}(b)$.*

Proof. From Theorem 3 we have that $M^a \cong M^b$. The boundary of M^a (w.r.t. M) is $f^{-1}(a)$ and likewise $\partial M^b = f^{-1}(b)$. The boundaries of two homeomorphic sets must themselves be homeomorphic. □

Proposition 5 can be applied to implicit surfaces, but must be restricted to non-intersecting implicit surfaces such that one implicit surface completely surrounds the other.

In order to show a homeomorphism between two implicit surfaces in general, we must define a family of implicit surfaces and define a height function on this family. Then the properties of the manifold due to the height function will also apply to the family of implicit surfaces.

Let $f : \mathbb{R}^n \times \mathbb{R}^m \to \mathbb{R}$ define a family of functions $f(\mathbf{x};\mathbf{q})$ parameterized by an m-vector $\mathbf{q}$ that defines a family of implicit surfaces as the collection of $(n-1)$-manifolds $f_{\mathbf{q}}^{-1}(0) = \{\mathbf{x} : f(\mathbf{x};\mathbf{q}) = 0\}$. Note that the domain of the instance $f_{\mathbf{q}} : \mathbb{R}^n \to \mathbb{R}$ differs from the domain of the family f. (The latter includes the parameter space.)

Consider two $(n-1)$-manifolds $M_0 = f_{\mathbf{q}_0}^{-1}(0)$ and $M_1 = f_{\mathbf{q}_1}^{-1}(0)$. Let $\mathbf{q}(t), t \in \mathbb{R}$, denote a linear interpolation of parameters such that $\mathbf{q}(0) = \mathbf{q}_0$ and $\mathbf{q}(1) = \mathbf{q}_1$. Let $\mathbf{q} : \mathbb{R} \to \mathbb{R}^m$ be parameterized such that $M \in \mathbb{R}^n \times \mathbb{R}^m = \{(\mathbf{x}, \mathbf{q}(t)) : f(\mathbf{x};\mathbf{q}(t)) = 0, t \in \mathbb{R}\}$ is an n-manifold. Define the height map $h : M \to \mathbb{R}$ as $h(\mathbf{x},\mathbf{q}) = t$.

Proposition 6. *Let $p = (\mathbf{x}_p, \mathbf{q}_p) \in M$. Then p is a critical point of h ($\nabla h(p) = \mathbf{0}$) if and only if $\nabla f_{\mathbf{q}_p}(\mathbf{x}_p) = (\partial f_{\mathbf{q}}/\mathbf{x}_1, \ldots, \partial f_{\mathbf{q}}/\mathbf{x}_n) = \mathbf{0}$.*

Proof. If p is a critical point of h, then it's value ($= t$) is locally constant along M and vice versa. Hence M is locally perpendicular to the t axis and orthogonal to the $\mathbf{q}$ hyperplane. The $\mathbf{x}$ coordinate system serves as a local coordinate system for M at p. □

Proposition 5 combines with this family of implicit surfaces to assert the following proposition that implicit surfaces do not change homotopy type if they do not intersect a critical point.

Proposition 7. *If the family of implicit surfaces $f_{\mathbf{q}(t)}^{-1}(0)$ is compact for every $t \in [t_0, t_1]$ and none contain a point $\mathbf{x}$ such that $\nabla f_{\mathbf{q}(t)}(\mathbf{x}) = 0$, then $f_{\mathbf{q}_0}^{-1}(0)$ is homeomorphic to $f_{\mathbf{q}_1}^{-1}(0)$.*

Proof. The surfaces $h^{-1}(t_0) = f_{\mathbf{q}_0}^{-1}(0)$ and $h^{-1}(t_1) = f_{\mathbf{q}_1}^{-1}(0)$. Proposition 6 asserts there are no critical values on M between t_0 and t_1, which allows Proposition 5 to show $h^{-1}(t_0) \cong h^{-1}(t_1)$. □

Proposition 8. *Let $\mathbf{x}_p$ be a non-degenerate critical point with index λ of $f_{\mathbf{q}(t_p)}$. If there exists some $\varepsilon > 0$ such that the set $\{\mathbf{x} : f(\mathbf{x};\mathbf{q}(t)) = 0, t \in [t_p - \varepsilon, t_p + \varepsilon]\}$ is compact and contains no other critical points than $(\mathbf{x}_p, \mathbf{q}_p)$, and assuming without loss of generality that $\partial f(\mathbf{x}_p; \mathbf{q}(t_p))/\partial t < 0$, then the n-manifold-with-boundary $f_{\mathbf{q}(t_p+\varepsilon)}^{-1}(-\infty, 0]$ has the same homotopy type as $f_{\mathbf{q}(t_p-\varepsilon)}^{-1}(-\infty, 0]$ with a λ-cell attached.*

The following proof follows the same logic as the proof of Theorem 4 but also uses a projection to show that the regions bounded by homeomorphic sets are also homeomorphic.

Proof. Following the proof of Theorem 4, choose a coordinate system such that $h = -\zeta + \eta$ in a neighborhood $U \subset M$ of p. Let $H = -\zeta + \eta - \mu(\zeta + 2\eta)$ inside U and $H = h$ outside U. As before, H has the same critical points as h, and the manifold-with-boundary $H^{-1}(-\infty, t_p + \varepsilon] = h^{-1}(-\infty, t_p + \varepsilon]$, but the critical point $(\mathbf{x}_p, \mathbf{q}_p)$ is in $H^{-1}(-\infty, t_0 - \varepsilon]$. Since there is no critical point in $H^{-1}[t_p - \varepsilon, t_p + \varepsilon]$, we have $h^{-1}(t_p + \varepsilon) \cong H^{-1}(t_p - \varepsilon)$ by Proposition 5.

Let $\pi : M \to \mathbb{R}^n$ be the projection $(\mathbf{x}, \mathbf{q}) \mapsto \mathbf{x}$. Proposition 6 shows us that near p, the manifold M is orthogonal to the $\mathbf{q}$ hyperplane, so ε can be set small enough such that the projection π is one-to-one in the neighborhood U.

Recalling the map φ from the proof of Theorem 3, we have

$$\begin{aligned} H^{-1}(t_p - \varepsilon) &\cong H^{-1}(t_p + \varepsilon), && (10)\\ &= h^{-1}(t_p + \varepsilon), && (11)\\ &\cong \pi \circ h^{-1}(t_p + \varepsilon), && (12)\\ &= f^{-1}_{\mathbf{q}(t_p+\varepsilon)}(0). && (13) \end{aligned}$$

Hence the homeomorphism $\pi \circ \varphi_{2\varepsilon}$ maps $H^{-1}(t_p - \varepsilon)$ to $f^{-1}_{\mathbf{q}(t_p+\varepsilon)}(0)$. The latter implicit surface is the boundary of the implicit solid $f^{-1}_{\mathbf{q}(t_p+\varepsilon)}(-\infty, 0]$. This region is mapped via the homeomorphism $\varphi_{-2\varepsilon} \circ \pi^{-1} : \mathbb{R}^n \to M$ into a subset of M with $H^{-1}(t_p - \varepsilon)$ as its boundary.

As before, the handle

$$\mathcal{H} = \text{closure}(H^{-1}(t_p - \varepsilon) - h^{-1}(t_p - \varepsilon)) \tag{14}$$

is the subset that creates the change in homotopy type, and this handle is homotopic to a λ-cell. Both the handle and the boundary of the λ-cell extend to $h^{-1}(t_p - \varepsilon)$ and hence their projections extend to $f^{-1}_{\mathbf{q}(t_p-\varepsilon)}(-\infty, 0]$. □

The disconnection direction $(\partial f(\mathbf{x}_0, q_0)/\partial q > 0)$ is not defined since there is no mechanism available to us to "remove a λ-cell." Instead, we must invert the t parameter about the critical point to treat the problem in the connection direction, or consider the closure of the complement of the implicit solid and attach an $(n - \lambda)$-cell.

These propositions allow us to classify changes in the topological type of implicit surfaces. The eight possible topological-type changes are listed in Table 1.

Index	Critical value $+ \to -$	Critical value $- \to +$
0	Create	Destroy
1	Connect	Cut
2	Spackle	Pierce
3	Burst	Bubble

Table 1. The eight possible homotopy equivalence class changes in 3-D at a non-degenerate critical point.

When a minimum value becomes negative, a new implicit surface component is created. This can be considered attaching a 0-cell to the empty set. When a minimum value becomes positive, the component is destroyed.

When an index 1 critical value becomes negative, a new connection is formed between two components. In terms of homotopy type, a 1-cell has been attached to the two solid components. When an index 1 critical value becomes positive, a connection is cut. These cases are shown in Figure 4.

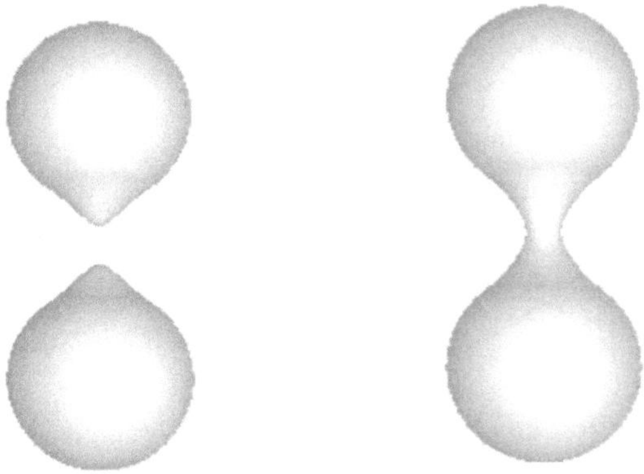

Fig. 4. An index 1 critical point with critical value positive (left) and negative (right).

When an index 2 critical value becomes negative, a hole in the solid is filled in. In terms of homotopy type, a 2-cell has been attached. When an index 2 critical value becomes negative, a new hole is pierced in the solid. These cases are shown in Figure 5.

Fig. 5. An index 2 critical point with critical value positive (left) and negative (right).

When a maximum value becomes positive, a hollow region is formed in a solid. Such an air bubble can be seen in the particle system rendering of the blobby cube in Figure 2. When a minimum critical value becomes negative, the bubble bursts. In terms of homotopy type, a 3-cell has been attached in place of the air bubble.

When the implicit solid changes homotopy type, simple algorithms exist to locally reconfigure the triangulation to reflect the new topology [11].

The only remaining problem for maintaining a triangulated version of a dynamic implicit surface is tracking all of the critical points of the function. Several techniques have been explored [11]. The most effective technique performs an interval Newton's method search across a given region of space over a given time interval for critical points that intersect the implicit surface. Such search methods based on interval analysis can be guaranteed not to miss any solutions, resulting in a guarantee that the triangulation is homotopy equivalent to the implicit surface it represents.

5 Conclusion

This document serves to provide a theoretical basis for the alteration of implicit surface topology in the presence of critical points. It ultimately shows that the topological type of an implicit surface before and after a non-degenerate critical value changes sign can be described through the attachment of an appropriate-dimension cell.

Morse theory might also add insight to current problems in shape transformation. The determination that the initial and final shapes share the same topological type would be the first step toward finding a possible topological-type preserving shape transformation. Likewise, the characterization of neighborhoods of critical points occurring during shape transformation may provide new insight into the maintenance of consistent texture coordinates through changes in homotopy type.

Thanks to Jules Bloomenthal, George Francis, John Hughes, Nelson Max and Bart Stander for informative discussions about Morse theory and implicit surfaces. This research was supported in part by the National Science Foundation under grants CCR-9309210 and CCR-9529809.

References

1. J. F. Blinn, *A generalization of algebraic surface drawing*, ACM Transactions on Graphics *1*:3 (1982), 235–256.

2. J. Bloomenthal, *An introduction to implicit surfaces*, Morgan-Kauffman, San Francisco, 1997.

3. J. Bloomenthal and K. Ferguson, *Polygonization of non-manifold implicit surfaces*, Computer Graphics (Annual Conference Series), Aug. 1995, pp. 309–316.

4. J. Bloomenthal and B. Wyvill, *Interactive techniques for implicit modeling*, Computer Graphics *24*:2 (1990), 109–116.

5. R. Bott, *Morse theory and its application to homotopy theory*, Universität Bonn, Germany, 1960.

6. J. C. Hart, *Sphere tracing: a geometric method for the antialiased ray tracing of implicit surfaces*, The Visual Computer *12*:10 (1996), 527–545.

7. ———, *Implicit formulations of rough surfaces*, Computer Graphics Forum *16*:2 (1997).

8. J. Milnor, *Morse theory*, Annals of Mathematics Studies, vol. 51, Princeton University Press, Princeton, NJ, 1963.

9. J. R. Munkres, *Topology: a first course*, Prentice-Hall, Englewood Cliffs, New Jersey, 1975.

10. B. T. Stander, *Polygonizing implicit surfaces with guaranteed topology*, Ph.D. thesis, School of EECS, Washington State University, May 1997.

11. B. T. Stander and J. C. Hart, *Guaranteeing the topology of an implicit surface polygonization for interactive modeling*, Computer Graphics (Annual Conference Series), Aug. 1997, pp. 279–286.

12. A. P. Witkin and P. S. Heckbert, *Using particles to sample and control implicit surfaces*, Computer Graphics (Annual Conference Series), July 1994, pp. 269–277.

Special Relativity in Virtual Reality

René T. Rau[1], Daniel Weiskopf[2], and Hanns Ruder[2]

[1] WSI/GRIS, University of Tübingen, Germany
[2] Theoretical Astrophysics, University of Tübingen, Germany

Abstract. The appearance of fast moving objects can be calculated according to the Theory of Special Relativity. In addition to the Lorentz contraction the effects of finite light speed and aberration play an important role. These phenomena were first discovered and described correctly in 1959 by Penrose and Terrel. Concerning the visualization of the phenomena there already exist systems with relativistic ray tracing and polygon rendering. Investigating these approaches in detail we found a reformulation of the problem which allows the treatment of acceleration in real-time. Therefore, user interaction could be integrated and a virtual reality for special relativity was possible.

1 Introduction

The Theory of Relativity is a fascinating topic in astrophysics. Most of the phenomena are beyond our normal experience and seem strange for a newcomer in the field. Especially for educational purposes, the visualization of many phenomena can give faster and better understanding of the transformations and effects. Visualization can create new virtual realities which make the effects part of our experience.

One aspect of the Theory of Special Relativity is the rendering of objects moving with relativistic velocities, i.e., velocities near the speed of light. The appearance of fast moving objects was already discussed in the beginning of this century with the first formulation of the theory. However, the early descriptions have been wrong, e.g. Einstein's statements in [4]. Even after the theory had been accepted the problem was not treated correctly, aside from a generally ignored article about the invisibility of the Lorentz contraction by Lampa in 1924. In 1959 Penrose [11] and Terrel [17] recognized the problem for the first time and gave correct solutions. Detailed descriptions were given later by Boas 1961 in [2], Scott and Viner 1965 in [15], and Scott and van Driel 1970 in [14].

We describe the relativistic transformations necessary for the computation of the correct appearance of relativistic objects and restrict our considerations to the correct display of the objects' shapes. Other effects, such as the Doppler effect, will not be considered but can be integrated as well.

The rendering of the objects can be considered as a generalization of already known rendering techniques used in computer graphics to the special relativistic situation. Mainly two approaches exist to obtain relativistic images: ray tracing and polygon rendering. We investigate these approaches with respect to efficiency and image quality.

Most of the literature does not treat the accelerated case or the authors suggest a method which is not suitable for real-time rendering on a workstation (cf. [6]). In this paper we show how the problem can efficiently be reformulated for the situation with acceleration and how a virtual reality environment can be built.

We proceed as follows: In the next section we give a brief introduction to the Theory of Special Relativity. Here we restrict ourselves to the situation with constant relative velocity. In Section 3 we investigate relativistic rendering as described in the literature. The reformulation in the accelerated situation in Section 4 allows the implementation of a virtual environment, which is described in Section 5. In the last section we discuss the results and the future work.

2 Special Relativistic Transformation

The Theory of Special Relativity usually considers inertial systems moving with constant relative velocity. However, accelerations can be treated within the same framework (cf. [8]).

For the moment, let us restrict ourselves to constant relative velocity. Figure 1 shows the situation for two given coordinate systems K and K', where K' is separating from K at a constant velocity v. On the assumption

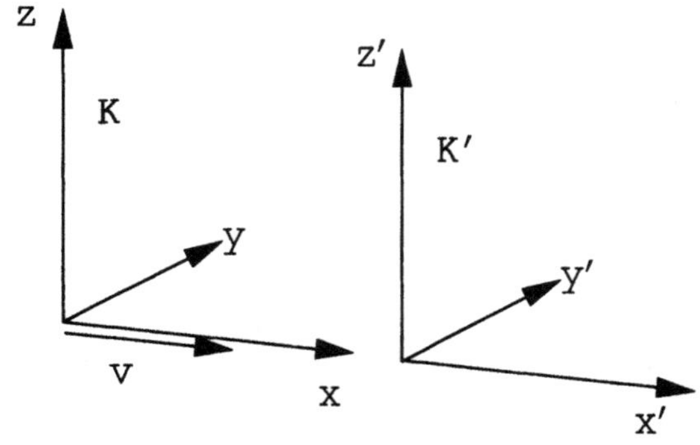

Fig. 1. Two coordinate frames moving with constant relative velocity.

that space is homogeneous and isotropic we can choose a special coordinate system without loss of generality. At time $t = t' = 0$ the origins of the coordinate systems coincide, i.e., $x = x' = 0$, $y = y' = 0$, and $z = z' = 0$. In addition, the axes x and x' are parallel to the relative velocity v, the x-y-plane coincides with the x'-y'-plane, and the x-z-plane coincides with the x'-z'-plane.

In classical mechanics the coordinate transformations are given by

$$x' = x + vt, \quad y' = y, \quad z' = z, \quad t' = t, \tag{1}$$

the so-called Galilean transformations. The time t remains invariant under these transformations and is in this sense "absolute". The laws of motion remain invariant as well.

The principle of special relativity states that the laws of nature are invariant under a particular group of transformations, the so-called Lorentz transformations. These transformations are given by

$$x' = \gamma(x + vt), \quad y' = y, \quad z' = z, \quad t' = \gamma(t + \frac{v}{c^2}x) \tag{2}$$

$$x = \gamma(x' - vt'), \quad y = y', \quad z = z', \quad t = \gamma(t' - \frac{v}{c^2}x'), \tag{3}$$

where $\gamma = \frac{1}{\sqrt{1-\beta^2}}$, $\beta = \frac{v}{c}$, and c denotes the velocity of light.

For a derivation of the Lorentz transformations from the Einstein Postulates see [13]. In contrast to the Galilean transformation the Lorentz transformation leaves the speed of light and Maxwell's equations invariant. For a more detailed presentation of the theory we refer, e.g., to [13] and [8].

3 Special Relativistic Rendering

The process of creating images from three-dimensional models is called rendering. Special relativistic rendering means the process of generating images of fast moving objects or image generation with a fast moving camera according to the equations given in the previous section. In addition, it means the visualization of relations given by these equations in order to obtain a better understanding of the image generation process.

Since the rendering of a static or slowly moving object can be considered as a special case in the relativistic situation, relativistic rendering should be a generalization of techniques already known in the field of computer graphics.

Many renderers for solid three-dimensional objects are based on a polygonal representation of the 3D objects and their surfaces. In many cases this representation is an approximation of the real object by a triangle mesh. In the rendering pipeline the vertices of the triangles are projected onto the image plane. The method is very fast and is supported by modern graphics hardware. For a detailed description of the pipeline we refer to [5].

Another well-known rendering technique is ray tracing. Here, for each pixel a ray is traced from the viewpoint into the scene. The rays are usually assumed to be infinitely thin and infinitely fast, and scattering is left out of consideration. The disadvantage of this technique are high computational costs.

In the relativistic situation Hsiung and Dunn (cf. [7]) formulated relativistic ray tracing. Gekelman et al. described the polygon rendering of a relativistically moving cube (cf. [6]). They also discussed the accelerated situation but their solution was not performable in real-time. In the most recent publication of Chang et al. [3] polygon rendering was used, but no acceleration was considered. Let us describe the two approaches in more detail.

3.1 Polygon Rendering

Let us first consider the situation with a single point light-source, which is at rest in K'. For the explanation of the effects we use spacetime coordinates (ct, x, y, z). In this framework the correct visualization can be derived by purely geometric operations. This geometric interpretation will allow us to treat the accelerated observer (cf. Section 4).

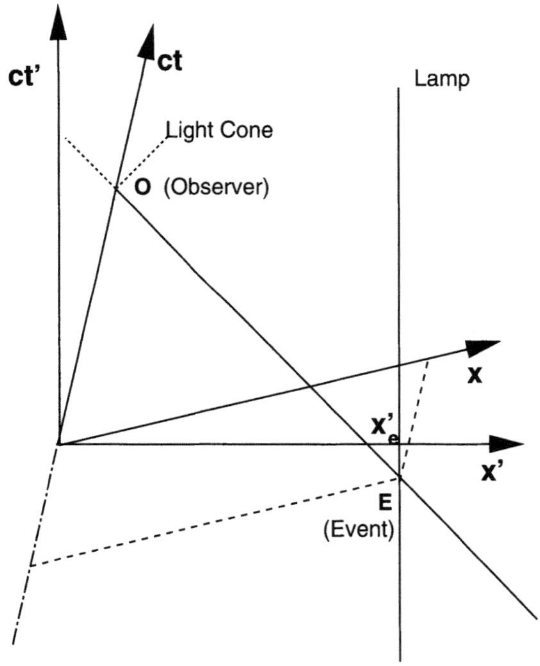

Fig. 2. Minkowski diagram with constant relative velocity.

In Figure 2 we show the Minkowski diagram, which is a spacetime diagram without the coordinates y and z. We use two coordinate systems. In the frame K' the object is at rest whereas the frame K is the observer's rest frame. Let (ct'_o, x'_o) denote the coordinates of the observer in K'. The light propagates on light cones. The light cone at the observer is outlined by the dotted lines. The line $\{(ct', x'_e)|t'\}$ denotes the world line of the point light-source. The intersection of the backward light cone with the world line of the light source is denoted by E, which is the emission position of the light reaching the observer at point O. Once this position is determined, we only have to compute the coordinates of E with respect to the frame K. In the diagram this can be done graphically as shown by the dashed lines, which corresponds to the Lorentz transformation from K' to K.

In four dimensions the coordinates of E can be computed by

$$(ct'_o - ct'_e) = \sqrt{(x'_e - x'_o)^2 + (y'_e - y'_o)^2 + (z'_e - z'_o)^2},$$

where $(ct'_e, x'_e, y'_e, z'_e)$ denote the coordinates of E and $(ct'_o, x'_o, y'_o, z'_o)$ the coordinates of O in K'. With the Lorentz transformation the coordinates of the emission event in K can then be determined. The space coordinates (x_e, y_e, z_e) determine the direction of the incoming light in the image plane and the emission time is not relevant. Light travels along straight lines in 4D space relative to every coordinate system and so does it with respect to a restriction to the three space coordinates. Therefore, standard computer graphics processing can be used for the correct projection onto the image plane.

In the polygonal representation the vertices hold the information such as color, material properties, surface normal etc., and, therefore, can be considered — after the evaluation of a suitable lighting model — as single light sources. We can then apply the transformation described above to all vertices and obtain a completely new object which approximates the emission positions of the objects surface in the coordinate frame of the camera. With Gekelman et al. [6] we call this virtual 3D object *photosurface.* This new object can be projected through a normal 3D projection with correct hidden surface elimination.

In addition to realistic rendering, this method offers another visualization of relativistic effects. The three-dimensional photosurface can be viewed from arbitrary view-points. To understand how the relativistic image was produced viewing the shape of the photo object from different view positions is helpful.

Figure 3 and 4 show the photosurface of a sphere and a cube. The observer is marked as a small sphere. The idea to transform only the vertices of a polygonal representation of a 3D object and to render the resulting object as shaded polygons is quite simple and seems to be very efficient. But due to the highly nonlinear transformation in the proximity of the observer for high velocities, new problems appear with the polygonalization.

The artifacts, which appear mainly at the boundaries, can be reduced by a fine triangulation of the objects. Although the fine triangulation of the moving object can slow down the rendering process the method remains still very fast for velocities up to $\beta = 0.99$.

Fig. 3. Photosurfaces of a sphere with $\beta = 0.95$ at different time steps. The observer is marked as a small sphere. The view is perpendicular to the direction of motion.

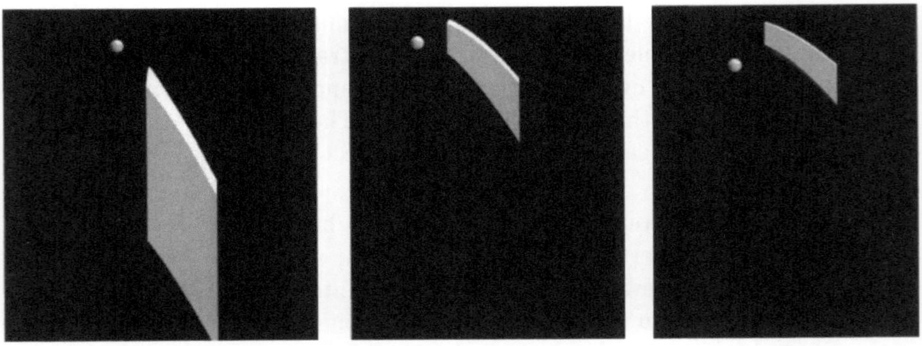

Fig. 4. Photosurfaces of a cube with $\beta = 0.9$ at different time steps. The observer is marked as a small sphere. The view is perpendicular to the direction of motion.

3.2 Ray Tracing

Ray tracing in computer graphics is normally performed in three-dimensional space and one may think of relativistic ray tracing as ray tracing in four-dimensional spacetime. This might be true for a general approach, but in many situations, e.g., the computation of the shape of the object, we can perform the task with "ordinary" ray tracing in three dimensions. This approach was proposed by Hsiung and Dunn in [7].

To explain this method we consider an eye-ray which is traced from the viewpoint through a pixel into the scene. The viewpoint, time, and direction of the ray are given in the coordinate system of the observer. This ray can be described by a four-dimensional start position (ct_0, x_0, y_0, z_0) and a direction $\boldsymbol{d} = (d_x, d_y, d_z)$ obtained from the camera parameters.

The transformation of the point (ct_0, x_0, y_0, z_0) is given by the Lorentz transformation described in Section 2 and results in a point $(ct'_0, x'_0, y'_0, z'_0)$. The direction $\boldsymbol{d}$ is transformed like a velocity and the transformed direction is obtained by

$$\boldsymbol{d}' = (\gamma(d_x + \beta), d_y, d_z)/\gamma(1 + \beta d_x). \tag{4}$$

For a fixed time t_0 we compute the direction for each pixel according to the viewpoint. The pixel coordinates in the observer frame and the time t_0 determine the start position of the ray. We transform the ray according to Equation (4) and send this ray through the scene performing classical ray tracing.

Since the transformation of the ray by Equation (4) is given analytically, a straight line which is bent due to the transformation, is displayed as a bent line in the raytraced image. Therefore, the fine triangulation or, in general, the subdivision of patches and lines is not necessary with ray tracing.

One disadvantage of ray tracing is the fact that the transformation given by Equation (4) is not very intuitive and the resulting image is frequently surprising. The method is well-suited for the generation of high quality images and less suited for educational purposes.

4 Virtual Reality for Relativistic Flights

User interaction being part of a virtual reality requires the consideration of the accelerated observer.

Again, as in the previous section, we use a Minkowski diagram for a fixed object coordinate system to reduce the visualization problem to a geometric one. Figure 5 shows the accelerated situation. The world line of the observer is no longer a straight line. For a given observer position O in K' the current relative velocity can be computed by the tangent to the world line. Once this velocity is known we can proceed as in the case of constant velocity. We obtain the coordinates of the emission event by computing the intersection of the backward light cone with the straight world line of the object. After a coordinate transformation into the current frame K of the observer we can again project the space coordinates onto the image plane.

The computation of the coordinates of the observer's world line can be done successively. Since the world lines of the light sources are not effected by the observer, we are able to render accelerated scenes with almost no delay.

4.1 Transformation Details

For a correct implementation of the accelerated observer it is necessary to parameterize its world line by the so-called *proper time*, which is defined as the time measured by a co-moving clock.

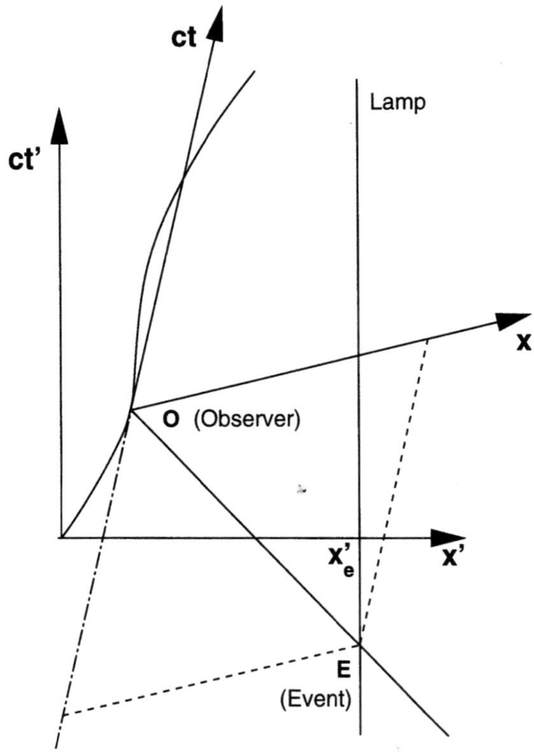

Fig. 5. Minkowski diagram with accelerated observer.

In order to show how this parameterization can be obtained we have to introduce the notion of spacetime and four-vectors and refer to, e.g., [8,9,12] for a detailed presentation. The time coordinate t and the three spatial coordinates (x, y, z) describe a point in spacetime and can be combined to form the position four-vector

$$x^\mu = (ct, x, y, z) = (x^0, x^1, x^2, x^3) \quad , \quad \mu = 0, 1, 2, 3.$$

In this framework a Lorentz transformation is a change of coordinate systems. A *four-vector* is defined as a quantity which has four components (b^0, b^1, b^2, b^3) relative to every coordinate system and which are transformed in the same way as the coordinates (x^0, x^1, x^2, x^3) (cf. [9]).

The *proper time* τ is a Lorentz scalar, i.e., it is independent of the reference frame. The differential proper time is given by

$$d\tau = \sqrt{1 - \beta^2}\, dt = \frac{dt}{\gamma},$$

where β and γ are defined by

$$\beta = \frac{v}{c} \quad , \qquad \gamma = \frac{1}{\sqrt{1-\beta^2}}.$$

Classical quantities such as velocity and acceleration can be extended to corresponding four-vectors.

The four-velocity is defined by

$$u^\mu = \frac{dx^\mu}{d\tau} \tag{5}$$

and the components of the four-velocity are obtained by

$$u^0 = \gamma\, c \quad , \quad u^1 = \gamma\, v_x \quad , \quad u^2 = \gamma\, v_y \quad , \quad u^3 = \gamma\, v_z .$$

The four-acceleration is given by

$$a^\mu = \frac{du^\mu}{d\tau} = \frac{d^2 x^\mu}{d\tau^2}. \tag{6}$$

The effect of the user interaction on the path of the observer, i.e., the observer's world line, is computed according to the equations for the four-velocity and four-acceleration. The user interaction determines the acceleration in the observer frame. From a given three-acceleration (a_x, a_y, a_z) we obtain a four-acceleration

$$a^\mu = (0, a_x, a_y, a_z)$$

in the observer frame, which is Lorentz transformed into the frame K' (see Equation (7)) and yields a coupled system of ordinary differential equations according to Equations (6). The system is numerically solved for the following time step using Euler's method. This way, we obtain the path of the observer in spacetime parameterized by the proper time.

The transformation of the photon emission event from the objects frame K' to the observer's frame K is performed according to the description in the previous section. This transformation can be divided into three parts.

- Translation of the origin of the K' frame to the current position of the observer.
- Lorentz transformation without rotation (Lorentz boost).
- Rotation according to the direction of motion.

The general Lorentz boost (cf. [8, p.69]) is given by

$$\begin{pmatrix} \gamma & -\beta\,\gamma\, n_x & -\beta\,\gamma\, n_y & -\beta\,\gamma\, n_z \\ -\beta\,\gamma\, n_x & (\gamma-1)\, {n_x}^2+1 & (\gamma-1)\, n_x\, n_y & (\gamma-1)\, n_x\, n_z \\ -\beta\,\gamma\, n_y & (\gamma-1)\, n_x\, n_y & (\gamma-1)\, {n_y}^2+1 & (\gamma-1)\, n_y\, n_z \\ -\beta\,\gamma\, n_z & (\gamma-1)\, n_x\, n_z & (\gamma-1)\, n_y\, n_z & (\gamma-1)\, {n_z}^2+1 \end{pmatrix} \tag{7}$$

where $\boldsymbol{n} = (n_x, n_y, n_z)$ is the normalized direction of motion and γ and β denote the velocity parameters defined above.

5 Description of the System

Our user interactive environment is approximately an airplane environment. We provide controls to accelerate and to decelerate in the current flight direction. Moreover, the user can change the direction by accelerating to the left or right side. Finally, a rotation around the current direction of motion is possible. We use keyboard, space-ball, and space-mouse as input devices.

We use textures, which are extremely useful for visualizing the apparent distortion of large scale objects due to relativistic effects. Our implementation accepts arbitrary scenes and we support VRML 1.0 file format (cf. [1]). In a preprocessing step we perform the fine triangulation of the surface patches. The meshing can be controlled interactively and individually for each surface in a radiosity program called RadioLab (cf. [16]). The rendering supports level-of-detail and is based on OpenInventor and OpenGL (cf. [18] and [10]).

An impression of the system can be obtained from color plates 33-33, p. 384. In Figure 33 the test scene with some geometric primitives is at rest. Besides the objects the parameters of the camera can be seen. In color plate 33-33, p. 384, images are displayed which have been taken during a flight with acceleration. The flight direction coincides with the viewing direction. Concerning the acceleration the most interesting effect is aberration of the incoming photons (cf. [8, p.68]). Some objects seem to move away from the observer although the movement is towards the objects.

This test scene with about 9,000 triangles was investigated on a SGI O2, R10000, workstation. With our current implementation we obtain a performance of about 25 frames per second for the non-relativistic and about 20 frames per second for the relativistic movement.

6 Conclusions and Further Work

According to the Theory of Special Relativity we described the transformation necessary for rendering. We described the two classical rendering techniques which are able to generate realistic images of fast moving objects. Polygon rendering is able to generate images very quickly and the image quality for lower velocities is sufficient for real-time applications. The ray tracing technique is much slower, but since the transformation is analytic, artifacts at the silhouettes do not appear.

Through the consideration of the spacetime the rendering equations can be obtained from intersection calculations. Our approach can be extended to the accelerated situation and allows rendering in real-time. Therefore, it was possible to build a virtual relativistic reality. The implemented system reveals surprising effects even for a user familiar with the theory.

We will integrate the Doppler effect and more realistic illumination into our future work. Finally, we plan to provide a freely distributed version of our environment based on OpenGL.

Acknowledgments

We would like to thank the referees for their constructive and detailed suggestions.

This work was supported by the German Science Foundation (DFG) and is part of the projects D1 and D4 within the Sonderforschungsbereich 382.

References

1. G. BELL, A. PARISI, AND M. PESCE, *VRML: Version 1.0 specification*, http://webspace.sgi.com/Archive/Spec1.0/index.html, 1995.
2. M. BOAS, *Apparent shape of large objects at relativistic speeds*, Amer. J. Phys. **29** (1961), 283.
3. M.-C. CHANG, F. LAI, AND W.-C. CHEN, *Image shading taking into account relativistic effects*, ACM Trans. Graphics **15** (1996), 265–300.
4. A. EINSTEIN, *Zur Elektrodynamik bewegter Körper*, Ann. d. Phys. **17** (1905), 891.
5. J. D. FOLEY, A. VAN DAM, S. K. FEINER, AND J. F. HUGHES, *Fundamentals of interactive computer graphics*, second ed., Addison-Wesley Publishing Company, 1990.
6. W. GEKELMAN, J. MAGGS, AND L. XU, *Real-time relativity*, Computers in Physics (1991), 372–385.
7. P.-K. HSIUNG AND R. H. P. DUNN, *Visualizing relativistic effects in spacetime*, Proceedings of Supercomputing '89 Conference, 1989, pp. 597–606.
8. C. MISNER, K. THORNE, AND J. WHEELER, *Gravitation*, Freeman and Company, 1973.
9. C. MØLLER, *The Theory of Relativity*, second ed., Clarendon Press, 1972.
10. J. NEIDER, *OpenGL programming guide*, Addison-Wesley, 1993.
11. R. PENROSE, *The apparent shape of a relativistically moving sphere*, Proc. Cambr. Phil. Soc. **55** (1959), 137.
12. W. RINDLER, *Introduction to Special Relativity*, second ed., Clarendon Press, Oxford, 1991.
13. H. RUDER AND M. RUDER, *Die Spezielle Relativitätstheorie*, Vieweg, 1993.
14. G. SCOTT AND M. DRIEL, *Geometrical appearances at relativistic speeds*, Amer. J. Phys. **38** (1970), 971.
15. G. SCOTT AND R. VINER, *The geometrical appearance of large objects moving at relativistic speeds*, Amer. J. Phys. **33** (1965), 534.
16. R. SONNTAG, *RadioLab: An object oriented global illumination system*, Dissertation, Universität Tübingen, Tübingen, Germany, 1997, Forthcoming.
17. J. TERREL, *Invisibility of the Lorentz contraction*, Phys. Rev. **116** (1959), 1041.
18. J. WERNECKE, *The Inventor Mentor*, Addison-Wesley, 1995.

Exploring Low Dimensional Objects in High Dimensional Spaces

Dennis Roseman

Department of Mathematics, University of Iowa, Iowa City, Iowa, USA

Abstract. We discuss general principles and a software implementation for visualizing low dimensional objects in high dimensional spaces. By high dimensional space we mean euclidian space of dimension greater than four. The low dimensional objects are modeled, mathematically, by simplicial complexes of dimension 4 or less. A particular software visualization project, named Hew, is discussed. An example of an embedded three dimensional projective space is featured in the figures.

1 Introduction

Computers can help visualize and mathematically explore low dimensional geometric objects in high dimensional spaces. Previous visualization efforts have been focused on the problem of 2-dimensional surfaces in 4-dimensional space [25,26,20]. These techniques and the associated software developed provide a starting point for these higher dimensional studies.

The objects studied are finite simplicial complexes of dimension 4 or less. These are piecewise linearly embedded in n-space R^n, also called the *ambient space*, where $4 < n$. (For mathematics of simplicial complexes, see [9], or most texts in algebraic topology.) To simplify discussion, we focus on the important special case where our objects are manifolds. The software we discuss can handle complexes. Modifications of our discussion from manifolds to complexes is straightforward. We will use the term "surface" to mean 2-dimensional manifold.

We have written a visualization tool called Hew. We describe Hew and discuss some general visualization strategies which utilize Hew.

2 Some Terminology for High Dimensional Viewing

For our viewing, the three dimensional space of computer graphics corresponds to some chosen 3-dimensional coordinate subspace of R^n we call the *viewing space*. Any of these three coordinates will be called a *viewing coordinate* or *v-coordinate*. The rest of the $n-3$ dimensions will be "unseen" and will be referred to as the *u-coordinates*. The three v-coordinates together with the i-th u-coordinate, determine a 4-dimensional subspace of R^n called *the i-th u-v-4-space*. The current version of Hew is implemented in terms of 4-dimensional viewing of such subspaces. Hew can examine higher dimensional spaces by examining an array of u-v-4-spaces.

(Equivalently, one might prefer to think of the three dimensions of computer graphics as first three coordinates of R^n and one rotates objects in R^n so that a given choice of 4 coordinates for the i-th u-v-4-space are sent to the first 4 coordinates.)

3 Four Sample Problems

For the purposes of focusing discussion we will consider four problems, concentrating on the first and third of these:

1. Visualizing a surface, S, in 8-dimensional space.
2. Visualizing a surface, S, in 31-dimensional space.
3. Visualizing a 3-manifold, M, in 5-dimensional space.
4. Visualizing a 4-manifold, N, in 6-dimensional space.

An ambient space dimension of 8 or less would seem sufficient for most problems involving surfaces. The admittedly large dimension of 31 for the ambient space for a surface was suggested to the author as being of a size for a certain problem in theoretical mechanics, and is briefly mentioned for the purpose of discussing the nature of the ensuing visualization problems. (See [11,12], for some examples of 2-dimensional data sets in high dimensions.)

For visualizing M, the primary problem is the 3-dimensionality of M. Dimension 5 is chosen for two mathematical reasons. There is some interest in 3-manifolds in R^4, such as [3], however only very simple 3-manifolds (such as the 3-dimensional sphere) embed in R^4. On the other hand, all 3-dimensional (closed) manifolds embed in R^5 [8,16,21]. Also, in dimension 5 one has a knotting phenomenon [17,10]. One goal of our visualization is to study this higher dimensional knotting.

Embeddings of 4-manifolds, in 6-dimensional space also are of interest since there is a rich knot theory in this situation. Since knotted 3-manifolds in R^5 have not yet been computationally explored, general investigation of this topic is premature. However, one special case merits note. The study of knotted 3-manifolds in R^5, is the study of isotopy classes that manifold in R^5. An isotopy, F_t, of M in R^5 (where $t \in I$, the unit interval) gives an embedding, called the *trace of the isotopy*, of the 4-manifold,$M \times I$, into 6-dimensional space, by the map $T(x,t) = F_t(x) \times t$. Hew is designed, in part, to investigate such an isotopy.

4 Hew

Hew is an interactive viewer for high dimensions. It is currently implemented as viewer of arrays of objects, in 4-space, each of which is a 2-dimensional. A basic method for Hew is slicing of surfaces in 4-space by hyperplanes, then viewing the result, see Section 4.2. In Section 4 we discuss this 4-dimensional viewing. In Section 5 we discuss methods for reducing higher dimensional problems to surfaces in 4-space, which can then be viewed in Hew.

4.1 Hew, the Basics

The basic structure of Hew is an array, H^{ij}, where each entry of the array is a surface (or a collection of surfaces) in (some) 4-dimensional space. At any time, a user selects one of these entries for interactive viewing. The viewing space is first three coordinates, denoted here as x, y, and z and the fourth coordinate will be the unseen coordinate, denoted u. The user will see the projection of H^{ij} into the x-y-z viewing space. The u-coordinate values are indicated by a color coding (Section 4.3). The user has access to real time 4-dimensional manipulations, 4-dimensional rotations and 4-dimensional slicing (Section 4.2), as well as the standard three dimensional manipulations of computer graphics. Additional features include use of marking (Section 4.4) and buoys (Section 4.5).

4.2 Real Time Slicing by Hew

A feature of Hew is its real time slicing of surfaces in 4-dimensional space. The user controls slice parameters but slicing is continually updated. In addition, the user can perform 4-dimensional rotations. Such rotations and slicing apply to each object of each H^{ij}.

At all times Hew slices objects using two (3-dimensional) hyperplanes parallel to the viewing space. The user selects two numbers: a "center" value, c and "thickness" value, ε. Letting $a = c - \varepsilon$ and $b = c + \varepsilon$, the two slicing hyperplanes have equations $u = a$ and $u = b$.

Thus each object of H^{ij} is divided into three pieces:

1. the *low zone*, consists of those points of H^{ij} whose u-coordinate is less than or equal to a.
2. the *mid zone*, or $[a, b]$-*slab*, consists of those points whose u-coordinate satisfies $a \leq u \leq b$.
3. the *high zone* consists of those points of H^{ij} whose u-coordinate is greater than or equal to b.

At any time, the user can view any combination of zones of any of the objects in view.

By using 4-dimensional rotations, one can, in effect perform slicing by any pair of parallel hyperplanes in 4-space. For example if one rotates the object by a 4-dimensional rotation of 90° in the x-u-plane, Hew's slicing then, in effect, slices using the x-coordinate.

4.3 Color Coding of a Fourth Dimension

Hew renders a point on a surface using color to indicate the value of the u-coordinate. This technique, we refer to as *color coding* has been used by all who work with surfaces in 4-space, for example [1,2,28,30]. Since slicing divides viewed objects into three zones, it is natural to have color codings that show this. Hew gives the user a variety of color schemes to choose from.

The color scheme for an object in Hew consists of 9 distinct colors, three for each zone, but it is not required to use all three. Typically bright colors are used for the slab zone, coordinated pastel colors for the high zones, and darker colors for the low zones.

Since one can view a number of objects at the same time, Hew provides a collection of such color schemes so that distinct objects can be viewed in contrasting colors, if desired. At any time the user can change the color scheme of any selected object from a built-in palate of choices.

For examples of this coloring, see, color plate 34, p. 385. In the top row is shown an object (an embedding of a 2-sphere in R^4). On the right only the mid zone, the slab, is shown, on the left all three zones are shown. In the lower row are examples where there are more than one object in view, each with a distinct color coding. In the lower left is seen three slabs of three objects, on the right there are two objects.

4.4 Marking Glazing

Hew has a *marking* mode in which the user can select any facet of a viewed surface and change its color to a contrasting marking color, thus leaving reference marks on the surface. Such marking colors are coordinated with color coding so that if one changes color coding, marked portions change to an appropriate contrasting color.

Glazing is a variation of marking with entirely different visualization uses. In *glazing* the user interactively marks facets of an object which are then rendered in a semi-transparent color (technically one uses α-blending with small values of α). The effect is that the user changes an opaque face into a window, and is able to see features otherwise hidden. This is very useful for investigating complex surface images. Marked faces can interactively be unmarked, thus one can use a glazing/unglazing change to briefly peek behind a face or group of faces.

This is illustrated in the middle row of color plate 34, p. 385. On the right, near the middle of the image, faces have been made semi-transparent, and reveal hidden detail. One could also see this detail by other means such as "flying through" the portion now glazed and getting a closer view. However, this glazing method helps to better relate this revealed detail to the whole of the object.

It is best not to use complete transparency (i.e. $\alpha = 0$) for a variety of visual reasons such as maintaining the illusion of three dimensions using (partial) occlusion. Also, with semi-transparent color, a user can distinguish edges which are created by slicing from those which border glazed areas.

4.5 Marking with Buoys

A *buoy* is a simple 2-dimensional object in R^4 that can be placed at a location in the space R^4, and once placed will appear at that same location when any other entry of the array is viewed. Buoys are subject to any of the rotations performed. The marking of Section 4.4, marks portions of objects viewed, a buoy, on the other hand marks a particular point in R^4.

Consider the case of viewing a 3-manifold where H^{ij} and $H^{(i+i)j}$ are views of close by slices M_i and M_{i+1}. As data sets (vertices and triangles) these objects are not easy to correlate, and may even be topologically distinct. A buoy can help keep visual track of changes as one views a sequence of slices.

In color plate 34, p. 385, in the lower right, we see two styles of buoys. The green one is simply a union of squares indicating the 2-dimensional coordinate subspaces of R^4. The second buoy replaces these squares by a frame. One might visually distinguish points using different styles of buoys. More elaborate buoys (not shown), have distinct less symetric shapes allowing a visual reminder of the current rotation in Hew.

Most often one wants to, interactively, place a buoy in a particular position relative to a viewed object. Color plate 35, p. 386, the top row shows annother use of buoys. In this case buoys have been placed at the corners of a unit four-dimensional cube. The objects viewed are two consecutive slices of a 3-manifold. This configuration can be useful in comparing views of different rotations.

4.6 History of Hew

Here is a short history of software developments leading up to the current Hew. The author's contribution has been largely in the development and design.

The first step was 4dview, a module of Geomview [28], coded largely by Daeron Mayer of the Geometry Center. It allowed slicing, projections and rotations of 2-dimensional objects in R^4. Next, is another module ndView of Geomview, ndView, was written at the Geometry Center which extended slicing, projections and rotations of to 2-dimensional objects in R^n. Although Geomview is a fairly portable program, it was not easy to port it to the CAVE (see [4]) virtual environment.

For these reasons the program, called slice [31], was written, largely coded by Glenn Chapell at NCSA and University of Illinois. The strengths of slice are real-time slicing and ability to run in the CAVE [5].

The Hew was written to combine features of slice and ndView. It was written at University of Iowa, largely coded by Josh Berdine. From its beginnings, Hew was designed to look at 3-dimensional objects (i.e. knotted 3-manifolds in 5-space), and associated isotopies.

5 Getting from High Dimensions Down to Four Dimensions

5.1 Surfaces

Consider a surface, S in R^8. Examination of each coordinate function for S, could be done using only two 4-dimensional views: view the first 4 coordinates and then, separately, the last 4. This has the obvious disadvantage that one has no opportunity to investigate possible relationships which involve these separated coordinates.

Let us assume for the moment that we wish to use the first three coordinates of R^8 as the viewing coordinates. We can then reduce our data to four dimensions by using one of the remaining coordinates for our u-coordinate. We have 5 choices for this u-coordinate. Thus we can obtain 5 objects, $U^{11}, \ldots U^{51}$, where U^{i1} is the projection of S into the i-th u-v-4-space.

However, there may not be any particular reason to prefer the first three coordinates of R^8 as the viewing coordinates. Suppose we want to look all the other possibilities for v-space. Assuming we do not care about the order of the axes (i.e. the orientation of the v-space) there are 56 such choices, which we will order in some way. This results in an array of $5 \times 56 = 280$ choices for viewing S, namely the H^{ij} where H^{ij} is the projection of S into the i-th u-v-4-space using the j-th choice of v-space. Most likely a user would scan all of these choices quickly when first examining a data set, and then narrow the choices to a more manageable number.

In the general case of surfaces in R^n, for a given 3-dimensional viewing space there are $n-3$ u-v-4-spaces to consider giving us a sequence of length $n-3$ of surfaces in a 4-dimensional space. The number of choices for v-space is $\binom{n}{3} = \frac{n!}{(n-3)!3!}$. For surfaces in 31 dimensions, there are 4495 v-space, and the array we might want to look at has $28 \times 4495 = 125,860$ entries! Of course this large number is a "worst case"—most likely particulars of a data set would help narrow these choices.

The above methods allow viewing of u-coordinate functions, one at a time. A different array of objects can be generated if one wishes to examine pairs of u-coordinates. One generates an array, H^{ij}, of size $(n-3)(n-4)$ where each each entry consists of two objects: the projection of S into the i-th u-v-4-space and the projection of S into the j-th u-v-4-space where $i \neq j$.

5.2 Three-dimensional Manifolds

In this section we will notate the coordinates of R^5 by x, y, z, u, and w. For a 3-manifold, M in 5-space, we slice M by a family of N hyperplanes, F_i given by $w = c_i$, where $\{c_i\}_{i=1}^{i=N}$ are appropriately chosen. Since each F_i is 4-dimensional, we can express M as a sequence of surfaces in a 4-dimensional space as follows.

Let $M_i = M \cap F_i$, and let U^{i1} denote the projection into the first four coordinates of M_i. (These will be generic slices as defined in Section 6). We may now view M as the sequence $U^{11}, \ldots U^{N1}$, with the viewing space taken to be the first three coordinates.

Alternatively, one might wish do our 5-dimensional slicing in any of the other coordinates, thus giving us an array of size $5N$ where H^{ij} is i-th 5-dimensional slicing along the j-th coordinate of R^5. There are 10 choices for viewing space for a 3-manifold in R^5. If we wish to examine all of these, one could generate and different sort of array of size $10N$ where H^{ij} is M_i Note: since viewing of these surfaces in Hew also uses slicing, we are "slicing the slices", a technique proposed for 3-spheres in R^5 in [10].

Of course one should not be restricted to only five types of 5-dimensional slicing. But certainly this is a minimum that one should consider, and even such limited investigation would involve considerable time on the part of the user.

6 Mathematics and Slicing

The various slicings that we have described involves computations that produce new objects. Besides their visualization utility, these have mathematical significance.

Morse theory deals with the study of non-degenerate critical points for smooth real-valued functions defined on manifolds [13]. Computationally, our objects are simplicial complexes and our coordinate functions are piecewise linear. At a vertex which has many edges, a real valued function might not be modeled by a quadratic, (as in the Morse Lemma in the smooth case). Nevertheless, we adopt the language and strategy of Morse theory.

If a slicing produces a manifold, we call it a *non-singular slice* or *generic slice*. Otherwise, the slicing is a *singular slice* and the non-manifold points are called *slice singularities*.

In practical terms, in most calculations, we expect that in all but a finite number of "bad" slicing directions, all but a finite number of slices will be generic. Furthermore in the singular slices, the singularities will consist of a finite number of points.

The manifolds obtained as non-singular slices are of potential interest in two ways: as submanifolds of the manifold being sliced and as submanifolds of the slicing hyperplane.

Singular slices are of interest since they determine useful information about the manifolds in much the way that Morse singularities can determine a combinatorial and topological description of the manifold on which they are defined. More than that, many of the "features" we might notice in our viewing correspond to such singularities.

6.1 Non-singular Slices

For any of the objects we view, the slabs produced by the 4-dimensional slicing of Hew, if made very thin, correspond to level curves of coordinate functions for that object. The topological changes that appear, as the slice midpoint is changed, correspond to singularities of this function. Also, these curves are subsets of the 3-dimensional slicing plane and correspond to slice knots and slice links, (see [6] for definitions). Color plate 34, p. 385, upper right shows a knot which is a slice knot.

When we slice a 3-dimensional manifold M, we generate a family of surfaces in M. A key to the topological study of 3-manifolds is the study of such subsurfaces. Our 5-dimensional slicing is an extrinsic way of generating such families. In addition, these surfaces, considered as subsets of the 4-dimensional slicing hyperplane, give examples of knotted surfaces, the focus of most of our recent research.

The study of knotted surfaces in R^4 is one of some interest. Our 5-dimensional slicing for 3-manifolds in R^5 provides a way of generating and viewing such examples. Techniques of [17] allow one to construct knottings of 3-manifolds in R^5. One of our motivations is to investigate these knottings.

6.2 Singular Slices for 3-Manifolds and Their Singularities

For surfaces, Morse type singularities are well known. They correspond to local maxima, local minima and saddle points. Here are the corresponding singularities in the next dimension.

In the case M is an embedded 3-dimensional manifold in R^n we are viewing M as a sequence of slices in the i-th direction. There are four kinds of Morse type singularities, a local minimum or local maximum, and two kinds of hyperbolic singularities. A local minimum (index 0) or local maximum (index 3)will correspond to the appearance or disappearance of a small sphere. In a hyperbolic singularities, one sees locally a hyperboloid of one sheet changing to a hyperboloid of two sheets (index 1), or the reverse of this (index 2). The hyperboloid of one sheet looks like a thin tube, as we pass through the singular point this "tube pinches off". This is visually fairly easy to spot. In color plate 34, p. 385, middle right, in the detail revealed, one sees such a "pinched" tube produced by a slice near a critical point of index 1.

6.3 A Concrete Example: an Embedding of Three Dimensional Projective Space in Five Space

The simplest closed 3-manifold (beyond the 3-sphere and the product of a surface and a circle) is three dimensional real projective space, P^3, see standard texts in topology such as [14,22]. It is classically defined as the space of all lines through the origin in R^4. It is homeomorphic to the space of rotations of R^3, and to the lens space $L(2,1)$. Another well-known equivalent description of P^3 is that it is the total space of the unit circle bundle of the tangent space of the S^2, the standard 2-sphere. In any manifold, M, a standard construction of a tangent disk bundle is as the regular neighborhood of the diagonal embedding of M in $M \times M$, [7]. The total space of unit tangent sphere bundle is then homeomorphic to the boundary of this regular neighborhood.

Besides the interest in viewing this object, the particular embedding is significant. In terms of the knot theory of 3-manifolds in R^5 it is a candidate for "a standard embedding of P^3 in R^5", or "the unknotted P^3 in R^5".

The embedding of P^3 investigated in color plate 35, p. 386, is obtained as follows: One begins with a copy, Σ of the standard 2-sphere in R^3. In our case we have chose an icosahedron. One then embeds the cartesian product $\Sigma \times \Sigma$ in R^5 analogous to the way one embeds the standard torus (a product of circles) in R^3. The embedding of P^3 shown is then constructed as the boundary of the regular neighborhood of the diagonal.

One problem with our 5-dimensional slicing is that we need to be able to relate one slice to the next. In color plate 34, p. 385, bottom left, we show three consecutive slices of P^3. Here we only show the slabs for these slices. This illustrates how one can get a detailed view of the changes as we vary the 5-dimensional slicing.

The bottom two rows of color plate 35, p. 386, show four slices of our embedding (the order is: middle left, middle right, lower left, lower right). To keep the computation small, we use an icosahedron to model the 2-sphere used in the construction. This accounts for the angularity of the projections as well as the pentagonality of the holes we see in the first and last slice. Slices shown are all torii (unknotted in 4-space) as one would expect from the construction since we are slicing the total space of a circle bundle (or also since we are slicing the lens space $L(2,1)$). The first and last in the seqence are shown from the point of view in the viewing space. The middle two in this squence have been separately rotated in the viewing space to more clearly show that they are torii.

7 Conclusions, Future Developments

There are many ways one can approach visualization in higher dimensions, and we have discussed here only a few simple tools. Our plan is to continue these developments. One thing that needs to be added is extension to add higher dimensional rotations and slicing to Hew. At the moment operations beyond 4-dimensional are done by pre-processing. Also we plan to incorporate use of texture coding as a replacement for color coding, see [27,19].

There are two basic ways to view any manifold–the intrinsic point of view and the extrinsic. In Hew, one is taking an extrinsic view for exploring 3-manifolds. Computer graphics exploration of an intrinsic nature is especially well done in the maniView module for Geomview, written by Charlie Gunn, and has been captured in the videos [23,24] Eventually one would like to, somehow, have the ability to consolidate these two points of view.

References

1. T. BANCHOFF, *Beyond the Third Dimension: Geometry, Computer Graphics and Higher Dimensions*, Scientific American Library (1990).
2. D. C. BANKS, *Interactive Manipulation and Display of Surfaces in Four Dimensions*, Symposium on Interactive 3D Graphics, Association for Computing Machinery (1992).
3. S. BERGMAN, *The Kernel Function and Conformal Mapping*, A.M.S. Math. Surv. **5** (1950).
4. C. CRUZ-NEIRA, D. J. SANDIN, T. A. DEFANTI *Surround-Screen Projection-Based Virtual Reality: The Design and Implementation of the CAVE*, Computer Graphics (Proceedings of SIGGRAPH '93), ACM SIGGRAPH (1993), 135–142.
5. G. K. FRANCIS, R. KUSNER, J. SULLIVAN, D. ROSEMAN, K. BRAKKE, *Laterna matheMagica*, Entry No. 6 in Virtual Environments and Distributed Computing Global Information Infrastructure Testbeds. ACM/IEEE Supercomputing Conference, San Diego, CA, (1995).
6. R. H. FOX, *A quick trip through knot theory*, Topology of 3-manifolds and related topics, Prentice Hall 120-167 (1961).
7. M. HIRSCH, *Differential Topology*, Grad. Texts in Math. No. 33, Springer Verlag, 1976.
8. M. HIRSCH, *Embeddings and compressions of polyhedra and smooth manifolds*, Topology **4** (1966), 361-369.
9. J. F. P. HUDSON, *Piecewise Linear Topology*, Benjamin, 1969.
10. S. J. JR. LOMONOCO, *Five dimensional knot theory*, Low Dimensional Topology, A.M.S. Contemp. Math **20** (1983).
11. J. H. MADDOCKS, R. F. MANNING, R. C. PAFFENROTH, K. A. ROGERS, J. A. WASRNER, *Interactive Computation, Parameter Continuation and Visualization*, preprint (1996).
12. MADDOCKS, J. H., SACHS, R. L. , *Constrained variational principles and stability in Hamiltonian systems*, Hamiltonian Dynamical Systems **63** I.M.A. Math. and Its Appl. pp. 231-264, Springer Verlag, 1995.
13. J. MILNOR, *Morse Theory*, Ann. of Math Study **51**, Princeton Univ. Press, 1963.
14. J. R. MUNKRES, *Elements of Algebraic Topology*, Benjamin/Cummings, 1984.
15. M. PHILLIPS, S. LEVY, AND T. MUNZNER, *Geomview — An Interactive Geometry Viewer*, Notices A.M.S., October 1993.
16. V. A. ROHLIN, *The embedding of non-orientable three-manifolds into five-dimensional Euclidean space*, Dokl. Akad. Nauk. SSSR **160** (1965), 549-551; English translation, Soviet Math. Dokl. **6** (1965), 153-156.
17. D. ROSEMAN, *Spinning knots about submanifolds; spinning knots about projections of knots,* Topology and its Applications **31** (1989), 225–241.
18. D. ROSEMAN, *Motions of Flexible Objects*, Modern Geometric Computing for Visualization (Eds. T.L. Kunii and Y. Shinagawa) Springer-Verlag, 1992, 91–120.
19. D. ROSEMAN, *Wiener's thought on the computer as an aid in visualizing higher-dimensional forms* A.M.S. Proc. of Symp. in Appl. Math. **52** (1997), 441-471.

20. D. ROSEMAN, *What Should a Surface in 4-space Look Like?*, Visualization and Mathematics, Springer Verlag, Edited by Hege and Polthier (1997).
21. C. T. C. WALL, *All 3-manifolds imbed in 5-space*, Bull. Am. Math. Soc. **71** (1965), 564-567.
22. N. STEENROD, *Topology of Fibre Bundles*, Princeton Univ. Press, 1951.

Listing of mathematical videos:

23. STAFF OF GEOMETRY CENTER, *Not Knot*, A.K. Peters, Wellesley MA.
24. STAFF OF GEOMETRY CENTER, *Shape of Space*, Geometry Center, Minneapolis, MN.
25. D. ROSEMAN, (with D. Mayer), *Viewing Knotted Spheres in 4-space*, video (8 mins) Produced at the Geometry Center, (June 1992).
26. D. ROSEMAN, (WITH D. MAYER AND O. HOLT), *Twisting and Turning in 4 Dimensions*, video (19 mins) produced at the Geometry Center, distributed by Great Media, Nicassio CA, (August 1993).
27. D. ROSEMAN, *Unraveling in 4 Dimensions*, video (18 mins) produced at the Geometry Center, distributed by Great Media, Nicassio CA, (July 1994)

Listing of software:

28. STAFF OF GEOMETRY CENTER, *Geomview*, software, Geometry Center, Minneapolis, Minn. Available via anonymous ftp from geom.umn.edu.
29. D. ROSEMAN, J. BERDINE, *Hew*, written University of Iowa (1996)
30. A.J. HANSON, ET AL, *MeshView 4D*, a 4d surface viewer for meshes for SGI machines, available via ftp from the Geometry Center (1994).
31. G. CHAPELL, G. FRANCIS, C. HARTMAN , *slice*, written at NCSA (1995)

Part V

Vector Fields and Flow Visualization

Fast LIC with Piecewise Polynomial Filter Kernels

Hans-Christian Hege and Detlev Stalling

Konrad-Zuse-Zentrum für Informationstechnik Berlin (ZIB), Germany

Abstract. Line integral convolution (LIC) has become a well-known and popular method for visualizing vector fields. The method works by convolving a random input texture along the integral curves of the vector field. In order to accelerate image synthesis significantly, an efficient algorithm has been proposed that utilizes pixel coherence in field line direction. This algorithm, called "fast LIC", originally was restricted to simple box-type filter kernels.

Here we describe a generalization of fast LIC for piecewise polynomial filter kernels. Expanding the filter kernels in terms of truncated power functions allows us to exploit a certain convolution theorem. The convolution integral is expressed as a linear combination of repeated integrals (or repeated sums in the discrete case). Compared to the original algorithm the additional expense for using higher order filter kernels, e.g. of B-spline type, is very low. Such filter kernels produce smoother, less noisier results than a box filter. This is evident from visual investigation, as well as from analysis of pixel correlations. Thus, our method represents a useful extension of the fast LIC algorithm for the creation of high-quality LIC images.

1 Introduction

Line Integral Convolution (LIC), introduced by Cabral and Leedom [5], is a particularly powerful and elegant method for synthesizing directional textures. Such textures are useful in visualization and computer art. The algorithm needs a texture and a vector field as input. The output image is computed by convolving the texture along the integral curves of the vector field. This causes anisotropic correlations of pixel intensities: the values are much higher correlated along individual integral curves than in directions perpendicular to field lines. The resulting image thereby clearly depicts the directional structure of the vector field.

The LIC technique can be used either to visualize a vector field or to impress a directional texture to an arbitrary image. In the first case inputs are the vector field one aims to visualize and a noisy, fairly arbitrary texture. In the second case the inputs are the image one wants to modify and some artificial vector field. This field may be generated, e.g., by taking the (rotated) gradient of the smoothed input image.

Elegance and usefulness of the LIC technique inspired many researchers to create variations and extensions of the original algorithm. Since the original algorithm is computationally rather expensive, a much faster algorithm has been developed which provides almost interactive speed [14]. Due to its significant lower computational complexity we call it "fast LIC". Another feature of this algorithm is that smooth zooms can be produced. This can be utilized to visualize details of vector fields. Furthermore texture animations with constant or spatially varying velocity can be created, in order to portray orientation and strength of the vector field in an intuitive manner. For interactive exploration of large vector fields further acceleration may be necessary. This can be achieved by parallelization. Designs of parallel algorithms for various kinds of computer architectures are presented in [4] and [15].

Another method, derived from LIC, for encoding field direction and orientation has been proposed in [17]. Here a low frequency input texture and a ramp like anisotropic convolution kernel are used. Several algorithmic approaches have been suggested for the generation of LIC images on curved surfaces [8,9,1,16,12]. In ref. [10] it is shown how surface shapes in volume data can be illustrated using principal direction-driven 3D LIC. By integrating the LIC algorithm into direct volume rendering, dye advection – as used in experimental flow visualization – has been simulated [13]. Strategies for effectively portraying 3D flow using volume line integral convolution are discussed in [11]. An overview on the current status of LIC algorithms and applications of LIC is given in [3].

In this paper we generalize the fastLIC algorithm [14] for use of piecewise polynomial kernels. We employ the new algorithm not only in the plane, but create LIC textures also on arbitrary surfaces, extending our work [1]. Furthermore, we investigate whether the use of higher order filters pays visually. This is done by analyzing the differences between LIC images being generated with different kernels. The examinations are supplemented by a statistical analysis of LIC images.

The material of this paper has been organized as follows. After an introduction in Sect. 2, providing the mathematical background of LIC, a suitable class of filter kernels is introduced, as well as a means for fast convolution (Sect. 3). The fast LIC algorithm for polynomial filters is described in Sect. 4. In the last section we provide a statistical analysis of LIC images based on white noise input textures.

2 Line Integral Convolution

Line integral convolution may be performed in flat and curved space of arbitrary dimension. For simplicity we describe the algorithm for the flat 2D case. It will be become obvious that the techniques and results can simply be generalized for higher dimensional and curved spaces. In fact we implemented the algorithm also for vector fields on curved surfaces.

2.1 Integral Curves

We are working on a domain $\Omega \subset \mathbb{R}^2$ with some vector field $\boldsymbol{v}: \Omega \to \mathbb{R}^2$. Let $\boldsymbol{\tau}: [t_0, t_1] \to \mathbb{R}^2$ denote the integral curves[1]

$$\frac{d}{dt}\boldsymbol{\tau}(t) = \boldsymbol{v}(\boldsymbol{\tau}(t)) \tag{1}$$

of the field $\boldsymbol{v}$ for initial conditions $\boldsymbol{\tau}(t_0) = \boldsymbol{x}_0$, $\boldsymbol{x}_0 \in \Omega$. For simplicity we assume that the right hand side locally obeys a Lipschitz-condition such that for all $\boldsymbol{x}_0 \in \Omega$ there is a unique solution. Furthermore, we assume that the field vanishes nowhere in Ω, i.e. that no critical points exist. All integral curves $\boldsymbol{\tau}(t)$ can then be reparametrized by arc-length s, where $s(t) = \int_{t_0}^{t} v(t')\, dt'$. Using s as parameter makes it particularly easy to step along a given field line with equi-distant steps. Both assumptions are taken to simplify the analysis. They are not imposed in our implementation, where special care is taken at critical and discontinuous points of the field $\boldsymbol{v}$. Applying the chain rule and the inverse function theorem, then using Eq. (1) and $ds/dt = v$, we obtain

$$\frac{d\boldsymbol{\tau}(t(s))}{ds} = \frac{d\boldsymbol{\tau}}{dt}\left(\frac{ds}{dt}\right)^{-1} = \frac{\boldsymbol{v}(\boldsymbol{\tau}(t(s)))}{v(\boldsymbol{\tau}(t(s)))}\,.$$

Introducing a new function $\boldsymbol{\sigma}(s) := \boldsymbol{\tau}(t(s))$, we get the alternative definition of integral curves

$$\frac{d}{ds}\boldsymbol{\sigma}(s) = \frac{\boldsymbol{v}(\boldsymbol{\sigma}(s))}{v(\boldsymbol{\sigma}(s))}\,, \tag{2}$$

with initial condition $\boldsymbol{\sigma}(s_0) = \boldsymbol{x}_0$. Note that the right hand side of Eq. (2) is just the normalized vector field.

2.2 Line Convolution

Now, in LIC an input texture T is convolved along the field lines $\boldsymbol{\sigma}$ of a given vector field. The intensity at point $\boldsymbol{x} = \boldsymbol{\sigma}(s)$ is defined as

$$I(\boldsymbol{x}) = \int_{s-L}^{s+L} T(\boldsymbol{\sigma}(s'))\, h(s-s')\, ds'\,, \tag{3}$$

where h is an arbitrary filter, normalized to $\int_{-L}^{L} h(s')\, ds' = 1$. Employing filter kernels h with finite support $supp\ h = [-L, L]$ and using the notation $I_{\boldsymbol{\sigma}}(s) := I(\boldsymbol{\sigma}(s))$, and $T_{\boldsymbol{\sigma}}(s) := T(\boldsymbol{\sigma}(s))$, we may also write

$$I_{\boldsymbol{\sigma}}(s) = \int_{-\infty}^{\infty} T_{\boldsymbol{\sigma}}(s')\, h(s-s')\, ds' = (T_{\boldsymbol{\sigma}} * h)(s)\,, \tag{4}$$

---

[1] We will use the terms 'integral curve', 'field line' and 'stream line' synonymously.

where $*$ means convolution.

In order to evaluate Eq. (4) at a point $\boldsymbol{x} = \boldsymbol{\sigma}(s)$, $\boldsymbol{x} \in \Omega$, we need a curve segment centered at $\boldsymbol{x}$ and extending a length L in both directions. For points $\boldsymbol{x}$ near $\partial\Omega$ such curve segments typically leave the domain Ω. Therefore we pad domain Ω by a sufficiently large boundary region B as shown in Fig. 1 and continue the vector field $\boldsymbol{v}$ arbitrarily but smoothly into region B. This allows us to calculate integral curves of sufficient lengths for all points $\boldsymbol{x} \in \Omega$. The input texture is also defined on the extended domain $\bar{\Omega} = \Omega \cup B$, in order to perform convolutions along these line segments. The output image lives on Ω.

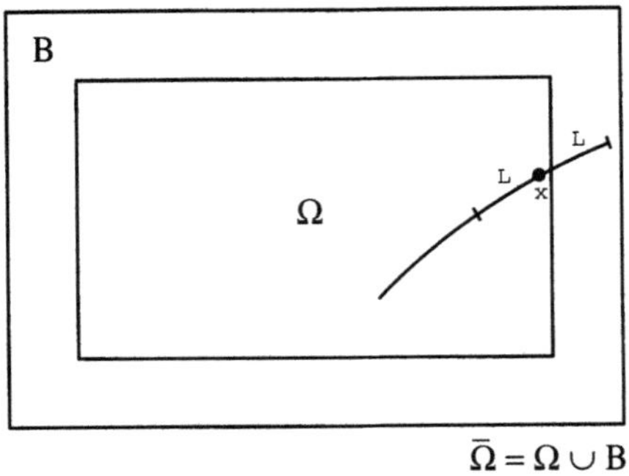

Fig. 1. Enlarged domain $\bar{\Omega} = \Omega \cup B$, used for calculation of convolution integrals for all points $x \in \Omega$.

2.3 Raster Images

The domains $\bar{\Omega}$ and Ω are partitioned into sets $\{\bar{\omega}_i\}$ and $\{\omega_j\}$ of rectangular pixels. The input texture T is a predefined raster image or is given procedurally. In either case it is a piecewise constant function $T : \bar{\Omega} \to \mathbb{R}$, assigning a texture value $T_{\bar{\omega}_i}$ to each pixel $\bar{\omega}_i$. The output is also a raster image[2], i.e. a piecewise constant function $O : \Omega \to \mathbb{N}_0$, assigning an integer grey value O_{ω_i} to each pixel $\omega_i \subset \Omega$.

First, let us focus on the computation of an intensity value I for a particular point $\boldsymbol{x} = \boldsymbol{\sigma}(s)$ on a field line, regarding the fact that T is a raster image. According to Eq. (3) all texture values along the curve segment $\boldsymbol{\sigma}[s-L, s+L]$ contribute to $I(\boldsymbol{x})$. Assume that the curve segment $\boldsymbol{\sigma}[s-L, s+L]$ passes n pixels $\omega_{j_1}, \ldots, \omega_{j_n}$, i.e. consists of n sub-segments $\boldsymbol{\sigma}[s_{k-1}, s_k]$ each crossing one pixel $\bar{\omega}_{j_k}$. Here $s_0 = s - L$ and $s_n = s + L$ are the limits of the convolution integral, whereas the values $s_1 \leq s_2 \leq \ldots \leq s_{n-1}$ correspond to all intersections

[2] Note, that the pixels $\bar{\omega}$ of the input texture contained in Ω may differ in size and location from the pixels ω of the output image.

between the curve segment $\boldsymbol{\sigma}[s-L, s+L]$ and the pixel boundaries[3]. The one-dimensional texture $T(\boldsymbol{\sigma}(s'))$ on interval $[s-L, s+L]$ then is a piecewise constant function with breakpoints $s_1, s_2, ...s_{n-1}$, that is $T(\boldsymbol{\sigma}(s')) = T_{\bar{\omega}_{j_k}}$, for $s' \in [s_{k-1}, s_k)$ and $k = 1, ..., n$. Therefore equation (3) becomes

$$I_{\sigma}(s) = \sum_{k=1}^{n} T_{\bar{\omega}_{j_k}} \int_{s_{k-1}}^{s_k} h(s - s')\, ds' . \tag{5}$$

Given all breakpoints s_k and a specific kernel h, e.g. a polynomial one, the intensity values $I_{\sigma}(s)$ can be easily calculated.

The output image must also be a raster image. Therefore a representative intensity value has to be obtained for each pixel ω_i. Ideally one should compute a spatial average,

$$O_{\omega_i} = \int_{\omega_i} I(\boldsymbol{x})\, d\mu \tag{6}$$

and, for quantization, cast the result value to the next integer. Here $d\mu$ defines a suitable measure with $\int_{\omega_i} d\mu = 1$.

For performance reasons both quantities, $I_{\sigma}(s)$ and O_{ω_i} will be computed only approximately, as shown in the next section.

2.4 Sampling and Aliasing

In a practical implementation the calculation of convolution integrals must be extremely fast, since for each pixel ω of the output image several intensity values have to be calculated. Computing the intersections of the curve segment $\boldsymbol{\sigma}[s-L, s+L]$ with the pixel boundaries and then evaluating Eq. (5) would be computationally too expensive. Trading some accuracy, the intensity $I_{\sigma}(s)$ can instead be approximated by a Riemann sum. Using the notation $T_l = T_{\sigma}(l\Delta s)$ and $h_l = h(l\Delta s)$ we get from Eq. (3) a discrete approximation

$$\hat{I}_{\sigma}(s) = \frac{1}{2m+1} \sum_{k=-m}^{m} T_{\sigma}(s + k\,\Delta s)\, h(-k\,\Delta s) = \frac{1}{2m+1} \sum_{k=-m}^{m} T_{j-k}\, h_k \tag{7}$$

where $m = L/\Delta s$, $j = s/\Delta s$, and $k = (s-s')/\Delta s$. Algorithmically this means that the integrand is sampled pointwise, e.g. starting at point $\boldsymbol{\sigma}(j\Delta s)$ and then stepping along curve $\boldsymbol{\sigma}$ with fixed step size Δs in both directions. Of course, the results computed with approximation (7) in general are correct only in the limit $\Delta s' \to 0$. Comparing the approximate expression with Eq. (5), the error sources for finite Δs are obvious: First, not all texture values $T_{\bar{\omega}_{j_k}}$ with $k = 1, ..., n$ are taken into account, and second, the integrals

[3] For this discussion we may disregard degenerate cases where the intersections are not single points.

in Eq. (5) are approximated only roughly. Rephrased in computer graphics language: using approximation (7), two signals are pointwise sampled, the 1D texture along a curve segment and the terms contributing to the integrals in Eq. (5). Since the distance of two subsequent pixel boundaries s_{k-1} and s_k may be arbitrarily small, the Nyquist frequency of the 1D random texture is unbounded – even if T would be continuous – and aliasing effects are in principle unavoidable. This is moderated by the fact that pixels being missed during sampling typically would receive only small weights according to Eq. (5).

Practical experience shows that using the approximation with a step size Δs of about a third or half a texture pixel width usually delivers visually pleasing results. Hence, this approximation is sufficient for practical purposes.

The pixel averages defined by Eq. (6) are approximated also by discrete sums

$$O_{\omega_i} \approx \frac{1}{n_c} \sum_{j=1}^{n_c} \alpha_j \, \hat{I}\left(\boldsymbol{x}_j\right) \qquad \text{where } \boldsymbol{x}_j \in \omega_i \tag{8}$$

with finite n_c. The sampling locations $\boldsymbol{x}_j$ and weights α_j can be chosen according to some anti-aliasing scheme. As will be discussed later, our LIC algorithm is able to produce multiple samples per pixel in a natural way. Although there is only limited control about the exact locations of these samples within a pixel (or optionally a subpixel), by adjusting a parameter min_{hit} the total number of samples per pixel (or subpixel) is guaranteed to be $n_c \geq min_{hit}$. It turns out that by averaging all samples with equal weights $1/n_c$ results of sufficient high quality are obtained.

A field line $\boldsymbol{\sigma}$ typically hits many pixels ω_i of the output image and therefore can be used to compute intensity values for a multitude of pixels. The fast LIC algorithm [14] avoids the redundancies during intensity calculations for neighbored points of a field line by relating these values mathematically and using such a discretization. For box filters the coefficients h_k are constant and the relation between $\hat{I}_{\boldsymbol{\sigma}}(s)$ and $\hat{I}_{\boldsymbol{\sigma}}(s + l\Delta s)$ is obvious. Therefore, calculating the LIC integral only at those locations which serve as sample points during integration, a fast algorithm can be designed. For non-constant kernels Eq. (7) is still is valid, but it is unfavorable to use it directly for calculating many intensity values along a field line.

In Sec. 3 and 4 and we will discuss how discrete convolutions can be performed very efficiently for certain types of filter kernels and how this procedure can be combined with the fast LIC concept.

3 Convolution Theorem, Filter Kernels

The original fast LIC algorithm utilizes the fact that for box-type filter kernels the convolution integrals are just differences of sums which can be easily updated while stepping along a field line. This fact is a special case of a certain convolution theorem stated below.

Piecewise polynomials provide a rather general function class that allows us to represent a wide variety of kernel shapes. Therefore, we will mainly aim at using this type of functions as filter kernels. We will define piecewise polynomial functions formally, introduce a basis of the corresponding linear space (following de Boor [2]), and then apply the convolution theorem to a linear combinations of these basis functions.

3.1 A Convolution Theorem

The single important theorem which our work is based on is that the convolution of two functions f and h, where h has finite support, is equal to the convolution of the integral F of f and the derivative h' of h, i.e.

$$f*h\,(x) = \int\limits_{-\infty}^{\infty} f(y)\,h(x-y)\,dy = \int\limits_{-\infty}^{\infty} F(y)\,h'(x-y)\,dy = F*h'\,(x) \tag{9}$$

for all $x \in \mathbb{R}$. To show this we consider a definite integral with finite bounds first, apply integration by parts, let the integration bounds move to infinity, and then use the fact that the filter kernel h has finite support:

$$\int_{-z}^{z} F(y)\,h'(x-y)\,dy = F(y)\,h(x-y)\Big|_{y=-z}^{y=z} + \int_{-z}^{z} f(y)\,h(x-y)\,dy \underset{z\to\infty}{\longrightarrow} f*h(x)$$

for every finite x. Assuming that f is (at least) n times integrable and h is (at least) n times differentiable, we may apply Eq. (9) n times repeatedly. In order to simplify the notation, we denote the nth integral of f by F_n,

$$F_n(x) = \int_{\infty}^{x} dx_n \int_{\infty}^{x_n} dx_{n-1} \ldots \int_{\infty}^{x_2} dx_1\, f(x_1) \qquad \text{for } n = 1,2,3,\ldots \tag{10}$$

Then we find the relation

$$f*h = F_n * h^{(n)}\,. \tag{11}$$

This is valid also if the n–th derivative $h^{(n)}$ is a delta distribution. Note that if $h^{(n)}(x) = c\,\delta(x-\xi)$ then

$$f*h\,(x) = c\int_{-\infty}^{\infty} F_n(y)\,\delta(x-\xi-y)\,dy = c\,F_n(x-\xi)\,. \tag{12}$$

Hence, convolution with a kernel h can be calculated by repeated integration if some derivative of h is a linear combination of delta distributions. A trivial case is a box filter, where $h^{(1)}$ consists of two delta functions and the convolution amounts to a difference of two integrals.

3.2 Piecewise Polynomial Functions

Given a strictly increasing sequence $\xi := (\xi_i)_{i=1\dots l}$ of knots $\xi_i \in \mathbb{R}$ and polynomials P_i, $i = 1...l$, each of order k (i.e., of degree $< k$), then we define a *piecewise polynomial function of order k* by

$$f(x) := \begin{cases} 0, & x < \xi_1 \\ P_i(x), & \xi_i \le x < \xi_{i+1}\,, \quad i = 1\dots l-1 \\ P_l(x), & x \ge \xi_l\,. \end{cases} \tag{13}$$

The function and its derivatives may or may not be continuous at the knots ξ_i, as illustrated in Fig. 2.

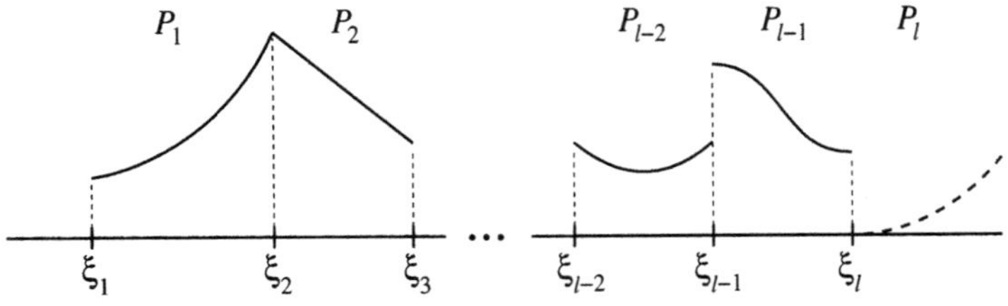

Fig. 2. A piecewise polynomial function, defined by a set of l knots ξ_i and l polynomials P_i.

It is easy to see that the set of piecewise polynomial functions of order k defined for a fixed knot sequence ξ generates a linear space. We will call this space $\mathbb{P}_{k,\xi}$.

For the time being we allow the function f to take non-vanishing values right from the last knot ξ_l. Later, when using the piecewise polynomial functions as filter kernels, we require the right-most polynomial P_l (dashed in Fig. 2) to be zero, since the kernels must have finite support.

3.3 The Truncated Power Basis

Let us now introduce a basis for the space of piecewise polynomial functions. We will built this basis from so-called truncated power functions, defined by

$$(x)_+^r := \begin{cases} 0, & x < 0 \\ x^r, & x \ge 0 \end{cases} \tag{14}$$

for $r \in \mathbb{R}_0$. Using this notation, we define a double sequence of functions corresponding to a knot sequence ξ by

$$\varphi_{ij}(x) = \frac{(x - \xi_i)_+^j}{j!} \qquad \text{for } i = 1\dots l \text{ and } j = 0\dots k-1. \tag{15}$$

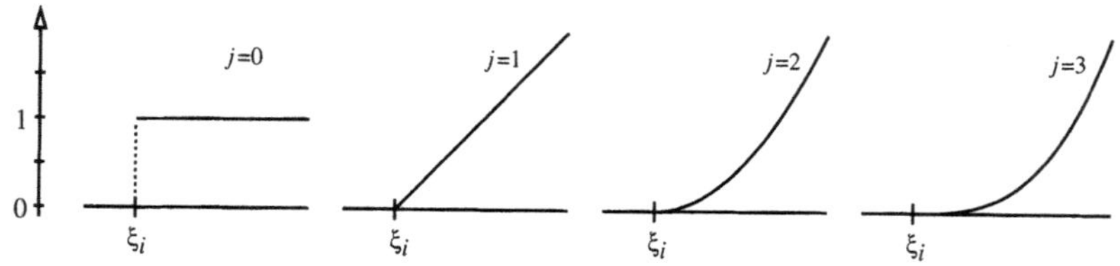

Fig. 3. Elements of the truncated power basis $\varphi_{ij} = (x - \xi_i)_+^j / j!$.

The φ_{ij} are piecewise polynomial functions of order $j+1$ with just one knot ξ_i. Obviously these functions are elements of $\mathbb{P}_{k,\xi}$. Note that this would not be the case if we had required $P_l(x)$ to be zero in Eq. (13). Plots of various φ_{ij} are shown in Fig. 3.

The derivatives of the functions φ_{ij} obey

$$\frac{d}{dx} \varphi_{ij} = \varphi_{i,j-1}, \qquad \text{for } j = 1 \dots k-1 \tag{16}$$

and

$$\frac{d^j}{dx^j} \varphi_{ij} = \varphi_{i,0}, \qquad \text{for } j = 0 \dots k{-}1 \tag{17}$$

where $\varphi_{i,0}$ is a step function which jumps at ξ_i from 0 to 1.

We will now show that the set of functions φ_{ij} is a basis of $\mathbb{P}_{k,\xi}$. Let us first look at the linear functionals λ_{ij} that take the difference of the jth derivative of f at a breakpoint ξ_i:

$$\lambda_{ij} f := \text{jump}_{\xi_i} f^{(j)} := f^{(j)}(\xi_i^+) - f^{(j)}(\xi_i^-), \tag{18}$$

with $i = 1 \dots l$ and $j = 0 \dots k$–1. Applying λ_{ij} to φ_{ij} and considering Eq. (17) yields

$$\lambda_{ij} \varphi_{rs} = \text{jump}_{\xi_i} \frac{d^j}{dx^j} \frac{(x - \xi_r)_+^s}{s!} = \delta_{ir} \delta_{js}. \tag{19}$$

The expression vanishes always except if the knot indices i and r as well as the exponents j and s are equal. We can use this result to show that the φ_{ij} are linear independent. From

$$\lambda_{ij} \sum c_{rs} \varphi_{rs} = \sum c_{rs} \lambda_{ij} \varphi_{rs} = \sum c_{rs} \delta_{ir} \delta_{js} = c_{ij} = 0. \tag{20}$$

it is obvious that $\sum c_{ij} \varphi_{ij} = 0$ implies $c_{ij} = 0$. The dimension of the space $\mathbb{P}_{k,\xi}$ is kl, which is equal to the number of functions φ_{ij}. Together with linear independence this shows that the φ_{ij} really comprise a basis for piecewise polynomial functions. Consequently, every $f \in \mathbb{P}_{k,\xi}$ has a unique representation of the form

$$f = \sum_{ij} (\lambda_{ij} f) \varphi_{ij} = \sum_{ij} (\text{jump}_{\xi_i} f^{(j)}) \frac{(x - \xi_i)_+^j}{j!}. \tag{21}$$

For many applications a more suitable and numerically advantageous basis of $\mathbb{P}_{k,\xi}$ is given by so-called B-splines [2,7]. However, as we will see the truncated power basis allows us to rewrite convolution integrals in an elegant way, thus facilitating the design of a general fast LIC algorithm. Therefore we will expand all filter kernels, even B-splines, in the truncated power basis.

We note that differentiating Eq. (17) yields the following distributional relationship for the basis functions φ_{ij} and all i and j:

$$\frac{d^{(j+1)}}{dx^{(j+1)}}\,\varphi_{ij}(x) \;=\; \delta(x-\xi_i)\,, \tag{22}$$

with δ representing the Dirac delta measure.

3.4 Applying the Convolution Theorem

Now suppose the filter kernel is given as a piecewise polynomial function, i.e. $h = \sum_{ij} c_{ij}\varphi_{ij}$. Differentiating φ_{ij} according Eq. (22) until a delta distribution is obtained and applying the convolution theorem (11), the convolution integral may be written as

$$\begin{aligned} f * h\,(x) = \int_{-\infty}^{\infty} f(y)h(x-y)dy &= \sum_{ij} c_{ij} \int_{-\infty}^{\infty} f(y)\varphi_{ij}(x-y)dy \\ &= \sum_{ij} c_{ij} \int_{-\infty}^{\infty} F_{j+1}(y)\delta(x-\xi_i-y)dy \\ &= \sum_{ij} c_{ij} F_{j+1}(x-\xi_i). \end{aligned} \tag{23}$$

Computing convolution integrals therefore amounts to calculating weighted averages of repeated integrals, or – after discretization – of repeated discrete sums.

3.5 Filter Kernels of B-Spline Type

For line integral convolution the filter kernel h should be as simple as possible. Usually there is no reason why to consider unsymmetric kernels. Furthermore, in order to obtain smoother results, the kernel should decrease or even approach zero at its boundaries. (The notion of "smoothness" will be made more quantitative by an intensity correlation analysis in Sect. 5.2.)

A nice class of filters that fulfill these conditions are B-splines with uniform knot sequences and centered around the origin. The most simple element in this class is a box filter b_1 with knots at $-L$ and L and normalized to $\int b_1 dx = 1$. Higher order filters can be obtained by repeatedly convolving box filters. Convolving two box filters we get a triangle filter $b_2 = b_1 * b_1$. In contrast to the box the triangle filter is continuous but still has a discontinuous derivative. Even smoother filters are obtained by convolving the triangle with a box again, and so on. In this way we get filter kernels of B-spline type.

Theorem. Let b_n be the $n-1$th convolution of a box filter with itself.

(i) Function b_n is given by

$$b_n(x) = \frac{1}{(2L)^n} \sum_{i=0}^{n} (-1)^i \binom{n}{i} \frac{1}{(n-1)!} \left(x + (n-2i)L\right)_+^{n-1} . \tag{24}$$

(ii) b_n has support $[-nL, nL]$.

(iii) b_n is normalized to $\int b_n dx = 1$.

(iv) In the limit $n \to \infty$ the functions b_n converge uniformly to a Gaussian.

Proof.

(i) We will show Eq. (24) by induction. For $n = 1$ we obtain the box filter itself:

$$b_1 = \frac{1}{2L}\left((x+L)_+^0 - (x-L)_+^0\right) .$$

Assuming that Eq. (24) is valid for b_n we get for b_{n+1}:

$$\begin{aligned} b_{n+1} &= b_n * b_1 \\ &= \frac{1}{(2L)^{n+1}} \sum_{i=0}^{n} (-1)^i \binom{n}{i} \int_{-\infty}^{\infty} \frac{(x+(n-2i)L)_+^{n-1}}{(n-1)!} \left((x-y+L)_+^0 - (x-y-L)_+^0\right) dy \\ &= \frac{1}{(2L)^{n+1}} \sum_{i=0}^{n} (-1)^i \binom{n}{i} \left(-\frac{(x+L+(n-2i)L)_+^n}{n!} + \frac{(x-L+(n-2i)L)_+^n}{n!}\right) \\ &= \frac{1}{(2L)^{n+1}} \sum_{i=0}^{n} (-1)^i \binom{n}{i} \left(-\frac{(x+(n+1-2(i+1))L)_+^n}{n!} + \frac{(x+(n+1-2i)L)_+^n}{n!}\right) \end{aligned}$$

Here we have used theorem Eq. (9) to replace the box filter by a delta distribution. Relabeling the index in the first term of the sum and using the equality $\binom{n}{i-1} + \binom{n}{i} = \binom{n+1}{i}$ we obtain

$$b_{n+1} = \frac{1}{(2L)^{n+1}} \sum_{i=0}^{n+1} (-1)^i \binom{n+1}{i} \frac{(x+(n+1-2i)L)_+^n}{n!}$$

in accordance with Eq. (24).

(ii) All filters b_n are normalized since $\int f\,dx = \int g\,dx = 1$ implies $\int f{*}g\,dx = 1$, and $\int b_1\,dx = 1$.

(iii) Elementary inspection of Eq. (24) shows that the functions b_n have support $[-nL, nL]$.

(iv) A proof that the functions b_n converge in the limit $n \to \infty$ uniformly to a Gaussian is given in Ref. [6]. □

In order to achieve $supp\, h = [-L, L]$ we re-scale $x \to \frac{x}{n}$. Furthermore we normalize the filters such that $\int h(x)\, dx = 1$. This yields the following sequence of bell-shaped filters of B-spline type:

$$\tilde{b}_n(x) = \frac{n}{(2L)^n} \sum_{i=0}^{n} (-1)^i \binom{n}{i} \varphi_{ij}(x) \tag{25}$$

where $\xi_i = \frac{2i-n}{n} L$ and $i = 1 \ldots n$. In Table 1 the first few functions of this type are given explicitly in the form of Def. 13. The expressions for $f * h$ contained in Table 1 result by applying relation (23).

Filter kernel h	Definition	Convolution $f * h$
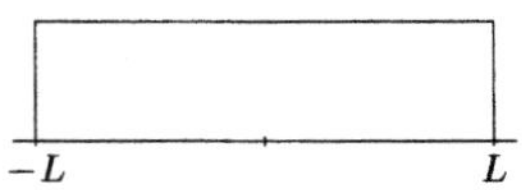	$P_1 = \frac{1}{2L}$	$\frac{1}{2L}(F(x+L) - F(x-L))$
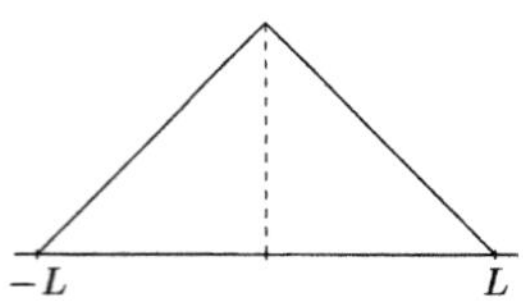	$P_1 = \frac{1}{L^2}(x+L)$ $P_2 = \frac{1}{L^2}(L-x)$	$\frac{1}{L^2}(F_2(x+L)$ $-2F_2(x) + F_2(x-L))$
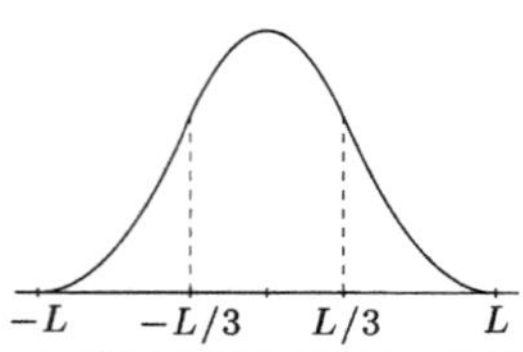	$P_1 = \frac{27}{16L^3}(x+L)^2$ $P_2 = \frac{9}{8L} - \frac{27}{8L^3}x^2$ $P_3 = \frac{27}{16L^3}(L-x)^2$	$\frac{27}{8L^3}(F_3(x+L) - 3F_3(x+L/3)$ $+3F_3(x-L/3) + F_3(x-L))$
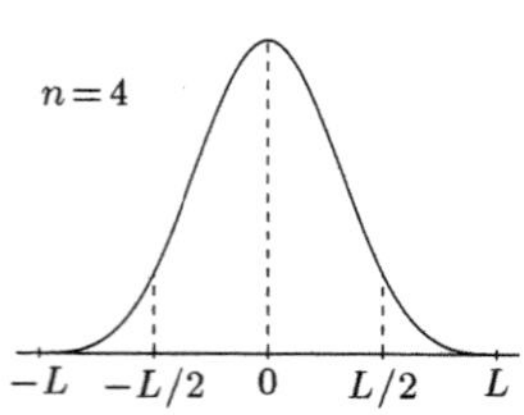		Convolution with a filter kernel of nth-order B-spline type: $f * h(x) = \left(\frac{n}{2L}\right)^n \sum_{i=0}^{n} (-1)^i \binom{n}{i} F_n(x + \frac{n-2i}{n} L)$

Table 1. Filter kernels of B-spline type obtained by recursively convolving box filters. All filters h in this table have support $[-L, L]$ and are normalized to one. Convolutions $f{*}h$ are represented as finite linear combinations of repeated integrals of f.

The construction of B-splines from repeated convolutions has first been described by Curry and Schoenberg in 1947 [6]. In modern spline theory B-

splines are usually defined in a more general context as divided differences on an arbitrary nondecreasing knot sequence, see e.g. [2].

A nice property of a nth order filter kernel of B-spline type is that all derivatives except of the $n-1$th one are continuous, i.e. the filter is built from exactly $n+1$ truncated power functions of degree $n-1$. Consequently, the evaluation of $f * h$ involves nth order integrals only. For general piecewise polynomial filter kernels usually other integrals have to be considered, too. The Catmull Rome Spline e.g. is given by

$$\begin{aligned} h(x) = 0.5(x+2)_+^3 - 2(x+1)_+^3 + 3(x)_+^3 - 2(x-1)_+^3 + 0.5(x-2)_+^3 \\ -0.5(x+2)_+^2 + (x+1)_+^2 - (x-1)_+^2 + 0.5(x-2)_+^2 \,. \end{aligned}$$

The occurrence of truncated power functions with different degrees makes the computation of $f * h$ slightly more expensive.

4 A General Fast LIC Algorithm

From Eq. (23) we can build a fast LIC algorithm in a straight-forward way. The results of the previous section allow us, like in the original fast LIC algorithm, to exploit the coherence of convolution values along a single field line. After the integrals F_k along a field line have been computed, we can use this information to quickly obtain the convolution values for a whole bunch of samples on that line.

4.1 Discretization

For numerical evaluation the integrals F_k have to be approximated by sums. For case of simplicity we use left-handed Riemann sums. Then the value of the first integral F_1 at locations $s = k\Delta s$, where $-m \le k \le m$ and $m = L/\Delta s$, is given by

$$F(k\Delta s) = \int_0^{k\Delta s} f(x)\,dx \approx \Delta s \sum_{i=0}^{k-1} f(i\Delta s), \tag{26}$$

while the higher order integrals correspond to the repeated sums

$$F_n(k\Delta s) = \int_0^{k\Delta s} F_{n-1}(x)\,dx \approx \Delta s \sum_{i=0}^{k-1} F_{n-1}(i\Delta s). \tag{27}$$

Note, that we cannot approximate the integrals by centered Riemann sums. These would have to be evaluated at $(i+\frac{1}{2})\Delta s$, which destroys the recursive relation in Eq. (27). However, we well might use trapezoid rule. Then the individual contributions would be given by $\frac{1}{2}(F(i\Delta s) + F((i+1)\Delta s))$. Since only function evaluations at integer multiples of Δs occur, the higher-order sums can again be computed recursively. However, in practice no difference between left-handed Riemann sums and trapezoid rule is visible. Therefore we use the simpler formulas (26) and (27).

In detail, our new fast LIC algorithm for piecewise polynomial filter kernels requires the following steps. First, the filter kernel h has to be expressed in terms of truncated power functions, i.e. $h = \sum_{ij} c_{ij}\varphi_{ij}$. Then the repeated sums F_n at positions $k\Delta s$ have to calculated up to the required order. Finally, a discrete approximation $\hat{I}$ of the convolution integral Eq. (23) is computed using

$$\tilde{I}(k\Delta s) = \sum_{ij} c_{ij} F_j(k\Delta s - \xi_i). \tag{28}$$

4.2 Outline of the Algorithm

Eq. (28) describes how to compute multiple samples on a single field line. In order to compute a full LIC image, we proceed as in the original fast LIC algorithm [14] . For each pixel of the output image we maintain two variables, a hit count and an accumulation variable. We then traverse all pixels in some order. Whenever we encounter a pixel (or optionally subpixel) with a hit count smaller than some user-defined limit min_{hit}, we start a field line calculation and compute multiple samples on that line. All samples are added to the accumulation variable of the corresponding pixel and the hit count of that pixel is incremented. After all pixels have been processed, the final image is computed by normalizing the accumulation variables against the number of hits per pixel. For the original fast LIC algorithm sophisticated seed point selection strategies or methods for determining the optimal number of samples per field line segment have been developed [15]. These techniques can be applied to the new algorithm without any modification.

In Table 1 various filter kernels of B-spline type are shown. The convolution value for the most simple filter kernel – the box filter – is given as the difference of the first-order sums at just two different locations. After evaluating these sums, only two operations – one subtraction and one addition – are needed per pixel (except for the first one). This is exactly the same as in the original fast LIC algorithm. Convolutions with piecewise polynomial filters can be evaluated at very little additional costs, by calculating a linear combination of repeated sums F_n and updating these sums while stepping along a field line (compare Table 1).

4.3 Implementation Issues

Some pitfalls of this new LIC techniques should be mentioned, too. The first one is a numerical aspect. The repeated (integer) sums quickly take on very large values, which can cause overflows. In our implementation we use unsigned 32 bit integers to represent these sums. For example, assuming an average value of the 8-bit input texture of 128 we can evaluate the fifth-order sum only for about 100 samples. Then an overflow would occur. However, as will be shown in Sect. 6, visual pleasant results can already be obtained by using the hat or triangle filter, which involves second-order sums only.

Another point which one has to take account of, concerns the inner control points of higher-order filter kernels. For example the third-order filter shown in Tab. 1 has two knots at $-L/3$ and $L/3$. In the discrete case it is necessary that the total number of samples used for approximating the kernel is divisible by 3. Using the nearest integer value of $L/3$ doesn't work.

5 Statistical Analysis of LIC Images

5.1 LIC viewed as a Stochastic Process

The input texture $T(\boldsymbol{x})$ for LIC can be arbitrarily chosen. In fact, every raster image can be used as input texture. Therefore, only global statements can be made about the statistical properties of LIC images. However, more detailed propositions are possible if the set of input textures is restricted to a class with certain statistical properties.

For vector field visualization the texture values $T_{\bar{\omega}_i} \in \mathbb{R}$ normally are chosen as a sequence of random variables T_i, i.e. as a (discrete) random process. We assume that the random variables T_i for distinct indices, i.e. for different texture pixels, are mutually independent. Then joint expectation values factorize, $\mathrm{E}(T_i T_j) = \mathrm{E}(T_i)\,\mathrm{E}(T_j)$ for any pair (i, j) with $i \neq j$, and the covariance therefore vanishes, $\mathrm{Cov}(T_i, T_j) = \mathrm{E}(T_i T_j) - \mathrm{E}(T_i)\,\mathrm{E}(T_j) = 0$. In accordance with the practice in scientific visualization let us assume that the random variables T_i are identically distributed with mean value

$$\mu_T = \mathrm{E}(T) \tag{29}$$

and finite variance

$$\sigma_T^2 = \mathrm{Var}(T) = \mathrm{E}(T^2) - \mathrm{E}(T)\,\mathrm{E}(T)\,. \tag{30}$$

We require $\sigma_T^2 > 0$, to exclude trivial cases like constant images. Since by definition $\mathrm{Cov}(T_i, T_i) = \mathrm{Var}(T_i)$, we get

$$\mathrm{Cov}(T_i, T_j) = \sigma_T^2\, \delta_{i,j} \tag{31}$$

for arbitrary pairs (i, j).

Some of the following calculations are more elegant, if the pixel position i is viewed as a continuous variable $\boldsymbol{x}$ (although a corresponding random process can physically not realized). Then many functions have to be considered as distributions; Eq. (31) for example becomes

$$\mathrm{Cov}(T_{\boldsymbol{x}}, T_{\boldsymbol{x}'}) = \sigma_T^2\, \delta(\boldsymbol{x} - \boldsymbol{x}')\,. \tag{32}$$

A random process $\{X_i\}$ is called (strictly) *stationary* if shifting in space has no effect on joint distributions, i.e. the distribution of $X_{i_1}, ..., X_{i_n}$ is the same as the joint distribution of $X_{i_1+k}, ..., X_{i_n+k}$. Utilizing this fact for $n = 2$ the autocovariance function

$$\gamma(i,j) := \mathrm{E}\{[X_i - \mathrm{E}(X_i)]\,[X_j - \mathrm{E}(X_j)]\} \tag{33}$$

obviously depends only on $\tau = j - i$ and may be written as

$$\gamma(\tau) = \mathrm{E}\{[X_t - \mu]\,[X_{t+\tau} - \mu]\} = \mathrm{Cov}[X_t, X_{t+\tau}]\,. \tag{34}$$

More useful is the standardized autocorrelation function

$$\rho(\tau) = \frac{\gamma(\tau)}{\gamma(0)}\,. \tag{35}$$

The random process $\{T_i\}$ obviously is stationary. For the autocovariance function we then get

$$\gamma_T(\tau) = \sigma^2\,\delta_{0,\tau} \tag{36}$$

and for the autocorrelation function

$$\rho_T(\tau) = \delta_{0,\tau}\,. \tag{37}$$

Random processes, constituted by a sequence of mutually independent and identically distributed random variables, are called a 'purely random process' or 'white noise'. Equations (29)-(37) represent the simple statistical properties of the random input textures typically used in vector field visualization with LIC. The LIC intensity variables I depend on these random variables T and therefore constitute another random process. Its statistical attributes are more interesting. The most characteristic feature of LIC images are the anisotropic correlations: While the intensity values I are strongly correlated along field lines due to the 1D convolution, they are almost uncorrelated in perpendicular directions.

5.2 Statistical Properties of LIC Images Along Field Lines

In order to obtain statistical properties of I values along a field line, we start with a simplifying assumption: We assume that the texture input grid $\{\bar{\omega}_i\}$ is fine enough such that distinct texture pixels $\bar{\omega}_i$ are sampled during the convolution according to Eq. (7).

Is this assumption realistic ? In practice one tries to choose a sampling distance along field lines of about one pixel width of the output image. Choosing identical locations, both for T and I samples, the condition is fulfilled to a rather good extend, if the spatial resolution of the input texture corresponds at least to that of the output image. In practice this condition usually is met.

We can therefore safely assume that the random variables $T_{\sigma,j-k}$, with $k \in \{-m, ..., m\}$, being used to calculate a specific I variable according to Eq. (7) are independent. For simplicity we drop the index $\boldsymbol{\sigma}$ in this equation and write for the discrete case

$$\hat{I} = \sum_{k=-m}^{m} T_{j-k}\, h_k \qquad \text{with} \qquad \sum_{k=-m}^{m} h_k = 1\,. \tag{38}$$

and for the continuous case:

$$I = \int_{-L}^{+L} T(s-s')\, h(s')\, ds \qquad \text{with} \qquad \int_{-L}^{L} h(s')\, ds' = 1\,. \tag{39}$$

Using these relations we find immediately:

$$\mathrm{E}(\hat{I}) = \mathrm{E}(I) = \mathrm{E}(T), \tag{40}$$

and, since the variables T are independent,

$$\mathrm{Var}(I) = \sigma_T^2 \sum_k h_k^2 \tag{41}$$

in the discrete case, or

$$\mathrm{Var}(I) = \sigma_T^2 \int_{-L}^{L} h^2(s')\, ds' = \sigma_T^2\, (h * h)(0) \tag{42}$$

in the continuous case. This shows that LIC leaves the average brightness of an image constant, but changes contrast. Using Eq. (31) for the autocovariance we have in the discrete case

$$\gamma_{\hat{I}}(\tau) = \mathrm{Cov}(\hat{I}_t, \hat{I}_{t+\tau}) = \begin{cases} \sigma_T^2 \sum\limits_{k=-m}^{m-\tau} h_k\, h_{k+\tau}, & \tau = 0, \ldots, 2m \\ 0, & \tau > 2m \end{cases} \tag{43}$$

and using Eq. (32)

$$\gamma_I(\tau) = \mathrm{Cov}(I_t, I_{t+\tau}) = \begin{cases} \sigma_T^2 \int\limits_{-L}^{L-\tau} h(s)\, h(s+\tau), & 0 \le \tau \le 2L \\ 0, & \tau > L \end{cases} \tag{44}$$

in the continuous case. Therefore the autocorrelation is

$$\rho_{\hat{I}}(\tau) = \frac{\sum\limits_{k=-m}^{m-\tau} h_k\, h_{k+\tau}}{\sum\limits_{k=-m}^{m} h_k^2} \qquad \text{for } \tau = 0, \ldots, 2m \tag{45}$$

and in the continuous case

$$\rho_I(\tau) = \frac{\int_{-L}^{L-\tau} h(s)\, h(s+\tau)}{\int_{-L}^{L} h^2(s)} \qquad \text{for } 0 \leq \tau \leq 2L\,. \tag{46}$$

If $h(s) = h(-s)$ the general relation $\rho_I(\tau) = \rho_I(-\tau)$ yields

$$\rho_I(\tau) = \frac{h * h\,(\tau)}{h * h\,(0)}\ . \tag{47}$$

Hence, with increasing τ the autocorrelation of LIC intensity values along a field line generated with a B-spline filter $\tilde{b}_n$ drops like a B-spline $\tilde{b}_{2n}$.

5.3 The Effective Filter Length

For a meaningful comparison between different filter kernels their lengths have to be adjusted. Usually, the larger the filter length L, the larger the feature size along a field line and the less the contrast of the resulting LIC image. In our case contrast can simply be defined as the variance of the image's overall intensity distribution.

When comparing e.g. a box filter and a triangle filter of equal length, it is clear that the triangle filter has smaller feature size and higher contrast. The reason is that pixels close to the boundaries of the triangle filter get smaller weights. Therefore the *effective* filter length is smaller for the triangle filter.

In a comparative study we want to choose equal effective filter lengths. We determine these filter lengths such that the variance of the resulting intensity distributions are equal. Using Eq. (42) and taking the ratio of the variances we get $L^{\text{eff}}_{\text{triangle}} = \frac{4}{3} L_{\text{box}}$ and $L^{\text{eff}}_{\text{3rd-order}} = \frac{33}{20} L_{\text{box}}$.

6 Results

Fig. 4 illustrates the effect of different filters kernels on LIC images. All images have been computed using the fast LIC algorithm outlined in the previous section. The differences between the box filter and higher-order filter kernels are well noticeable on a computer monitor. In order to make these differences visible in the printed book, i.e. after a complex reproduction process, the images are magnified.

In Color Plates 13 and 14 on page 374 in the Appendix we depict solutions of ODEs, using LIC with triangular filter kernels. Furthermore, contrast enhanced difference images of LIC images with triangular respectively box filter kernels are shown. Note that the error structures are much smaller than the characteristic feature length of the LIC texture. Therefore, box filter images appear noisier on the screen than triangular filter images.

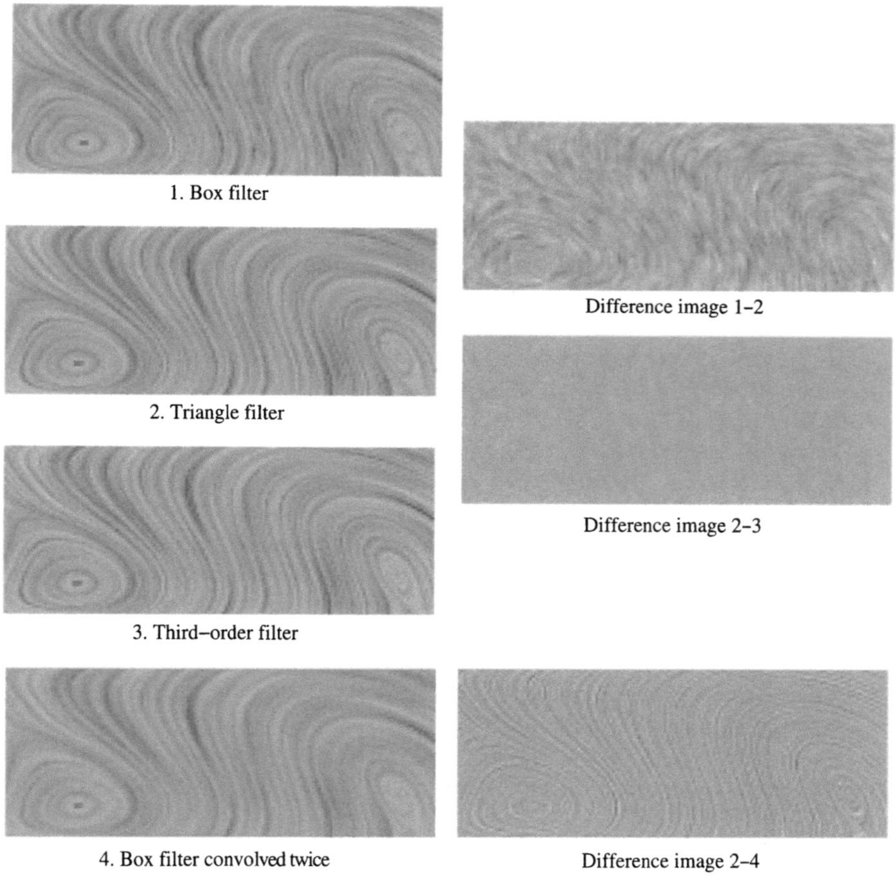

Fig. 4. LIC images obtained with different filter kernels. The contrast of the three difference images has been increased equally for better visibility. On a computer monitor the differences between the images on the left are much more obvious.

We conclude that higher order filter kernels can be combined with the fast LIC algorithm and therefore are well suited for interactive visualization. Statistical analysis of pixel correlations as well as visual investigation show that higher higher order filter kernels lead to smoother, less noisy images. For practical purposes LIC images produced with triangular or quadratic B-spline filters are sufficient. The differences between these and higher order filters in general are hardly visible.

References

1. H. Battke, D. Stalling, and H. Hege, *Fast line integral convolution for arbitrary surfaces in 3D*, Visualization and Mathematics (H. Hege and K. Polthier, eds.), Springer, 1997, pp. 181–195.

2. C. de Boor, *A practical guide to splines*, Applied Mathematical Sciences, vol. 27, Springer, New York, Heidelberg, Berlin, 1978.
3. B. Cabral, H.-C. Hege, V. Interrante, K.-L. Ma, and D. Stalling, *Texture synthesis with line integral convolution – course notes*, Siggraph 97, ACM Siggraph, Cambridge, 1997.
4. B. Cabral and C. Leedom, *Highly parallel vector visualization using line integral convolution*, Seventh SIAM Conference on Parallel Processing for Scientific Computing, February 1995, pp. 802–807.
5. B. Cabral and L. C. Leedom, *Imaging vector fields using line integral convolution*, Computer Graphics (SIGGRAPH '93 Proceedings) (J. T. Kajiya, ed.), vol. 27, August 1993, pp. 263–272.
6. H. Curry and I. Schoenberg, *On spline distributions and their limits: the pólya distribution functions*, Bull. Amer. Math. Soc. **53** (1947), 1114.
7. P. Deuflhard and A. Hohmann, *A First Course in Scientific Computation*, Verlag de Gruyter, Berlin, 1995.
8. L. K. Forssell, *Visualizing flow over curvilinear grid surfaces using line integral convolution*, Visualization '94, IEEE Computer Society, 1994, pp. 240–247.
9. L. K. Forssell and S. D. Cohen, *Using line integral convolution for flow visualization: Curvilinear grids, variable-speed animation, and unsteady flows*, IEEE Transaction on Visualization and Computer Graphics **1**:2 (1995).
10. V. Interrante, *Illustrating surface shape in volume data via principal direction-driven 3D line integral convolution*, Computer Graphics Proceedings, Annual Conference Series, ACM SIGGRAPH, Addison Wesley, August 1997, held in Los Angeles, California, 3-8 August 1997, pp. 109–116.
11. V. Interrante and C. Grosch, *Strategies for effectively visualizing a 3D flow using volume line integral convolution*, ICASE, Technical Report TR-97-35, July 1997.
12. X. Mao, M. Kikukawa, N. Fujita, and A. Imamiya, *Line integral convolution for arbitrary 3D surfaces through solid texturing*, Proceedings of Eighth Eurographics Workshop on Visualization in Scientific Computing, 1997 (to appear).
13. H. Shen, C. R. Johnson, and K. Ma, *Visualizing vector fields using line integral convolution and dye advection*, 1996 Volume Visualization Symposium, IEEE, October 1996, pp. 63–70.
14. D. Stalling and H. Hege, *Fast and resolution independent line integral convolution*, Annual Conference Series, ACM SIGGRAPH, Addison Wesley, August 1995, held in Los Angeles, California, 6-11 August 1995, pp. 249–256.
15. D. Stalling, M. Zöckler, and H. Hege, *Parallel line integral convolution*, Parallel Computing **23** (1997), 975–989.
16. C. Teitzel, R. Grosso, and T. Ertl, *Line integral convolution on triangulated surfaces*, Proceedings of WSCG '97, The Fifth International Conference in Central Europe on Computer Graphics and Visualization '97 (N. M. Thalmann and V. Skala, eds.), vol. 3, University of West Bohemia Press, 1997, pp. 572–581.
17. R. Wegenkittl, E. Gröller, and W. Purgathofer, *Animating flowfields: Rendering of oriented line integral convolution*, Vienna University of Technology, Institute of Computer Graphics, Technical Report TR-186-2-96-23, December 1996.

Visualizing Poincaré Maps together with the Underlying Flow

Helwig Löffelmann, Thomas Kučera, and Eduard Gröller

Institute of Computer Graphics, Vienna University of Technology,
Wien, Austria

Abstract. We present a set of advanced techniques for the visualization of 2D Poincaré maps. Since 2D Poincaré maps are a mathematical abstraction of periodic or quasi-periodic 3D flows, we propose to embed the 2D visualization with standard 3D techniques to improve the understanding of the Poincaré maps. Methods to enhance the representation of the relation $x \leftrightarrow P(x)$, e.g., the use of spot noise, are presented as well as techniques to visualize the repeated application of P, e.g., the approximation of P as a warp function. It is shown that animation can be very useful to further improve the visualization. For example, the animation of the construction of Poincaré map P is a very intuitive visualization. During the paper we present a set of examples which demonstrate the usefulness of our techniques.

1 Introduction

Poincaré sections are an important tool for the investigation of dynamical systems in theory as well as in applications. They are used for models (usually in 3D) that exhibit periodic or quasi-periodic behavior. In addition to mathematical descriptions—there is a lot of theory about dynamical systems [3,4,7,15]—periodic or quasi-periodic dynamical systems can be found in many fields, e.g., in physics, chemistry, biology, ecology. Especially chaotic systems are often examined by the use of Poincaré sections [13–15].

Roughly speaking a 2D Poincaré section through a periodic 3D flow is a planar cross-section transverse to the flow such that a periodic orbit intersects it at its center. The corresponding Poincaré map is defined as a map correlating consecutive intersections of flow trajectories with the Poincaré section. The Poincaré map is a discrete dynamical system of one dimension less than the continuous flow which it is constructed from. Since many of the most important flow properties are inherited by the Poincaré map and its analysis is usually more simple due to its reduced dimensionality, it is often used for analysis instead of the 3D flow. See Sect. 2 for a little bit more detailed discussion of some basics on Poincaré sections and Poincaré maps.

Since their introduction to dynamical system analysis by Henri Poincaré in 1899 [12] the visual representation of Poincaré maps has been always a very important part of this research technique. Hand-drawn sketches of the

Poincaré map were used for a long time to guide or illustrate the mathematical analysis [1]. Due to the ability of integrating a dynamical system numerically by the use of a computer, visualization techniques that are based on the numerical approximation of the Poincaré map have become popular [11,14]. There are a number of programs that calculate Poincaré maps [6]. See Sect. 3 for a brief review of previous work in this field.

Most visualization techniques for Poincaré maps suffer from rather severe limitations. One problem is, that with the use of 2D visualization techniques the context of the 3D flow is lost. Certain features, e.g, the number of windings of a Möbius band, can not be derived from the 2D Poincaré map alone. Another problem with these techniques is that the temporal correlation between points of the Poincaré map is not encoded within the 2D image. Most limitations of the 2D techniques in this field are usually not due to a weakness of the software or method, but rather due to an inherent difficulty with dimension reduction approaches.

We therefore propose a set of advances within this rather untouched field of visualization. First, we suggest to adapt some well-known visualization techniques as, e.g., spot noise, to Poincaré maps to improve the visual representation of the 2D map. See Sect. 4 for a discussion of these ideas. Furthermore we present an embedding of these techniques within a 3D visualization of the underlying flow. This approach allows to significantly reduce some limitations of previously known techniques. Please refer to Sect. 5 for a description of this approach. Our ideas are implemented within DynSys3D [9], an experimental workbench for the development of advanced visualization techniques in the field of dynamical systems. The capabilities of the implementation are illustrated with some examples in Sect. 7. We draw some conclusions in Sect. 8 and give ideas for future work.

2 About Poincaré Maps

A *Poincaré section* is used to construct a $(n-1)$-dimensional discrete dynamical system, i.e., a *Poincaré map*, of a continuous flow given in n dimensions. This reduced system of $(n-1)$ dimensions inherits many properties, e.g., periodicity or quasi-periodicity, of the original system. We will concentrate in the following on the important case of n being equal to three.

Poincaré maps are used to investigate periodic or quasi-periodic dynamical systems. Often these systems exhibit a periodic cycle or a chaotic attractor. A Poincaré section $\mathcal{S}$ is now assumed to be a part of a plane, which is placed within the 3D phase space of the continuous dynamical system such that either the periodic orbit or the chaotic attractor intersects the Poincaré section. The Poincaré map is now defined as a discrete function $P : \mathcal{S} \to \mathcal{S}$, which associates consecutive intersections of a trajectory of the 3D flow with $\mathcal{S}$ (see Fig. 1).

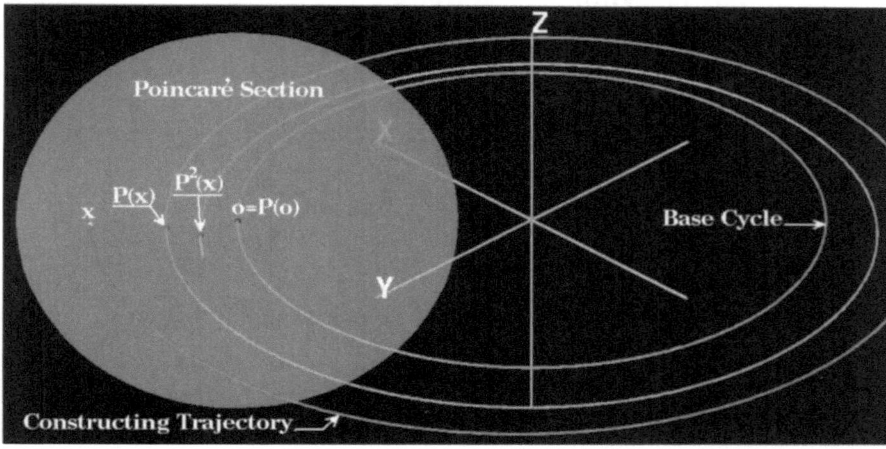

Fig. 1. An illustration of the Poincaré map definition (cf. color plate 20, p. 378)

There are some important relations between a 3D flow and the corresponding Poincaré map: Cycles $\mathcal{C}$ of the 3D system which intersect the Poincaré section $\mathcal{S}$ in q points ($q \geq 1$) are related to periodic points $o = \mathcal{C} \cap \mathcal{S} = P^q(o)$ of Poincaré map P, i.e., o is a fixed point of the map P^q. Furthermore stability characteristics of the cycle are inherited by the fixed point: stable, unstable, or saddle cycles result in stable, unstable, or saddle nodes, respectively. Therefore many characteristics of periodic or quasi-periodic dynamical systems can be derived from the corresponding Poincaré map.

3 Previous and Related Work

Visualization techniques have been used for the illustration of Poincaré maps since they were introduced by Henry Poincaré. Most of these imaging methods are 2D or 1D plots that are calculated by numerically integrating the underlying flow.

In addition to these traditional 2D plots a few more general techniques can be found in the literature. Hand-drawn images in a book by Abraham and Shaw [1] demonstrate that a combination of the Poincaré section and the underlying 3D flow within within a single image convey a better understanding of the underlying flow characteristics. The book by Abraham and Shaw is in effect quite an inspirational source. We think that a number of artistic techniques used for the hand-drawn images in the book are well-suited for computer-supported visualization techniques. Streamarrows [10] are another approach to automate techniques presented in this book.

4 Visualizing Poincaré Maps

The contribution of this paper can be divided into two parts. First, we suggest a set of advanced visualization techniques directly applicable to 2D Poincaré maps. We describe these techniques in subsections 4.1 and 4.2. Secondly, we demonstrate how an embedding of our techniques within a 3D flow visualization improves the expressiveness of images and animations. See Sect. 5 for a discussion of this part of our work.

In this section we focus on a 2D Poincaré section $\mathcal{S}$ and how the mathematical method of analysis by the use of Poincaré map P can be supported by visualization. Different aspects of $\mathcal{S}$ or P are of interest and might therefore be visualized. The transformation of $\mathcal{S}$ under the application of P is certainly the most important issue to visualize (see Sect. 4.1). Additionally the repeated application of the Poincaré map, namely P^n, is also often investigated. This is done, for example, when the asymptotical behavior of the map is studied (see Sect. 4.2).

4.1 Visualizing Poincaré Map *P*

As Poincaré map P maps points x onto $P(x)$ both lying in $\mathcal{S}$, a visualization based on (directed) strokes connecting x and $P(x)$ has been done. We implemented a module named FLOW that generates a set of arrows on the Poincaré section, which all start in some point x_i and end in correlated $P(x_i)$. See Fig. 2 for a visualization of a non-linear saddle cycle where this technique was used. $\mathcal{S}$ is represented as a semi-transparent disk. A set of light-grey arrows is placed on $\mathcal{S}$ to visualize P^1. Sequences of consecutive applications of P are visualized by the use of small red spheres, representing $\{P^j(x_i)\,|\,j \geq 0\}$, whereby the sphere depicting $P^0(x_i)$ is colored differently from the others. The visualization of sequence $\{P^j(x_i)\,|\,j \geq 0\}$, which starts near the origin of phase space, is furthermore enhanced by combining it with a visualization of the constructing flow trajectory.

We also adapted spot noise [16] to Poincaré maps. We place elliptic spots onto $\mathcal{S}$ such that the focal points of the ellipses coincide with x_i and $P(x_i)$, respectively. This choice is due to the fact that no directional information should be encoded, when $P(x_i) = x_i$. In this case both focal points coincide and the elliptic spot degenerates to a circle. Images rendered with this method are well suited to visualize the entirety of $P(\mathcal{S})$ within one still image. See color plate 23, p. 379, for a visualization of a non-hyperbolic saddle cycle (3 stable and 3 unstable manifolds), where spot noise was used for visualization. Similar as in Fig. 2 white and red spheres are used to visualize certain sequences $\{P^j(x_i)\,|\,j \geq 0\}$.

The results of the previous techniques are embedded into a 3D visualization of the underlying flow. We therefore represent Poincaré section $\mathcal{S}$ as

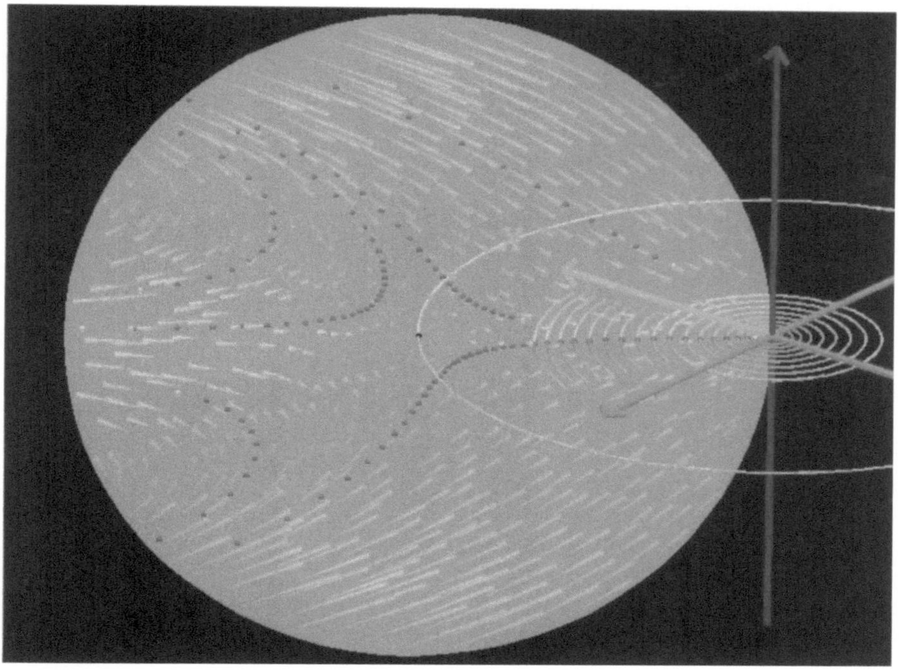

Fig. 2. Visualizing a non-linear saddle cycle (see also color plate 22, p. 379)

a semi-transparent disk placed within the flow and realized the arrows and spot noise as a texture of this disk (cf. Fig. 2 & color plate 23, p 379). Semi-transparency was used for the map to allow the viewer to see through. This improves the understanding of the context of map P.

4.2 Visualizing the Repeated Application P^n

There are several reasons to investigate the repeated application of Poincaré map P, i.e., P^n. Probably the most important one is the analysis of the asymptotic behavior induced by the (iterated) Poincaré map. Periodic systems near a cycle can exhibit different asymptotical behavior, e.g., convergence or divergence with respect to the cycle. This aspect is due to different possible cycle characteristics, namely stable, saddle, or unstable behavior. A stable cycle attracts near trajectories, whereas unstable and saddle cycles repel near trajectories. A saddle cycle $\mathcal{C}$ separates its Poincaré section $\mathcal{S}$ into regions of attraction and such of repulsion. Almost all trajectories near $o = P(o)$ emerge into the repelling parts of $\mathcal{S}$ and thus are finally repelled from $\mathcal{C}$. Figure 2 and color plate 23, p. 379, show two Poincaré maps of different saddle cycles.

We implemented a module TRAJECTORY which produces a visualization of the set $\{P^j(x_i) \mid j \geq 0\}$ by generating a sphere for each $P^j(x_i)$. See, e.g, the small red spheres in Fig. 2 & color plate 23, p. 379, for examples where this module was used. The sphere representing $P^0(x_i) = x_i$ is colored differently from all the others such that the origin of a sequence can be identified (cf. color plate 23, p. 379).

Sometimes P^q, $q > 1$ is more interesting to investigate than P. This is the case, for example, when base cycle $\mathcal{C}$ itself pierces $\mathcal{S}$ q times during one complete loop (see Fig. 3). In the given example $\mathcal{C}$ intersects $\mathcal{S}$

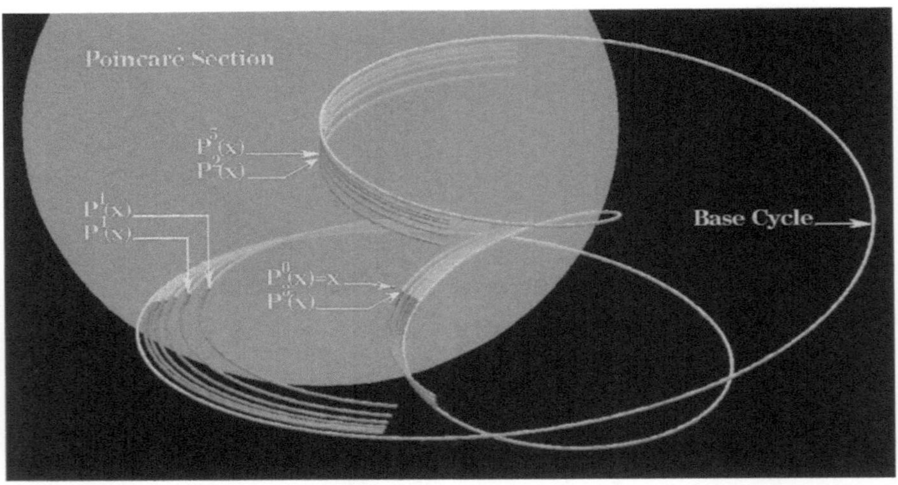

Fig. 3. Visualizing why P^q is sometimes more expressive than P

two more times before it returns to the initial intersection point and thus closes the cycle. The behavior of trajectories $\mathcal{T}$ near $\mathcal{C}$ are better described by one arbitrary $x \in \mathcal{T} \cap \mathcal{S}$ & $P^3(x)$ rather than x & $P(x)$. In fact any pair $(P^j(x),\, P^{j+q}(x))$, $0 \leq j < q$, can be chosen for this improved analysis.

The user can change the default value of q, i.e., 1, such that all the previously discussed visualization techniques, e.g., spot noise, are adapted to this new parameter setting. Moreover we allow the user to specify that only those intersections $y = \mathcal{T} \cap \mathcal{S}$ are considered where $f(y) \cdot f(\mathcal{C} \cap \mathcal{S}) > 0$ (f being the underlying 3D flow). Only those points on trajectory $\mathcal{T}$ are of interest where $\mathcal{T}$ crosses the Poincaré section with the same orientation. This means that $\mathcal{T}$ crosses $\mathcal{S}$ in both cases either in front-to-back or back-to-front orientation. In Fig. 4 the left image was rendered with $q = 1$ and the right one with $q = 3$. In this case the base cycle pierces $\mathcal{S}$ three times during one complete loop.

Although there are still some artifacts in the right image which are due to the limited size of $\mathcal{S}$, it is more expressive than the left one. The egg-shaped intersection of an invariant torus (containing the base cycle) and Poincaré section $\mathcal{S}$ can be clearly seen as dark line around the center of this image. Furthermore the unstable cycle within this torus cross-section can be distinguished as a fixed point of the Poincaré map P. The radial repulsion away from this fixed point towards the torus is well represented by the star like spot noise texture.

The visualization of P^n, $n > 1$, is more difficult than visualizing P itself. A technique we investigated for the representation of P^n, n increasing, is image warping [5]. We implemented a module WARP that approximates P by a warp function W on the basis of $\mathcal{L}$ and $P(\mathcal{L})$ where $\mathcal{L}$ can be chosen to be either a jittered or regular set of line segments spread over $\mathcal{S}$. WARP loads an initial texture TEX onto $\mathcal{S}$ and then applies W n times, where n is specified via a parameter of WARP. The resulting texture $TEX \circ W^{-n}$ (cf. Fig. 5) placed on $\mathcal{S}$ gives a good impression of the main characteristics of P^n. See Fig. 7 for an image series rendered using this technique. The reason why we approximate Poincaré map P by the use of a warping function instead of using P directly for the transformation of the texture is that P is usually rather costly to compute. Warp function WARP on the other hand is capable of approximating P quite good if P is sufficiently smooth and warping parameters are chosen appropriately. At least an idea of P^n is gained using this technique.

5 Embedding the Visualization of Poincaré Maps within the 3D Flow

The simultaneous visualization of a Poincaré map and the underlying flow allows to overcome some limitations of 2D visualization techniques of Poincaré

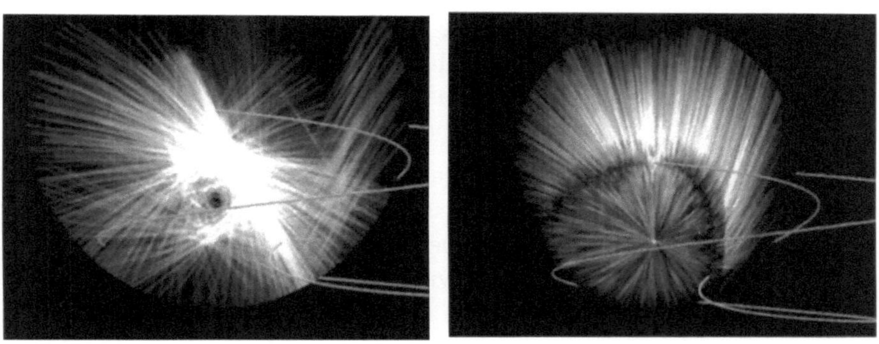

(a) (b)

Fig. 4. Visualizing $\{(x_i, P(x_i))\}$ (a) vs. $\{(x_i, P^3(x_i))\}$ (b)

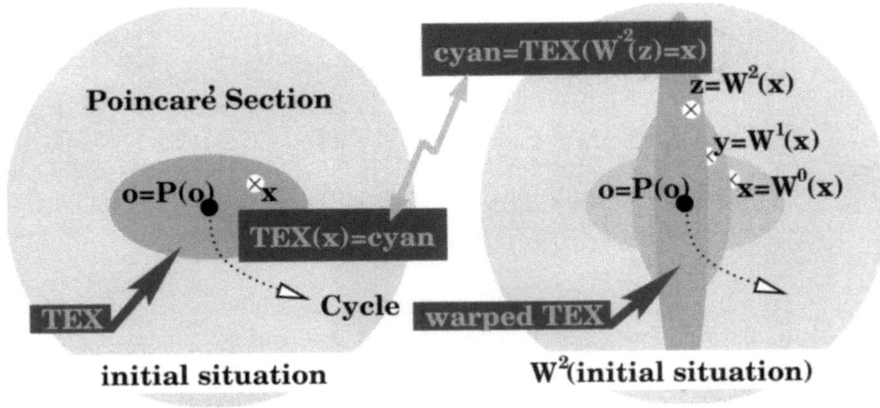

Fig. 5. Evaluating the initial texture after two applications of W (cf. color plate 24, p. 380)

maps. Flow characteristics which cannot be derived from 2D Poincaré maps alone as, e.g, the relation between consecutive intersections, can be made visible and thus enrich the capabilities of this visualization technique.

A Poincaré section $\mathcal{S}$ is represented as a circular patch that is rendered semi-transparently. Therefore visualization icons before as well as behind $\mathcal{S}$ are visible. Our current implementation is based on the existence of a base cycle $\mathcal{C}$ within the 3D flow. Cycle $\mathcal{C}$ defines the center of Poincaré section $\mathcal{S}$. The cycle $\mathcal{C}$ is rendered as an opaque tube through 3D phase space together with a sphere at $\mathcal{C} \cap \mathcal{S}$ where the base cycle intersects the Poincaré section. See Fig. 2 or color plate 23, p. 379, for an image where $\mathcal{S}$, $\mathcal{C}$, and the intersection $\mathcal{C} \cap \mathcal{S}$ can be easily detected.

The module which generates this Poincaré map visualization takes care that initially a useful view point is chosen. $\mathcal{S}$ is viewed under an angle which is little bit less than $\pi/2$ so that both the Poincaré map and the intersecting base cycle are easily recognizable (see Fig. 2). We also found it very useful to provide a relative placement capability such that the user can move the Poincaré section easily around the base cycle $\mathcal{C}$. The actual position of the map on the cycle is specified by a value between 0 and 1.

Additionally we suggest some more elaborated 3D extensions to Poincaré map visualization. Module TRAJECTORY, for example, generates either the entire trajectory which constructs the sequence $\Omega = \{P^j(x) \,|\, j \geq 0\}$, only short parts of this trajectory in the vicinity of $\mathcal{S}$, or just the spheres representing the consecutive intersections.

Drawing the entire trajectory that generates sequence Ω, for example, helps to relate consecutive points on $\mathcal{S}$ mentally (cf. Figs. 1 & 3). Long trajectories, however, may clutter the image.

Using only short parts of the trajectory near $\mathcal{S}$ avoids the problem of visual clutter (see Fig. 6). Assume a periodic 3D system which exhibits some high frequency oscillation parallel to the rotational axis of the flow (see Fig. 6, left image). If the frequency (measured in oscillations per one total revolution of the carrying periodic system around the rotational axis of the flow) is an integer number, the resulting Poincaré map is not affected by this oscillation at all. Results are the same as for a system without the modulated frequency. Compare left and right image in Fig. 6.

Both images in Fig. 6 are generated with techniques already discussed previously in this paper. Spot noise is used to represent the entirety of P, whereas white and red spheres are used to visualize three sequences $\{P^j(x_i) \mid j \geq 0\}$. The one sequence starting near the origin of phase space is enhanced by short parts of the constructing flow trajectory. This enhancement is necessary to visualize the differences between both systems.

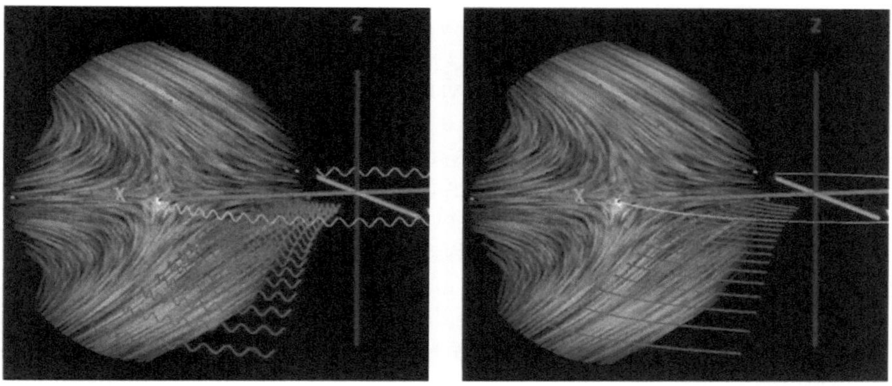

Fig. 6. Visualizing supplementary information in 3D (cf. color plate 26, p. 380)

Another technique for the investigation of Poincaré maps is realized as module SEEDLINE. By parameters r, φ, and *dist* a line segment $\mathcal{Y}$ of length $2 \cdot \mathrm{R}$ is specified within $\mathcal{S}$ that is perpendicular to the vector connecting $\mathcal{C} \cap \mathcal{S}$ and the mid-point of $\mathcal{Y}$. The length of this vector is specified by parameter *dist*. Parameters φ and r are the polar coordinates of one end-point of this line segment with respect to its mid-point. SEEDLINE can be used to generate a stream surface [8] or a rag of stream lines alternatively (see color plate 21, p. 378). In addition to the supplementary stream surface, the flow trajectory, which constructs a certain sequence of consecutive applications of P, was visualized by the use of a green tube. Other techniques discussed previously in this paper have been used for visualizing Poincaré section $\mathcal{S}$.

6 Animation Aspects

Animation is a powerful approach to increase the dimensionality of visualization results. We found the following parameters of our modules suitable to be animated:

TRAJECTORY parameters *no* and *len* – Module TRAJECTORY generates a sequence Ω of consecutive intersections of trajectory $\mathcal{T}$ and Poincaré section $\mathcal{S}$, i.e., $\Omega = \mathcal{T} \cap \mathcal{S}$. Parameter *no* specifies how many intersections should be calculated. *len* can be used to control $|\Omega|$ via the spatial length of $\mathcal{T}$.
Animating one of these two parameters, the construction of the Poincaré map P can be visualized. Furthermore the asymptotical behavior of $P^n(x)$, $n \to \infty$ can be investigated. This specific application of animation is capable of representing the *inherent* nature of Poincaré map P.

TRAJECTORY parameters r and φ – Another pair of TRAJECTORY parameters, which is very well suited for animation, is (φ, r). It encodes the initial condition x of solution $\mathcal{T}(x)$ in polar coordinates with respect to some arbitrary local coordinate system on $\mathcal{S}$. In other words (φ, r) specifies the starting point of sequence $P^j(x)$, $j \geq 0$.
Animating these two parameters allows to investigate the development of arbitrary curves $\mathcal{Q}$ within $\mathcal{S}$. Such a curve $\mathcal{Q} = (\mathcal{Q}_\varphi(s), \mathcal{Q}_r(s))$ should be given as a parameterized subset of $\mathcal{S}$ with s as the parameter. Given such a curve $\mathcal{Q}$, parameter s can be animated: The module TRAJECTORY takes $x(s) = (\mathcal{Q}_\varphi(s), \mathcal{Q}_r(s))$ as an initial condition for the generation of $\{P^j(x(s)) \,|\, j \geq 0\}$. Initial condition $x(s)$ moves along curve $\mathcal{Q}$, and simultaneously its long-term future $\{P^j(x(s)) \,|\, j \geq 0\}$ is visualized.

WARP parameter n – The animation of WARP parameter n (number of applications) improves the expressiveness of the image warping approach. A sequence of images with consecutive applications of warping function W shows efficiently the overall behavior of P^n, $n \to \infty$. Refer to Fig. 7 for three images resulting from consecutive applications of the warping function W.

7 Implementation Issues

All the visualization techniques proposed in this paper have been implemented within DynSys3D [9], which is a flexible development workbench for advanced visualization techniques of three-dimensional dynamical systems. DynSys3D itself is based on AVS [2], which is a general purpose visualization system. A main feature of DynSys3D is the treatment of *dynamical system*, *numerical integrator*, and *visualization technique* as separate components.

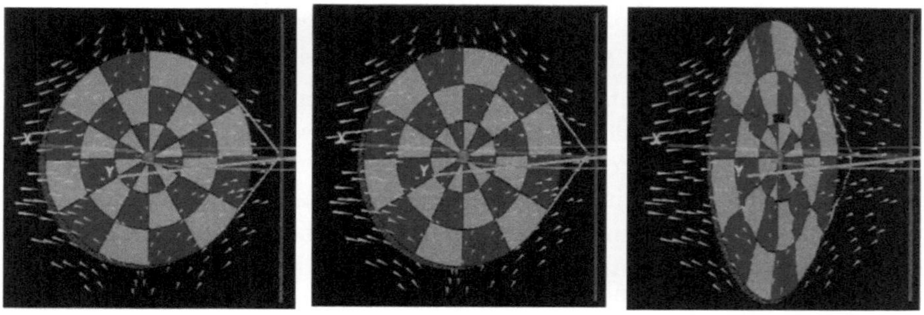

Fig. 7. Images resulting from one, two, and eleven applications of warp function W, i.e., $W(\mathcal{S})$, $W^2(\mathcal{S})$, and $W^{11}(\mathcal{S})$ (color plate 25, p. 380)

Thereby it is very easy to extend this system by a new visualization technique or an additional dynamical system. Dynamical systems are specified analytically within DynSys3D.

A Poincaré map is realized as a new *visualization technique* named MAP. It provides the core functionality of all the techniques presented here. Additionally modules TRAJECTORY, FLOW, SEEDLINE, and WARP have been implemented as separate modules which are connected to basis module MAP in case one of the previously described visualization mappings should be instantiated. These modules allow a flexible composition of an arbitrary number of visualization mappings.

Additionally we investigated a set of dynamical systems to test our techniques. We constructed five simple systems that exhibit most of the characteristics usually obtained in periodic systems. See the Appendix for a brief note on the construction of these periodic dynamical systems.

8 Conclusions

The use of Poincaré sections for the investigation and analysis of periodic or quasi-periodic dynamical systems has become an established technique quite a long time ago. The investigation of, e.g., chaotic dynamical systems together with informative pictures of Poincaré maps strengthened the importance of this technique. The vast majority of visualization techniques for Poincaré maps are based on 2D plots of 3D systems.

We propose several advances to this field in our paper. On the one hand we suggest some improvements of the planar visualization of Poincaré maps. We place arrow-like icons on the Poincaré section that connect x and $P(x)$. Thereby the dynamics induced by the Poincaré map is visualized. Alternatively a technique similar to Wijk's spot noise can be used to visualize the entirety of P. The visualization of a repeated application of P is also very important. We use spheres to depict series $\{P^j(x) \mid j \geq 0\}$.

In addition to the enhanced planar visualizations of Poincaré map P we discussed how an embedding of these techniques within the underlying dynamical system can be done. Drawing the entire trajectory which constructs series $\{P^j(x) \mid j \geq 0\}$ or just parts of it can give important information on the relation between points of the Poincaré map. Additional information about the characteristics of the underlying flow as, e.g., the number of windings of a Möbius band, is visualized, which cannot be derived from planar visualization. Using stream surfaces to enhance the visualization of the underlying flow improves the representation of the combined flow and Poincaré map.

Together with both approaches animation techniques must be revisited carefully. Meaningful applications can be found, for example, the consecutive construction of the Poincaré map by animating the underlying flow. Animation can be used as an additional dimension of visualization and thus provides additional possibilities. For example, the visualization of the entirety of repeated application of Poincaré map P^n can be visualized using a warp function and animation.

We implemented a set of dynamical systems to demonstrate the usefulness of our techniques. They exhibit most of the characteristics which can be found in 'real world' models. See URL `http://www.cg.tuwien.ac.at/research/vis/dynsys/Poincare97/` for a webpage with images and animations generated during this project. Due to the limited set of advanced visualization techniques in this field, we think that future work on the visualization of Poincaré maps, especially in higher dimensions ($n > 2$) is required.

9 Acknowledgements

The authors would like to thank Helmut Doleisch and Rainer Wegenkittl for helpful comments on preliminary versions of this paper.

References

1. R. H. Abraham and C. D. Shaw, *Dynamics – the geometry of behavior*, 2nd ed., Addison-Wesley, 1992.
2. Advanced Visual System Inc, *Avs developers guide, release 4*, May 1992.
3. V. I. Arnold, *Ordinary differential equations*, M.I.T. Press, Cambridge, MA, 1973.
4. D. K. Arrowsmith and C. A. Place, *An introduction to dynamical systems*, Cambridge University Press, 1990.

5. T. Beier and S. Neely, *Feature-based image metamorphosis*, Computer Graphics (SIGGRAPH '92 Proceedings) (E. E. Catmull, ed.), vol. 26, July 1992, pp. 35–42.

6. E. Doedel, *Software for continuation and bifurcation problems in ordinary differential equations (auto 86 user manual)*, 2nd ed., February 1986, See also URL `http://ute.usi.utah.edu/software/math/automan.html`.

7. J. H. Hubbard and B. H. West, *Differential equations: A dynamical systems approach: Ordinary differential equations*, Texts in Applied Mathematics, vol. 5, Springer, 1991.

8. J. P. M. Hultquist, *Constructing stream surfaces in steady 3d vector fields*, Proceedings Visualization '92, October 1992, pp. 171–177.

9. H. Löffelmann and E. Gröller, *DynSys3D: A workbench for developing advanced visualization techniques in the field of three-dimensional dynamical systems*, Proceedings of The Fifth International Conference in Central Europe on Computer Graphics and Visualization '97 (WSCG '97) (N. Thalmann and V. Skala, eds.), Plzen, Czech Republic, February 1997, pp. 301–310.

10. H. Löffelmann, L. Mroz, and E. Gröller, *Hierarchical streamarrows for the visualization of dynamical systems*, Accepted for publication at the 8th EG Workshop on Visualization in Scientific Computing, Boulogne sur Mer, France, April 1997.

11. H.-O. Peitgen, H. Jürgens, and D. Saupe, *Chaos and fractals – new frontiers of science*, Springer, 1992.

12. H. Poincaré, *Les méthodes nouvelles de la mécanique céleste (3 vols.)*, Gauthier-Villars, Paris, 1899.

13. S. H. Strogatz, *Nonlinear dynamics and chaos: with applications to physics, biology, chemistry, and engineering*, Addison-Wesley, 1994.

14. A. A. Tsonis, *Chaos – from theory to applications*, Plenum Press, 1992.

15. S. Wiggins, *Introduction to applied nonlinear dynamical systems and chaos*, 3rd ed., Texts in Applied Mathematics, vol. 2, Springer, 1996.

16. J. J. van Wijk, *Spot noise*, Proceedings SIGGRAPH '91, July 1991, pp. 309–318.

Appendix: Dynamical Systems – Examples

All the sample systems are 3D flows and exhibit at least one periodic cycle. As a basic example (used for Figs. 1, 5, & 7) we constructed system RealCyc which contains a saddle cycle with a linear Poincaré map, i.e., $P(x) = A \cdot x + b$, that evolves around the z-axis. It is given as follows:

$$\begin{aligned} \dot{x} &= a \cdot x - y \quad \text{with } a = A \cdot (r - 1/r) \text{ and } r = \sqrt{x^2 + y^2} \\ \dot{y} &= a \cdot y + x \\ \dot{z} &= C \cdot z \end{aligned}$$

The other sample systems were all together constructed along the same principle: First a 2D system is specified by $\dot{u} = f(u,v)$ and $\dot{v} = g(u,v)$. Functions f and g are modeled such that desired properties are guaranteed (see below). As the next step the 2D system is transformed into half-space $\{(a,b) \mid a \geq 0\}$ by transformation $a = e^u$ and $b = v$. Finally the 3D system is constructed by $x = a \cdot \cos\varphi$, $y = a \cdot \sin\varphi$, and $z = b + h(\varphi)$. By differentiation—assuming $\dot{\varphi} = \omega$ const.—the 3D flow is derived as follows:

$$\left.\begin{array}{l} \dot{x} = f(u,v) \cdot x - \omega \cdot y \\ \dot{y} = f(u,v) \cdot y + \omega \cdot x \\ \dot{z} = g(u,v) + \omega \cdot \dot{h}(\varphi) \end{array}\right\} \text{ with } \left\{\begin{array}{l} u = \ln a \text{ and } a = \sqrt{x^2+y^2} \\ v = z - h(\varphi) \text{ and } x \cdot \tan\varphi = y \\ \text{and } \varphi \text{ given by } x \cdot \tan\varphi = y \end{array}\right.$$

To complete the specifications of NLCYC1, STAR, and RTORUS, we list functions f, g, and h (assume $h(\varphi) \equiv 0$ for all but NLCYC1) that were used for our systems together with a brief explanation of the system parameters:

NLCYC1 – A 3D flow with a non-linear Poincaré map P containing one cycle. Stability characteristics of this cycle depend on A and B. If $C \neq 0$ an oscillation of frequency D is added collinear with the z-axis.

$$\begin{array}{ll} f(u,v) = A \cdot u + (2B - A) \cdot v^2 & \\ g(u,v) = B \cdot v & \text{and } h(\varphi) = C \cdot \cos(D \cdot \varphi) \end{array}$$

NLCYC1 is visualized in Figs. 2 & 6 .

NLCYC2 – A non-linear saddle cycle, constructed from a linear saddle by applying a rotational shear transformation: $(r,\varphi) \to (r, \varphi + D \cdot r)$.

$$\begin{array}{rl} \alpha\text{:} & u \cdot \tan\alpha = v \text{ and } r = \sqrt{u^2+v^2} \text{ and } \bar{r} = D \cdot r \\ f(u,v) = & (u - \bar{r} \cdot v) \cdot \cos\beta + r \cdot v \cdot \sin\beta \quad \text{with } \beta = 2 \cdot (\alpha - \bar{r}) \\ g(u,v) = & (v + \bar{r} \cdot u) \cdot \cos\beta - r \cdot u \cdot \sin\beta \end{array}$$

Refer to color plate 21, p. 378, for an image of NLCYC2.

RTORUS – A 3D flow which exhibits a torus-like invariant set of 'radius' $\sqrt{R}$ within 3D phase space. Its stability characteristics depend on A.

$$\begin{array}{l} f(u,v) = s \cdot u - v \quad \text{with } s = A \cdot (r - R/r) \text{ and } r = \sqrt{u^2+v^2} \\ g(u,v) = s \cdot v + u \end{array}$$

RTORUS was used for Figs. 3 & 4.

STAR – A flow exhibiting a non-hyperbolic saddle cycle, i.e., $\nabla P|_o = 0$. Parameter A controls the number of stable (instable) manifolds. Ω may be used to rotate Poincaré map P around its fixed point $(u,v) = (0,0)$.

$$\begin{array}{rl} \alpha\text{:} & u \cdot \tan\alpha = v \text{ and } r = \sqrt{u^2+v^2} \\ f(u,v) = & u \cdot \cos\beta + r \cdot v \cdot \sin\beta \quad \text{with } \beta = A \cdot (\alpha - \Omega) \\ g(u,v) = & v \cdot \cos\beta - r \cdot u \cdot \sin\beta \end{array}$$

See color plate 23, p. 379, for a visualization of dynamical system STAR.

Accuracy in 3D Particle Tracing

Adriano Lopes and Ken Brodlie

School of Computer Studies, University of Leeds, United Kingdom

Abstract. This paper presents a novel way of identifying and illustrating the accuracy in the particle tracing method for flow visualization. We make use of explicit Runge-Kutta methods for particle tracing in steady velocity fields, and describe three approaches to estimate the accuracy of the calculated path. These approaches are: re-integration (with smaller tolerance or in a backward direction), global error estimators and residuals in the velocity field. Visualization paradigms are also presented to convey data accuracy information and these ideas are implemented in an Open Inventor / IRIS Explorer environment.

1 Introduction

In scientific visualization, there is a responsibility to deliver some measure of reliability associated with the visualization that is created. This measure affects the validity of subsequent analysis, and therefore the credibility of conclusions drawn by the scientist. Errors in data visualization can arise from the data itself, or from the visualization process that turns numbers to images. Yet, accuracy is often taken for granted, maybe because it is quite difficult to define, characterize and even control its propagation throughout the visualization pipeline.

The GIS (Geographical Information Systems) community has been devoting considerable attention to this problem, particularly in the field of cartography. Their concern has mostly been with accuracy of the data, rather than the visualization process. General concepts like data quality and derived elements have been studied [5,14] but nonetheless the main emphasis still remains on visualization of positional accuracy in maps.

The scientific visualization community has belatedly taken an interest in the topic, and the work of Wittenbrink et al. [15,8] has illustrated a number of ways of representing the quality of the original data within the visualization.

In this paper we look at the accuracy of the visualization process, when a flow field is viewed as a trace of particles. Our aim will be not to create a new visualization system for this purpose, but rather try to extend an existing commercial dataflow system — IRIS Explorer.

The layout of the paper is as follows: in Sect. 2 the concept of accuracy is introduced within the context of the dataflow model. Then particle tracing and related accuracy work is presented in Sect. 3, while in Sect. 4 we describe in detail an architecture to visualize accuracy information. Some tests carried out are presented in Sect. 5 and in Sect. 6 conclusions and future work directions are presented.

2 Accuracy and the Dataflow Model

The dataflow paradigm [6] is the underlying model for most current visualization systems. It turns out we can readily extend the model to accommodate accuracy information.

We shall work with the variation of the Haber-McNabb model described by Brodlie [2]. Here the input "raw data" is regarded as samples of some underlying physical phenomenon to be visualized. Thus the visualization pipeline consists first of a modeling stage where an estimate, or "model", of the underlying field is created; then an abstract geometric representation is created in a mapping stage; and finally the "geometry" is turned to "image" in a rendering stage (see Fig. 1).

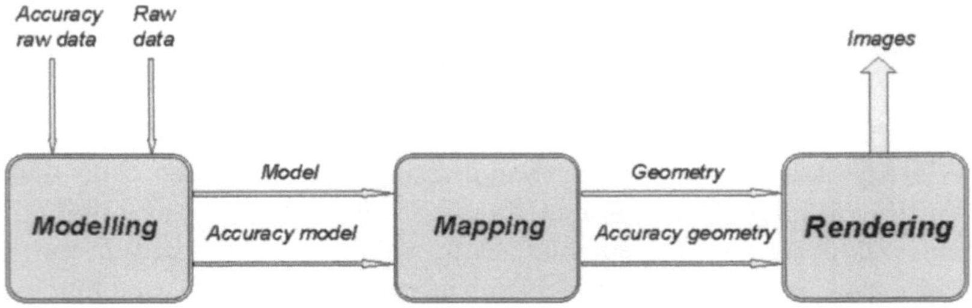

Fig. 1. Variation of dataflow model to accommodate accuracy information.

However, the model created may not be accurate; it is very likely that errors will be introduced at this stage — some due to the input data itself, some due to the model creation process (usually an interpolation). It is then desirable to have at the end of the modeling stage not only the assumed "model" but also information about its accuracy. We call it the "accuracy model". Thus, "model" and "accuracy model" better represent the true underlying phenomenon than the "model" itself. Similarly we include "accuracy raw data" as input to the modeling stage, for example to include metadata about the way the raw data was simulated or collected.

Next comes the mapping stage where again errors are introduced. If we want to be able to describe the geometric representation of the underlying physical model, we have to provide the assumed "geometry" information plus a measure of its accuracy. For example, if the computed geometry is a sphere defined by its center (x_c, y_c, z_c) and radius r then we can define its accuracy by giving a variation in the center $(\Delta x_c, \Delta y_c, \Delta z_c)$ and in the radius Δr.

Finally the same principles can also be applied to the rendering stage. The final picture will embody both the assumed "image" data and its related accuracy, as a result of all the transformations in the pipeline, in order to reflect as faithfully as possible the underlying physical model we are trying to understand.

It is worth noting that by accuracy we mean the closeness of the computed value to the "true" value. Of course in most (real) cases the true value is unknown and therefore we normally replace the "true value" by a value of higher accuracy than the one we are assessing. This will motivate us to study ways of gaining sources of higher accuracy.

This extended reference model will be the basis of our implementation of an accurate visualization of particle traces.

3 Particle Tracing

3.1 The Problem

Particle tracing is a common visualization technique used in many scientific areas, particularly in Computational Fluid Dynamics (CFD). In this context, the motion of a particle (massless) is based on the physical velocity from the flow field, and so the problem is to find out particle positions $\boldsymbol{p}$ given velocity data $\boldsymbol{V}(\boldsymbol{p}, t)$ and initial position at time t_0. That is, the numerical problem is stated as:

$$\frac{d\boldsymbol{p}}{dt} = \boldsymbol{V}(\boldsymbol{p}, t), \qquad \boldsymbol{p}(t_0) \text{ given.} \tag{1}$$

Errors are made at each integration step as (1) is solved. Even worse, they are enhanced as the integration proceeds. We consider a "local error" as the error made at the integration step while "global error" as a measure of the cumulative effects of errors from every time step including any start-up procedures. Therefore the global error can be defined as $\varepsilon_G = ||\boldsymbol{p}(t_i)^* - \boldsymbol{p}(t_i)||$, where $\boldsymbol{p}(t_i)^*$ is the exact position at step i and $\boldsymbol{p}(t_i)$ is the computed solution. Indeed, users are concerned with global error rather than local error.

There are a range of methods for the solution of Ordinary Differential Equations (ODEs); methods frequently used for particle tracing include Euler's method (only first order accurate), and Runge-Kutta methods of second and fourth order, with fixed or variable step size.

Another important aspect of particle tracing is the use of interpolation techniques to obtain velocities since the velocity field is usually given only at grid points. Different methods can be used but, as Buning mentions in [3], errors due to interpolation might be in some cases more significant than errors due to integration. For more details about particle tracing, see [10,4].

3.2 A Runge-Kutta Solver

Runge-Kutta methods with an adaptive time-stepping algorithm are widely used for steady flows. A typical particle tracing code proceeds as follows:

the input is taken as a set of velocity values on a regular grid. Seed point locations within a probe are generated, and all these particles are advected until they are out of domain or reach zero velocity. The calculation also stops after a specific time period, or if a pre-set maximum number of iterations has been reached [9].

Since step size is adjusted from integration point to integration point while particle positions are required to be output at a pre-set sample frequency, interpolation among integration points must be performed. We call this task "sampling" and it has no effect on the integration itself.

The computed sequence of closely spaced points are then joined by line segments to produce a piecewise-linear graphical representation of the particle trace.

3.3 Related work

The majority of related work looking at accuracy of particle tracing concentrates on a comparison of different methods and normally uses analytical models to assess the accuracy of these different methods. Darmofal and Haimes in [4] present an interesting numerical analysis of accuracy and stability of various particle tracing algorithms, particularly for unsteady flows. Lodha et al. in [8] visualize differences in particle traces obtained from different integration methods and/or different step sizes for each method. Walton compares results from two different methods in [13] and Knight and Mallison compare their own method with widely used ones [7].

In contrast, we are more interested in the analysis of a particular particle tracing algorithm rather than a comparison of different ones. Moreover, we concentrate on practical rather than theoretical data sets in order to meet the demands of scientists.

4 Accuracy Assessment in a Runge-Kutta Method

The primary issue is to establish how to assess accuracy of a given particle trace, say $\mathcal{P}$. If the exact solution is known in advance as in the case of analytical data sets, then the answer is quite obvious: the difference between the two solutions gives a measure of accuracy for $\mathcal{P}$. However, since we are interested in practical cases with no analytic solution, alternative approaches are needed.

Looking at the visualization pipeline, we realize there are many sources of errors. In the case of particle tracing, the model creation is the interpolation of a continuous field of velocity values from the gridded data. For now we shall assume that the model (created by either nearest neighbour or tri-linear interpolation) is accurate and therefore we are concerned here only with errors due to the Runge-Kutta integration and the associated sampler.

Taking into account error control strategies that are commonly used in the solution of ODEs, we use three approaches to derive accuracy measurement for particle tracing, which are explained in detail in Sect. 4.2. They are:

Re-integration: the problem is integrated at least twice and the discrepancy between solutions gives a measure of accuracy.

Global error estimators: as the problem is integrated, the magnitude of the global error at each integration point is estimated.

Velocity residuals: as the main trace is computed, a residual of velocity is also calculated at sampling times. That is, the difference is computed between an inferred first derivative computed by the solver and the velocity obtained from the data input. The bigger the residual the less the accuracy.

4.1 Architecture

The architecture we envisage to identify and visualize accuracy information is shown in Fig. 2. The main core is the numerical solver which has the task of computing particle traces. It can deliver:

- the main trace $\mathcal{P}$;
- other paths $\mathcal{Q}$ obtained by re-integration;
- inferred velocities to the unit responsible for computing the velocity residuals $\mathcal{R}$;
- global error estimators $\mathcal{E}$, directly from the integrator.

As explained above, the velocity field is calculated by interpolation from the input grid values — this field is assumed accurate for our purposes. Further down the pipeline, data values have to be properly delivered to the mapping units that subsequently produce geometries to be rendered. For instance, information contained in $\mathcal{Q}$ must match that in $\mathcal{P}$, in terms of time output. This is the responsibility of the synchronizer. It behaves like a filter but, very importantly, it does not perform any additional interpolation and therefore this unit is not itself a source of errors.

To implement this architecture we have built two new modules in the IRIS Explorer environment. They are:

GenPathAccuracy, which mainly deals with the generation of particle tracing data and measurement of its accuracy. It is able to deliver data synchronized in time according to a IRIS Explorer lattice format.

VisPathAccuracy, which takes the information $\mathcal{P}, \mathcal{Q}, \mathcal{R}$ and $\mathcal{E}$ from GenPathAccuracy and creates a visualization by passing geometries to a Render module.

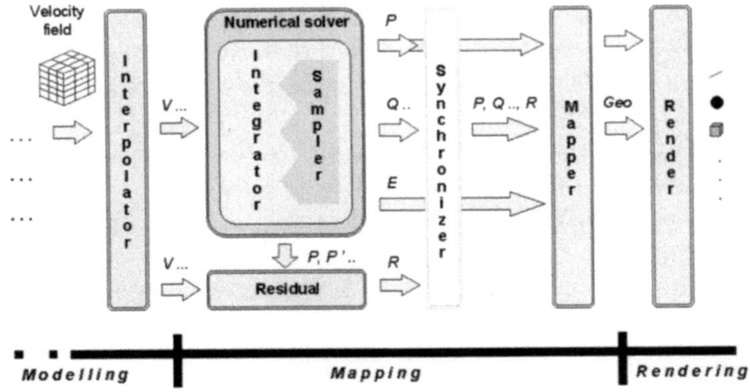

Fig. 2. Architecture to identify and visualize accuracy of particle tracing data.

4.2 GenPathAccuracy Module

This module is an extension of the IRIS Explorer NAGAdvectSimple module in order to incorporate accuracy measurements. The numerical solver is unchanged but extra functionalities are offered. Figure 3 shows the user interface of our new module, GenPathAccuracy, alongside that of NAGAdvectSimple. The similarities are so obvious that it will make it easy for a user to replace the standard module with our new accuracy module.

In respect to output, not only the geometry of $\mathcal{P}$ and lattices describing $\mathcal{P}, \mathcal{Q}, \mathcal{R}$ and $\mathcal{E}$ can be generated but also geometries concerning re-integrations with a sequence of different tolerances. This last feature gives the user additional insight into accuracy as the tolerance changes.

In the following we will give details about the numerical solver, the three techniques (re-integration, global error estimators and velocity residuals) and the data synchronizer. Notice that is up to the user to decide which technique(s) is (are) to be used in a particular trial.

Numerical Solver. We have used the routine *nag_ode_ivp_rk_onestep* for initial value problems in ODEs, from the NAG C library. There are two pairs of Runge-Kutta formulae available, with error orders of (2,3) and (4,5) respectively. At each step the integration result is from the higher-order formula (local extrapolation). The step size is adjustable and it is the difference between results of formulae in the pair that gives a local error estimator [1].

The local tolerance requirement followed in this module is a recipe based on the average magnitude of the vector over the field and the minimum values of the components of the vector throughout the field. To be more precise, the (positive) threshold values for each direction i are defined by a pre-set relative error tolerance multiplied by the global minimum of the ith component of the

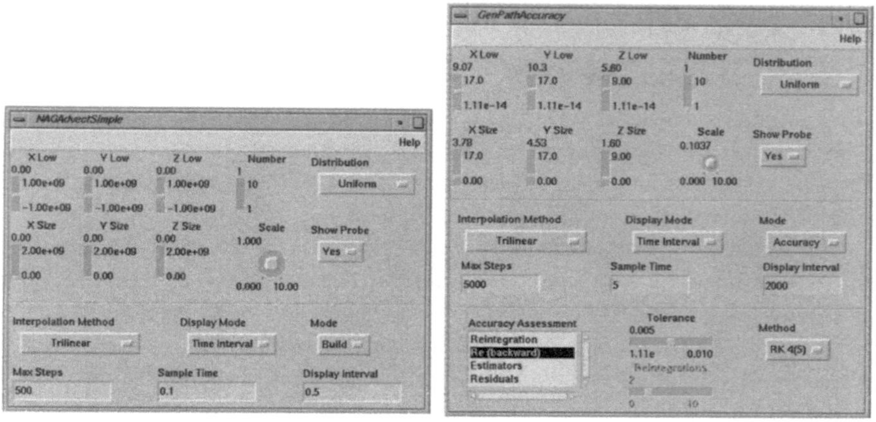

(a) (b)

Fig. 3. GenPathAccuracy (b) as an extension of NAGAdvectSimple (a).

vector throughout the field. However, if the absolute value of a component is less than the respective threshold value, then this component will be ignored in the error control.

Moreover, since particle positions are required at specific sampling times, the solver uses internal interpolants to do this. These interpolants are related to each Runge-Kutta formulae pair and have the same orders of error. All the computation is done in double precision.

Re-Integration. Here the problem is integrated again, looking for (different) solutions to be compared. So far we have implemented three variants. They are:

1. Performing a second integration with smaller relative error tolerance. The chosen reduction factor was 1/10.
2. Performing a second integration in backward direction; in this case the only difference from the first integration lies in the seed point which is the last particle position from the main trace — and of course the time direction which is reversed.
3. Performing additional integrations with different relative error tolerances, say reducing by 1/10, 1/100, 1/1000 and so on. This information can give the user a valuable insight about the trade-off between accuracy of solution and the cost in terms of computing time.

All the solutions obtained consist of particle positions and associated times.

Global Error Estimators. In contrast to local error assessment, used to control the integration process, global error assessment aims to look at the cumulative effects of local errors at all integration steps. It appears a natural way to assess accuracy.

If required, the NAG solver can provide the Random Mean Square (RMS) of the true error at any integration point. The RMS is taken from the beginning of the integration to the point of interest.

The underlying procedure is as follows: after each integration step, two half steps are taken over the same interval in a subsidiary integration. The assessment is related to the difference of results between the primary integration and the (more accurate) subsidiary integration [12].

However, according to Shampine and Gladwell in [11], the theory behind such an approach is no longer valid when the code is tried on problems with discontinuities in the derivative function, or even when the tolerances used are at the extremes of the range. As a result, if at one integration point the assessment is clearly implausible (for instance because the estimated local error in the subsidiary integration is not sufficiently smaller than that of the main integration) then it is sensible to assume that any assessment onwards is not valid as well.

As a note, the reported solution is slightly different when global error assessment is required. This is due to the fact that reported solutions are from the subsidiary integration, which is supposed to be more accurate.

Velocity Residuals. Unlike global error estimators, this technique is mainly used to provide accuracy information at sampling points. The assessment is made in the velocity data space and the deviations in velocity are returned as output. This emerges from the idea that an obtained solution can be regarded as the exact solution of a different problem and so it is acceptable to use the difference between the problems as a measure of accuracy.

In our case, the sampler of the numerical solver can compute not only positions between integration steps but also corresponding first derivatives, all with the same order of accuracy. Therefore we can make use of this information and define the residual $\boldsymbol{r}$ at each point as the difference between this first derivative and the value we obtain if we interpolate the velocity field at that position. Thus, the defining equation is of the form:

$$\boldsymbol{r}(t) = \frac{d\boldsymbol{p}}{dt} - \boldsymbol{V}(\boldsymbol{p}(t)) \,. \tag{2}$$

We regard this as a check on the accuracy of the interpolation carried out by the sampling process.

Data Synchronizer. The aim of this filter is to make sure data will be suitable for visualization. It is worth noting for example that there is no absolute guarantee solutions obtained by re-integration are returned at exactly the same sampling times as in the main trace.

All data obtained by re-integration and velocity residuals techniques have to be filtered in order to include only those whose times are explicitly available in the main trace, say the set of times $\mathcal{T}$. The main trace does not require filtering at all and is defined as 3D positions and times at constant intervals. However, the last position might not follow this rule as would be the case when a particle leaves the domain. In this case the exact time and position has to be recorded as well. With respect to error estimators, this data should be passed through untouched because it has not necessarily been obtained at sampling times.

As a result data is delivered according to the following format:

$$\begin{array}{lll} \mathcal{P}_{\;x,\; y,\; z,\; (t)}, & t \in \mathcal{T} \cup \{t_{n+1}\}, & t_{n+1} - t_n \leq \alpha, \\ \mathcal{Q}_{\;x,\; y,\; z,\; (t)}, & t \in [\; t_{q_i} \dots t_{q_k} \;] \subseteq \mathcal{T}, & \Delta t = \pm\alpha, \\ \mathcal{E}_{\;x,\; y,\; z,\; |\delta x|,\; |\delta y|,\; |\delta z|,\; (t)}, & t \in [\; t_{e_0} \dots t_{e_k} \;] \subseteq \mathcal{T}, & \\ \mathcal{R}_{\;\delta v_x,\; \delta v_y,\; \delta v_z,\; (t)}, & t \in [\; t_{r_0} \dots t_{r_k} \;] \subseteq \mathcal{T}, & \Delta t = \alpha, \end{array} \tag{3}$$

where $\mathcal{T} = [\; t_0 \dots t_n \;]$, $\Delta t = \alpha$, and $\mathcal{P}$, $\mathcal{Q}$, $\mathcal{E}$ and $\mathcal{R}$ are respectively the particle trace, re-integration, global error estimators and velocity residuals solutions.

4.3 VisPathAccuracy module

The VisPathAccuracy module takes data generated by the GenPathAccuracy module and performs appropriate visualization according to what the user selects. The former module is shown in Fig. 4.

Features. The data sets mentioned in (3) are input as lattices. Only $\mathcal{P}$ is enforced to be input; the others including the color map are optional. Based on the input data, at any time the user can:

- select the group of particles to be analyzed;
- scale visualization primitives, one for each type of data set;
- select a visualization paradigm for each accuracy data type;
- adjust visualization attributes like transparency, complexity, etc.
- query accuracy data and then visualize textual information.

All the visual primitives are generated as Open Inventor nodes and then passed to the IRIS Explorer environment.

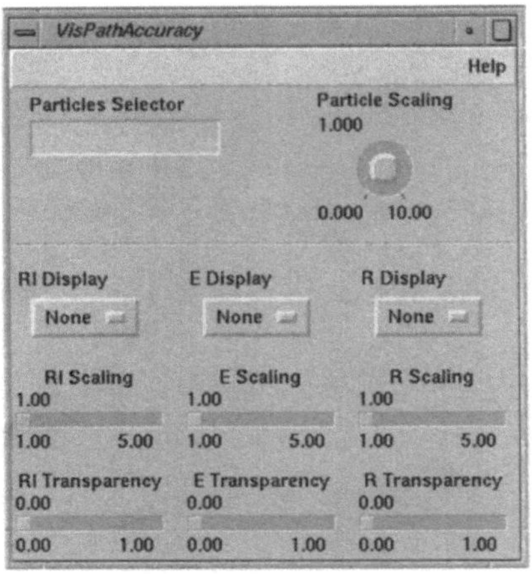

Fig. 4. VisPathAccuracy control panel.

Visualization Paradigms. Our aim is to provide meaningful pictures as far as the accuracy information is concerned. In addition we aim to offer a useful set of options; users may well have different opinions about which visualization technique is the most effective. For consistency, the options are provided within a common framework for visualizing data accuracy information.

The visual relation between the particle trace and its accuracy is combined in one picture. However, visual attributes for each set of graphical primitives can be independently adjusted to enhance the comparison. The accuracy analysis can be indeed an interactive process.

The particle trace itself is represented in a traditional way: a polyline through particle positions and a sphere located at each of these positions. As a rule, consecutive spheres should imply constant variation in time.

For each accuracy data type, there are visualization techniques based on 1D, 2D, and 3D primitives. In general, in re-integration we want to represent distances and direction in space whereas in estimators the aim is to represent the magnitude of the distances along each axis. Finally, in the case of velocity residuals we are interested on variations of velocity. The latter is quite different from the others in the sense that accuracy is shown in a different data space and so the user should be aware of that.

The basic visualization technique is based on *cuboids* aligned according the x, y, z axes, with edges proportional to the accuracy of data in each direction. It is therefore possible to acknowledge differences from axis to axis. If that information is not so important then a set of *lines* can be used. Both of these two approaches are meant to represent a discrete world. If the representation of a continuum domain is required, then we offer respectively *tube* and *strip* techniques. Notice that drawing styles can always be changed at render stage, for example representing the primitives as wireframes. These techniques are shown in the Appendix, p. 381.

In addition to these visualization techniques the user is offered the possibility to query accuracy information at sampling points in the particle trace. The user might feel in some cases that the accuracy data is so small — it is why scaling options are also offered — that a efficient way to acknowledge it is by means of textual representation.

5 Results

The main tests carried out so far were based on a realistic data set representing the air flow in the space between two parallel sheets of glass — one sheet cold, one sheet warm — a simulation of the gap in double glazing. The data set was obtained from the solution of the Navier-Stokes equation using a multigrid code. This dataset was input as a uniform lattice.

The examples shown in the color plates were obtained using the pair Runge-Kutta 4(5) and tri-linear interpolation.

Plate 1, p. 381, shows three paths of one particle using different relative tolerances — 0.005 and then reduced by 1/10 and 1/100 — and colored according to the work required for integration. The red one corresponds to the smaller tolerance whereas the green one to the bigger tolerance. The region where differences become very large deserves particular attention. We then might conclude that velocity in that area changes more dramatically.

Plate 2 is a variation of the former, but with different seed point and sampling interval. As in Plate 1, the *multi-strip* is colored regarding the cost of each integration.

Plates 3 and 4 correspond to the same problem as Plate 2. Plate 3 shows a yellow *tube* representing a particle path and its accuracy, derived from re-integration with tolerance reduced by 1/10. Again, at some point differences start to increase significantly.

Plate 4 represents accuracy information gathered from re-integration in a backward direction, shown as a yellow *strip*. This is cheaper in terms of computational time (using smaller tolerance implies smaller integration step sizes and so more steps are required). However, this should only be taken as an indicator of accuracy and not as a possible solution for the problem. When differences are small we can accept that there is no need for additional tests in order to confirm the accuracy of the main solution.

6 Conclusion and future work

It is well known that it is important to have a measure of reliability associated with a final picture. We believe that the ideas used to identify and measure the accuracy of particle tracing data have proven to be valid, particularly if taken as a whole set of options.

Some techniques have clearly shown what the differences can be if we solve the same problem in a different way. In the case of re-integration it is not even necessary to change the formulae; only a change in a scalar parameter like relative tolerance can be enough. These differences are very much dependent upon the data.

Our focus has been on providing the user with an indication of the error associated with the main path, $\mathcal{P}$. In cases where error estimation involves the calculation of other path $\mathcal{Q}$, supposed to be more accurate, then the user can obviously take this path as the solution.

In respect to global error estimators technique, the values obtained were not large. Also, some tests carried out lead to situations where these estimators were not available from some point onwards. As mentioned in Sect. 4.2, this happens because the integration stops when there is no guarantee about the reliability of such estimators. It is worth remembering that a good understanding of the details of the techniques is always crucial.

As far as visualization paradigms is concerned, the techniques available are so far sufficient for our purposes. Future research will focus on visual perception issues.

This work so far has focussed on errors due to integration. There is also a modeling error, due to interpolation of the input data. We are currently studying the inter-relation of these two sources of error.

Both modules are publicly available at the module repository of the IRIS Explorer Center of Excellence, University of Leeds, United Kingdom (URL http://www.scs.leeds.ac.uk/iecoe).

7 Acknowledgements

The authors would like to thank Jeremy Walton of NAG Ltd for supplying the NAGAdvectSimple code and Andy Sleigh and Tim David for his practical data sets. Adriano Lopes is also grateful to his sponsors, the Portuguese Programme PRAXIS XXI under contract BD / 2756 / 94 and the Department of Mathematics of the University of Coimbra, Portugal.

References

1. R. Brankin, I. Gladwell, and L. Shampine, *Rksuite: a suite of explicit runge-kutta codes for the initial value problem for odes softreport 91-s1*, Tech. report, Department of Mathematics, Southern Methodist University, Dallas, TX 75275, USA, 1991.

2. K. BRODLIE, *A typology for scientific visualization*, Visualization in Geographical Information Systems (H. HEARNSHAW AND D. UNWIN, eds.), John Wiley & Sons, 1994, pp. 34–41.

3. P. BUNING, *Sources of error in the graphical analysis of cfd results*, Journal of Scientific Computing *3*:2 (1988), 149–164.

4. D. DARMOFAL AND R. HAIMES, *An analysis of 3-d particle path integration algorithms*, Journal of Computational Physics *123* (1995), 182–195.

5. M. GOODCHILD, B. BUTTENFIELD, AND J. WOOD, *Introduction to visualizing data validity*, Visualization in Geographical Information Systems (H. HEARNSHAW AND D. UNWIN, eds.), John Wiley & Sons, 1994, pp. 141–149.

6. R. HABER AND D. MCNABB, *Visualization idioms: A conceptual model for scientific visualization systems*, Visualization in Scientific Computing (G. NIELSON, B. SHRIVER, AND L. ROSENBLUM, eds.), IEEE Computer Society Press, 1990, pp. 75–93.

7. D. KNIGHT AND G. MALLINSON, *Visualizing unstructured flow data using dual stream functions*, IEEE Transaction on Visualization and Computer Graphics *2*:4 (1996), 355–363.

8. S. LODHA, A. PANG, R. SHEEHAN, AND C. WITTENBRINK, *Uflow: Visualizing uncertainty in fluid flow*, Proceedings Visualization '96 (R. YAGEL AND G. NIELSON, eds.), ACM Press, 1996, pp. 249–254.

9. NAG, *Iris explorer*, URL http://www.nag.co.uk.

10. F. POST AND T. VAN WALSUM, *Fluid flow visualization*, Focus on Scientific Visualization, Springer Verlag, 1993, pp. 1–40.

11. L. SHAMPINE AND I. GLADWELL, *The next generation of runge-kutta codes*, Computational Ordinary Differential Equations (J. CASH AND I. GLADWELL, eds.), Oxford University Press, 1992, pp. 145–164.

12. L. SHAMPINE AND H. WATTS, *Global error estimation for ordinary differential equations*, ACM Transactions on Mathematical Software *2*:2 (1976), 172–186.

13. J. WALTON, *Visualization benchmarking: A practical application of 3d publishing*, 14th Eurographics UK Chapter Conference Proceedings, vol. 2, Imperial College of London, March 1996, pp. 339–351.

14. F. V. D. WEL, R. HOOTSMANS, AND F. ORMELING, *Visualization of data quality*, Visualization in Modern Cartography (A. MACEACHREN AND D. TAYLOR, eds.), Modern Cartography, vol. 2, Elsevier Science Ltd, 1994, pp. 313–331.

15. C. WITTENBRINK, E. SAXON, J. FURMAN, A. PANG, AND S. LODHA, *Glyphs for visualizing uncertainty in environmental vector fields*, SPIE & IS&T Conference Proceedings on Electronic Imaging: Visual Data Exploration and Analysis, vol. 2410, SPIE, February 1995, pp. 87–100.

Clifford Algebra in Vector Field Visualization

Gerik Scheuermann[1], Hans Hagen[1], and Heinz Krüger[2]

[1] Computer Science Department, University of Kaiserslautern, Germany
[2] Physics Department, University of Kaiserslautern, Germany

Abstract. The visualization of vector fields is still based on piecewise linear approximation. This is fast and good enough in large areas but has drawbacks if the non-linear behavior of a field has local topological implications like close simple critical points or higher order singularities. This article introduces the concept of Clifford algebra into the visualization of vector fields to deal with these difficulties. It derives a close relationship between the description of some polynomial 2D vector fields in Clifford algebra and their topology, especially the index and the position of critical points. This is used to develop an algorithm for vector field visualization without the problems of conventional methods.

1 Introduction

There is a big interest in methods for the visualization of vector fields because of their ability to describe a large area of problems in mathematics, natural sciences and engineering. The reason is simply that they are the geometric equivalent of differential equations.

The role of visualization is usually to show the results of a numerical simulation or an experiment on a screen with the help of computer graphics. The emphasis has been more and more on structural information like topology in the last years compared to simply showing arrows or streamlines in the beginning. In most cases one starts with discrete data on a structured or unstructured grid and uses piecewise linear interpolation to get a vector field in the whole space. In the view of topology this is a good and fast approach in all areas where only well separated simple critical points occur. If there are close simple critical points or critical points of higher complexity, this is not a good idea because these topological patterns are impossible in a linear field and a piecewise linear approach produces errors as we shall see in the last section.

Another interesting point is that nearly all approaches use Cartesian coordinates in an Euclidean vector space to describe the geometry ignoring the fact that there are sometimes good reasons for using different mathematical theories describing geometry.

We want to show a solution to the problems of linear approximations in two dimensions. This is done by using Clifford algebra to describe the vector field. This allows a direct relation between topology and coordinates. For special polynomial fields we derive this in an explicit way in section 3. This gives the basis for the algorithm and the results in the remaining sections.

2 Clifford Algebra

We need Clifford algebra only in two dimensions, so for simplicity we should stay there. Let $\{e_1, e_2\}$ be an orthonormal basis of R^2 with standard scalar product. The Clifford algebra G_2 is then the R-algebra of maximal dimension containing R and R^2 such that for each vector $x \in R^2$ holds $x^2 = ||x||^2$.

This implies the following rules

$$e_j^2 = 1 , \qquad j = 1,2$$
$$e_1e_2 + e_2e_1 = 0 .$$

We get a 4-dimensional algebra with the vector basis $\{1, e_1, e_2, e_1e_2\}$. We now have the real vectors

$$xe_1 + ye_2 \in R^2 \subset G_2$$

and the real numbers

$$a1 \in R \subset G_2$$

both in the algebra.

We may also define the *projections* $< \cdot >_k$, $k = 0, 1, 2$ by

$$\begin{aligned} < \cdot >_0 &: G_2 \to R \subset G_2 \\ a1 + be_1 + ce_2 + de_1e_2 &\mapsto a1 \\ < \cdot >_1 &: G_2 \to R^2 \subset G_2 \\ a1 + be_1 + ce_2 + de_1e_2 &\mapsto be_1 + ce_2 \\ < \cdot >_2 &: G_2 \to Re_1e_2 \subset G_2 \\ a1 + be_1 + ce_2 + de_1e_2 &\mapsto de_1e_2 \end{aligned}$$

For two vectors one can then describe the new product by already known products. Let $v = v_1e_1 + v_2e_2$, $w = w_1e_1 + w_2e_2$ be two vectors, $v_1, v_2, w_1, w_2 \in R$. Then

$$\begin{aligned} vw &= (v_1w_1 + v_2w_2)1 + (v_1w_2 - w_1v_2)e_1e_2 \\ &= < vw >_0 + < vw >_2 \\ &= v \cdot w + v \wedge w \end{aligned}$$

where $\cdot$ denotes the scalar (inner) product and $\wedge$ denotes the outer product of Grassmann. This unification of inner and outer product is the starting point of the geometric interpretation. We will not need it here, so see [3] for a good introduction.

More important for us is that the complex numbers can also be canonically embedded by recognizing $(e_1e_2)^2 = -1$, so set

$$i := e_1e_2$$

and it is

$$a1 + bi \in C \subset G_2$$

a subalgebra.

The next section describes a vector field in Clifford algebra and how one can use this for better insights in their behavior in some cases.

3 Clifford Analysis

Here we need again only the two-dimensional case, so we limit our definitions to that case to avoid technical overload.

Our basic maps will be multivector fields

$$\begin{aligned} A : R^2 &\to G_2 \\ r &\mapsto A(r) \,. \end{aligned}$$

A Clifford vector field is just a multivector field with values in $R^2 \subset G_2$

$$\begin{aligned} v : R^2 &\to R^2 \subset G_2 \\ r = xe_1 + ye_2 &\mapsto v(r) = v_1(x,y)e_1 + v_2(x,y)e_2 \,. \end{aligned}$$

The directional derivative of A in direction $b \in R^2$ is defined by

$$A_b(r) = \lim_{\varepsilon \to 0} \frac{1}{\varepsilon}[A(r + \varepsilon b) - A(r)], \quad \varepsilon \in R \,.$$

This allows the definition of the vector derivative of A at $r \in R^2$

$$\begin{aligned} \partial A(r) : R^2 &\to G_2 \\ r &\mapsto \partial A(r) = \sum_{k=1}^{2} g^k A_{g_k}(r) \,. \end{aligned}$$

This is independent of the basis $\{g_1, g_2\}$ of R^2.

The integral in Clifford analysis is defined as follows: Let $M \subset R^2$ be an oriented r-manifold and $A, B : M \to R^2$ be two piecewise continuous multivector fields. Then one defines the integral

$$\int_M AdXB$$

as the limes

$$\lim_{n \to \infty} \sum_{i=0}^{n} A(x_i) \Delta X(x_i) B(x_i)$$

where $\Delta X(x_i)$ is a r-volume in the usual Riemannian sense. This allows the definition of the Poincaré-index of a vector field v at $a \in R^2$ as

$$ind_a v = \lim_{\varepsilon \to 0} \frac{1}{2\pi i} \int_{S^1_\varepsilon} \frac{v \wedge dv}{v^2}$$

where S^1_ε is a circle of radius ε around a.

For our result it is necessary to look at $v : R^2 \to R^2 \subset G_2$ in a different way. Let $z = x + iy$, $\bar{z} = x - iy$. This means

$$x = \frac{1}{2}(z + \bar{z}) \qquad y = \frac{1}{2i}(z - \bar{z}).$$

We get

$$\begin{aligned} v(r) &= v_1(x, y)e_1 + v_2(x, y)e_2 \\ &= [v_1(\frac{1}{2}(z + \bar{z}), \frac{1}{2i}(z - \bar{z})) - iv_2(\frac{1}{2}(z + \bar{z}), \frac{1}{2i}(z - \bar{z}))]e_1 \\ &= E(z, \bar{z})e_1 \end{aligned}$$

where

$$\begin{aligned} &E : C^2 \to C \subset G_2 \\ &(z, \bar{z}) \mapsto v_1(\frac{1}{2}(z + \bar{z}), \frac{1}{2i}(z - \bar{z})) - iv_2(\frac{1}{2}(z + \bar{z}), \frac{1}{2i}(z - \bar{z})) \end{aligned}$$

is a complex-valued function of two complex variables. The idea is now to analyze E instead of v and get topological results directly from the formulas in some interesting cases.

We assume that E is a polynomial splitting into linear factors in the following way :

$$E(z, \bar{z}) = a \prod_{j=1}^{n} (z - z_j)^{l_j} (\bar{z} - \bar{z}_j)^{m_j}$$

with $a, z_1, \ldots, z_n \in C$. Then we have

Theorem 1. *Let $v : R^2 \to R^2 \subset G_2$ be the vector field $v(r) = E(z, \bar{z})e_1$. Then the only zeros of v are $z_1, \ldots, z_n$ and the Poincaré-index at z_j is $m_j - l_j$.*

Proof. The z_j are the only zero because $Ee_1 = 0 - E = 0$. For the indices we write E as

$$E(z, \bar{z}) = a \prod_{j=1}^{N} F_j(z, \bar{z})$$

with

$$F_j(z, \bar{z}) = \begin{cases} z - z_j \\ \bar{z} - \bar{z}_j \end{cases}$$

For the derivatives we get

$$\frac{\partial E}{\partial z} = a \sum_{j=1}^{N} \frac{\partial F_j}{\partial z} \prod_{k=1,k\neq j}^{N} F_k$$

$$\frac{\partial E}{\partial \bar{z}} = a \sum_{j=1}^{N} \frac{\partial F_j}{\partial \bar{z}} \prod_{k=1,k\neq j}^{N} F_k \,.$$

For the computation of the Poincaré-index we assume $z_{j_0} = 0$ after a change of the coordinate system and that ε is so small that there are no other zeros inside S^1_ε. We get

$$\begin{aligned}
ind_{z_{j_0}} v &= \frac{1}{2\pi i} \int_{S^1_\varepsilon} \frac{v \wedge dv}{v^2} \\
&= \frac{1}{2\pi i} \int_{S^1_\varepsilon} \frac{1}{v^2} < a \prod_{j=1}^{N} F_j(z,\bar{z}) e_1 [dza \sum_{j=1}^{N} \frac{\partial F_j}{\partial z} \prod_{k=1,k\neq j}^{N} F_k + \\
&\quad d\bar{z} a \sum_{j=1}^{N} \frac{\partial F_j}{\partial \bar{z}} \prod_{k=1,k\neq j}^{N} F_k] e_1 >_2 \\
&= \frac{1}{2\pi i} \int_{S^1_\varepsilon} \frac{1}{v^2} < a\bar{a} \sum_{j=1}^{N} F_j \overline{(dz \frac{\partial F_j}{\partial z} + d\bar{z} \frac{\partial F_j}{\partial \bar{z}})} \prod_{k=1,k\neq j}^{N} F_k \bar{F}_k >_2 \\
&= \sum_{j=1}^{N} \frac{1}{2\pi i} \int_{S^1_\varepsilon} \frac{1}{F_j \bar{F}_j} < F_j e_1 (dz \frac{\partial F_j}{\partial z} + d\bar{z} \frac{\partial F_j}{\partial \bar{z}}) e_1 >_2 \\
&= \sum_{j=1}^{N} ind_{z_j} F_j e_1 \\
&= l_j - k_j \,.
\end{aligned}$$

The last equation is easily seen by figure 1 showing the vector fields $v(r) = ze_1$ and $v(r) = \bar{z}e_1$, or by simple computation.

The importance of Clifford algebra lies in our eyes in a direct relation between the complex numbers and the real vectors describing the vector field. It gives a canonic way to come from one description to the other by pure algebra without defining an isomorphism allowing several choices. The proof of the theorem shows the elegance of this approach.

4 Vector Field Visualization Using Clifford Algebra

The last section opens the way for a visualization of vector fields without topological restrictions coming from piecewise linear approaches.

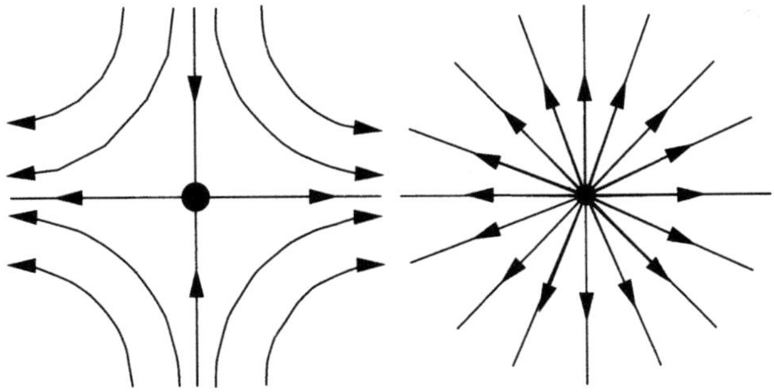

Fig. 1. $v(r) = ze_1$ and $v(r) = \bar{z}e_1$.

Our central point is the fact that the distance between critical points in conventional approaches depends on the grid because each linear cell has usually only one critical point. In two dimensions one can see this very easily in an unstructured grid consisting of triangles. If one approximates the triangles by linear interpolation then each triangle contains only one critical point but in reality there may be more inside. The key for a solution is to analyze the data to get information about the number and index of critical points and to choose an approximation in the light of the theorem to allow several critical points if necessary. Outside the areas with more than one critical point we still use linear interpolation to keep the algorithm fast.

The basic idea is that a critical point has topological implications into the field if its Poincaré-index is different from 0. One may look at figure 2. There are two close saddles in one triangle, but in a piecewise linear approximation there will be two different triangles containing one saddle each.

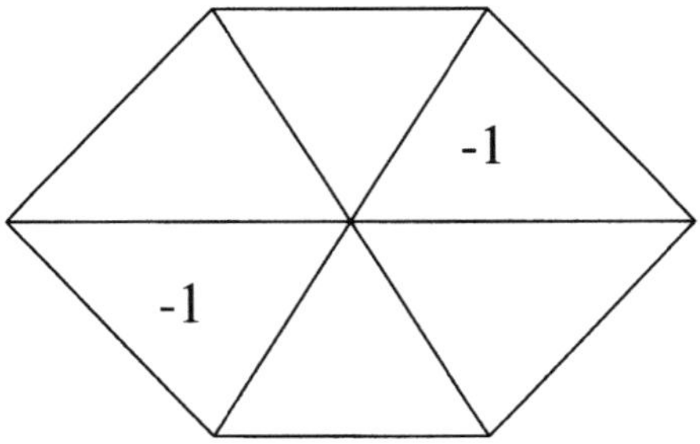

Fig. 2. Indices of triangles around a triangle with two saddles.

This behavior tells how to find such situations. If several critical points are in the same cell and have the same index, one will notice close cells with critical points of that index in the linear approach. This areas can be found in a first step and then one can approximate with polynomials like the ones in the last section. In our example above one could use

$$v(r) = b(z - a_1)(z - a_2)e_1$$

in the area and then one can get the saddles in the same triangle.
Our algorithm includes therefore the following steps

(1) Compute the Poincaré-index around each triangle assuming linear interpolation along the edges. One gets -1, 0 or +1.
(2) Build the regions of close triangles with possible higher order critical points.
   (a) If there are two triangles with a common edge and opposite index as on the left of figure 3, mark them and save the neighboring connection.
   (b) If there are unmarked triangles with the same index and a common vertex, put all the triangles with that vertex in a region as on the right of figure 3. If one of the triangles was marked in (a), put its neighbor also in the region as on the left of figure 4. If any of the triangles is already in a region, do not build this region. Otherwise mark all the triangles in this new region.
   (c) If there are unmarked triangles A and B with the same index and a triangle C with a common edge with A and a common vertex V with B as on the right of figure 4, build a region consisting of A and all the triangles having V as vertex. Like (b) look for triangles which have been marked in (a) and make sure to put the neighbors always also in the region. Again, if any triangle in this new region is already in a region, do not build this region.
(3) Compute the index of each region by just adding the index of all triangles in that region. Then set up a polynomial approximation of the type

$$v(r) = b(\bar{z} - \bar{a_1})(\bar{z} - \bar{a_2}) \cdots (\bar{z} - \bar{a_n})e_1$$

for index $+n$, $n \in N$, or

$$v(r) = b(z - a_1)(z - a_2) \cdots (z - a_n)e_1$$

for index $-n$, $n \in N$ and fit $a_1, \ldots, a_n, b$ by a least square fit.

We are not maintaining continuity across the boundaries at the moment but are thinking about some kind of blending in an area close to the boundary of our regions to solve this problem.

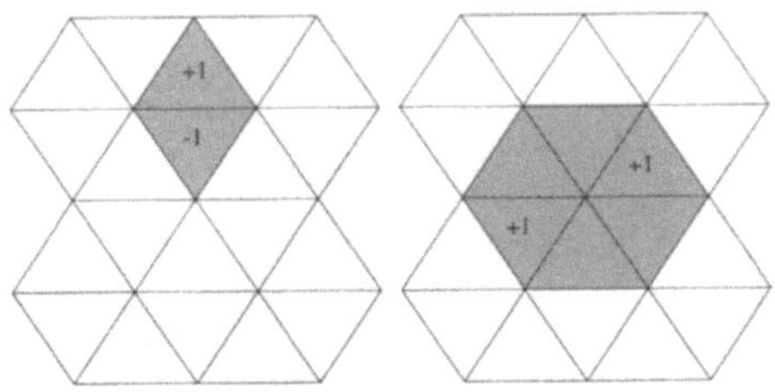

Fig. 3. Triangles with opposite indices and a region with index +2.

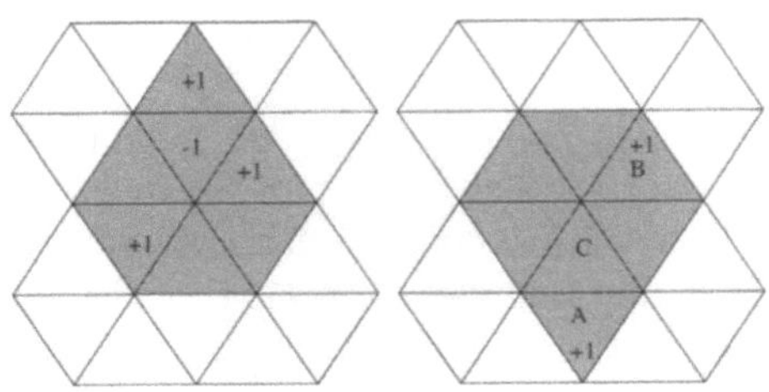

Fig. 4. A +2-region with +1/-1-pair and a more complicated region.

5 Results

We want to show two examples of the algorithm.

The first example is a monkey saddle on a 20×20 quadratic grid with coordinates $[-1,1] \times [-1,1]$. Plates 36 and 37 on page 387 show the piecewise linear approximation and a zoom into the interesting part. Plates 38 and 39 show the results obtained by using our algorithm. The second example consists also of a 20×20 quadratic regular grid with Cartesian coordinates $[-1,1] \times [-1,1]$. The field contains a source and a sink close together and two saddles and a second source scattered over the remaining area. The linear approximation produces again two extra critical points as can be seen in plate 40 and 41 on page 388. The Clifford algorithm removes this artifact and puts the source and the sink closer together as can be seen in plate 42 and 43. These two examples show the limits of piecewise linear approaches and how methods based on Clifford algebra can help to do better.

6 Acknowledgement

This work was partly made possible by financial support by the Deutscher Akademischer Auslandsdienst (DAAD). The first author got a "DAAD-Doktorandenstipendium aus Mitteln des zweiten Hochschulsonderprogramms" for his stay at the Arizona State University from Oct. 96 to Jan. 97. We also want to thank Alan Rockwood, Greg Nielson and David Hestenes from Arizona State University for many comments, suggestions and inspiration.

References

1. V. I. ARNOLD, *Gewöhnliche Differentialgleichungen*, Deutscher Verlag der Wissenschaften, Berlin, 1991.

2. J. L. HELMAN, L. HESSELINK, *Visualizing vector field topology in fluid flows*, IEEE Computer Graphics and Applications *11*:3 (1991), 36–46.

3. D. HESTENES, *New Foundations for classical mechanics*, Kluwer Academic Publishers, Dordrecht, 1986.

4. H. KRÜGER, M. MENZEL, *Clifford-analytic vector fields as models for plane electric currents*, Analytical and Numerical Methods in Quaternionic and Clifford Analysis (W. SPRÖSSIG, K. GÜRLEBECK, eds.), Seiffen, 1996.

5. G. SCHEUERMANN, H. HAGEN, H. KRÜGER, R. ROCKWOOD, *Examples of Clifford vector fields in two dimensions*, Technical Report, Arizona State University, 1997.

6. G. SCHEUERMANN, H. HAGEN, H. KRÜGER, M. MENZEL, R. ROCKWOOD, *Visualization of higher order singularities in vector fields*, submitted to IEEE Visualization 1997.

Visualization of Complex ODE Solutions

Laurent Testard

LMC–IMAG, Tour des Maths, Grenoble, France

Abstract. This paper presents a visualization method for Complex Ordinary Differential Equations (*CODEs*) based on real extended phase portraits. This technique enables to visualize functions of classical ODE theory and supports the visual interpretation of numerical integration results. The paper comprises three parts: firstly a presentation of the problem and related existing methods, secondly an introduction to extended phase portraits and their application to complex equations, and thirdly two applications of this visualization technique.

1 Introduction

1.1 Complex Ordinary Differential Equations

Complex Ordinary Differential Equations (CODEs) arise in many domains, such as:

- Mathematical studies of real differential equations. The behavior of the solutions of real ordinary differential equations may be affected by complex phenomena, like the vicinity of arbitrarily near complex singularities. It is to be noted that isolated (in the complex plane) real singularities can be avoided by choosing a complex path around the singularity and performing the integration on this complex path [1].
- Studies of physical phenomena, such as liquid flows on surfaces [9], or many other physical processes.

Our aim is to visualize the solutions of differential equations

$$y' = f(x, y) \tag{1}$$

where x belongs to a parameterized path $\mathcal{P}$ in the complex domain, given an initial condition (x_0, y_0) with $x_0 \in \mathcal{P}$, and $y_0 \in \mathbb{C}$.

If f is analytic in a domain $D \subset \mathbb{C}^2$, and if $(x_0, y_0) \in D$, then the problem has exactly one solution y, analytic in the vicinity of x_0, such that $y(x_0) = y_0$. This existence and uniqueness theorem is the complex version of the Cauchy-Lipschitz theorem, see e.g. [5].

Many methods exist for the computation of differential equation solutions, some are numerical (mainly adaptations of real numerical methods, like complex Runge-Kutta or spectral methods) and others are formal (e.g. consisting in the computation of a formal series that obeys term-to-term the differential equation). This paper focuses on the problem of visualizing the solution y, not on the problem of computing $y(x)$. We assume that we can numerically compute the values of the solution y for every point x of a path $\mathcal{P}$.

When interpreting the data, one of the problems is their visualization: the results are both complex functions and solutions of differential equations. Visualization of these solutions as complex functions is often not sufficient. Additionally data from numerical CODE integration itself are of interest.

1.2 Visualization of Complex Functions

Complex functions generally need *four* dimensions to be completely visualized. Several methods exist:

1. Visualization as a *pair of real functions* (real and imaginary parts, or modulus and argument pair). But this method does not provide a comprehensive visualization of the complex function.
2. Visualization as a *colored graph* by depicting a real function (the modulus) and colors (encoding the argument), see Fig. 1. This graphical representation may be improved by plotting the values of the function not only for x belonging to a curve, but also for x belonging to a complex domain, which leads to a surface visualization.

The two previous methods are implemented in many software packages (numerical and symbolical software).

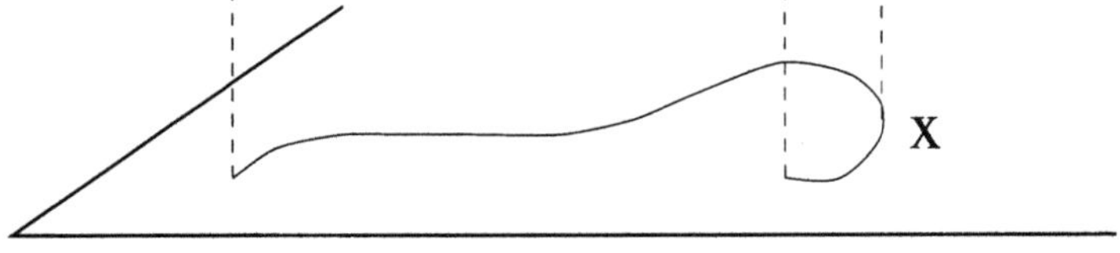

Fig. 1. The usual representation of complex functions

1.3 Visualization of CODE Solutions

Can this general representation model be useful for CODEs ? If we restrict ourselves to the study of integration in the complex domain, that is, computing the values of a solution of a CODE with x in a given subset of the complex plane such as a complex path, then the previous visualization techniques can generally be applied.

The second method in the above list was used in [8] to study the behavior of *linear homogeneous* CODE solutions in the vicinity of their singularities. It provided good comparisons between numerical and formal solution methods. This method does not allow to study several solutions resulting from different initial conditions. Such solutions generally intersect, rendering the interpretation of the picture very difficult. There is not much gained from studying several solutions as far as linear equations are concerned. But concerning the solutions of general equations such studies may be helpful.

Uniqueness theorems can be used in the complex plane, but the visualization methods described in Sect. 1.2 are not able to use this property in order to separate the different solutions resulting from different initial conditions. Furthermore, this visualization technique is dependent on the path $\mathcal{P}$.

The main principle in order to provide an intrinsic visualization technique for the solutions of CODEs is to consider the functions values of the solution as a parameterized path (as in [8]), with the same real parameter as the complex path $\mathcal{P}$. By doing so, only three dimensions are used to depict the graph of the given solution. It will be drawn as a three-dimensional curve, with the horizontal sections being the solution's values, and the value of the parameter being drawn on the z-axis, as shown on Fig. 2.

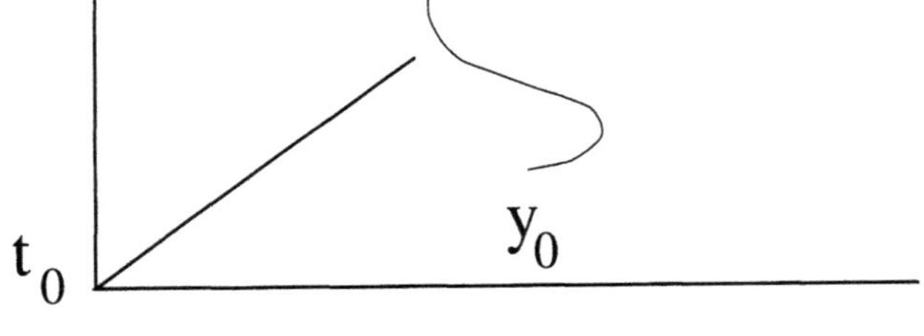

Fig. 2. Parameterized solutions curves

This visualization model has three main advantages:

- The visualization is more intrinsic, since the mathematical object being visualized *is* the solution of the CODE.
- A color scheme may be used to provide an encoding of additional data, like the estimated local error during numerical integration.

- The curve no longer depicts path $\mathcal{P}$ and thus can be projected on Riemann surfaces (for instance the Riemann sphere, which allows a compactification of the complex plane).

Furthermore, different solutions resulting from distinct initial conditions can be visualized on the same graph as a surface. If we take two solution curves, obtained after integrations starting from two "nearby" initial conditions, these curves can be used to build a mesh of a surface by connecting points with the same parameter's value. The resulting quadrangles lead to a surface representation of the solutions, taken as functions of t, the time parameter of the integration, with an additional parameter, the initial condition. The resulting function is a complex version of the real *flow function*, as defined in [3]. This function can provide a dynamical interpretation of the CODE, as we will see in 2.3.

1.4 Notes on the Experiments

The graphical environment that produced the three-dimensional images in this paper is called GANJ, and was developed by the author. This tool allows end-users to specify and program the visualization modes, whatever type of graphic subsystem is being used for the experiment, Furthermore, it can be distributed in a non-heterogeneous network of computers. The tool provides standard APIs (C,C++, Maple, C++/Parallel, ...) and the way the API commands are performed is completely transparent to the user.
Different C++ classes were used to carry out the experiments:

- A complex number class, which can treat these numbers as logarithmic Riemann surface elements, and thus allows end-users to use complex functions such as square-roots and logarithms.
- A numerical-formal mixed path class, which allows programmers to formally manipulate the paths and evaluate the points on demand.

(See [11] for further details.)

2 Extended Phase Portraits

2.1 Real Extended Phase Portraits

Extended phase portraits are a classical notion in the area of two-dimensional dynamical systems. We consider a system in the real variable t, e.g. system (2) of two real differential equations, that meets the conditions of the Cauchy-Lipschitz theorem [3] in a subset D of the plane:

$$\begin{cases} \dfrac{dy_1}{dt} = f_1(t, y_1, y_2) \\[2ex] \dfrac{dy_2}{dt} = f_2(t, y_1, y_2)\,. \end{cases} \tag{2}$$

Given an initial condition $(t_0, y_1^0, y_2^0) \in \mathbb{R} \times D$, we can define an *orbit* [7] of this system as the two-dimensional curve given by

$$\omega(t_0, y_1^0, y_2^0) = \{(y_1(t), y_2(t)), t \in \mathbb{R}\} \tag{3}$$

where $\begin{bmatrix} y_1(t) \\ y_2(t) \end{bmatrix}$ is the unique solution of the system such that

$$\begin{bmatrix} y_1(t_0) \\ y_2(t_0) \end{bmatrix} = \begin{bmatrix} y_1^0 \\ y_2^0 \end{bmatrix} .$$

We then define the phase portrait of system (2) in D', a given subset of D, by

$$\mathcal{PP}(D', t_0) = \{\omega(t_0, y_1^0, y_2^0), (y_1^0, y_2^0) \in D'\} . \tag{4}$$

If the system is autonomous, i.e. if the functions f_1 and f_2 do not depend on t, two orbits corresponding to two different initial conditions cannot intersect. However if the system is non-autonomous, then two such orbits may intersect. Since these intersections can obscure the comprehension of the phase portrait, we define the *lifted orbits* by

$$\Omega(t_0, y_1^0, y_2^0) = \{(y_1(t), y_2(t), t), t \in \mathbb{R}\} , \tag{5}$$

and the *extended phase portrait* (called `txy-space` in [7]) by

$$\mathcal{EPP}(D', t_0) = \{\Omega(t_0, y_1^0, y_2^0), (y_1^0, y_2^0) \in D'\} . \tag{6}$$

These extended phase portraits can be used to visualize many solutions of a given dynamical system, since different lifted orbits issued from different initial conditions *cannot* intersect. We can now apply this principle to the complex equations under study.

2.2 Complex Extended Phase Portraits

Suppose path $\mathcal{P}$ is parameterized by a diffeomorphism Φ. We can write

$$x = \Phi(t)$$

and then perform the substitution in the Eq. (1) (since Φ is supposed to be a diffeomorphism)

$$\frac{dy}{dx} = \frac{1}{\Phi'(t)} \frac{d\tilde{y}}{dt}$$

where $\tilde{y}(t) = y(\Phi(t))$. Doing so, we can rewrite Eq. (1) as

$$\frac{1}{\Phi'(t)} \frac{d\tilde{y}}{dt} = f(\Phi(t), \tilde{y}(t)) . \tag{7}$$

If we now write $\tilde{y}(t) = y_1(t) + iy_2(t)$ and

$$\Phi'(t)f(\Phi(t), \tilde{y}(t)) = f_1(t, y_1, y_2) + if_2(t, y_1, y_2)$$

we obtain a 2×2 real dynamical system

$$\begin{cases} \dfrac{dy_1}{dt} = f_1(t, y_1, y_2) \\ \\ \dfrac{dy_2}{dt} = f_2(t, y_1, y_2) \,. \end{cases} \tag{8}$$

Generally, this system is not autonomous, even if Eq. (1) is autonomous, because the term in $\Phi'(t)$ is present in both equations of the system. Nevertheless, we can build the lifted orbits of this system as defined in (5). The horizontal sections of these curves represent the two-dimensional points

$$(y_1(t), y_2(t)) \tag{9}$$

which are equivalent to $\tilde{y}(t)$ in $\mathbb{C}$. So $y(\Phi(t))$ is the solution of Eq. (1) such that $y(\Phi(t_0)) = y_0$. If we draw the extended phase portrait of system (8), we obtain a set of curves, where the horizontal sections provide the complex values of the solutions of the system, and the z-value represents the parameter of the integration path. The idea presented in Sect. 1.3 is therefore valid, and we can use the fact that lifted orbits don't intersect if they arise from different initial conditions. Furthermore, we saw in Sect. 1.3 that only three dimensions are needed to draw these curves, so that we can encode related information by using color. Finally the embedding of the curves in spaces other than $\mathbb{C} \times \mathbb{R}$ is made possible. This last operation can provide topological methods to study the behavior of CODEs solutions.

2.3 The Flow Function

In classical dynamical theory, the *flow function* provides a way of estimating the initial conditions' dependence on a set of solutions of a dynamical system. The flow $\Psi_{x_0,y_0}(t)$ associated to a dynamical system like Eq. (2) is defined by the following two conditions :

- $\Psi_{x_0,y_0}(t) = \begin{bmatrix} x(t) \\ y(t) \end{bmatrix}$ is a solution of the dynamical system,
- $\Psi_{x_0,y_0}(t_0) = \begin{bmatrix} x_0 \\ y_0 \end{bmatrix}$.

The dependence on initial conditions can be appreciated using this function: if two initial conditions (x_0, y_0) and (x'_0, y'_0) are "close" in the plane, but if, after a given integration time T, the values $\Psi_{x_0,y_0}(T)$ and $\Psi_{x'_0,y'_0}(T)$ are radically different, the system is said to be initial condition dependent. Furthermore, some dynamical system may be area-preserving, which means the image of a subset of $\mathbb{R}^2$ by the flow function has the same area as the initial subset (some applications may be found in Hamiltonian systems theory).

With the aforementioned method, we can visualize the flow function of a complex equation as a bi-parameterized (t, u) surface, where t is the time parameter, and u is a parameterization of a sub-domain (of complex dimension 1) of $\mathbb{C}$ where different values of y_0 may be taken.

An example of a visualization of a flow function is shown on the color plates 46a and 46b in the appendix.

Color plate 46a, p. 391, visualizes the flow function of

$$y' = xy^3 \tag{10}$$

where x is taken on a circle of radius 1 and centered at 2. The initial conditions are the pairs $(1, y_0)$ where the y_0s were taken on a circle of low radius (0.3), centered at 0. If we only consider the points for $t = 0$, the figure shows the initial condition path. The projection of the surface on the plane $z = 0$ shows the complex values of the solutions. On the figure, the real axis is labeled R. The solutions curves on the figure can be traced in integration time, since the color of a particular curve is the same as the color of the point representing the initial condition.

Color plate 46b, p. 391, depicts the flow function of Eq. (10) (rotated by 90°). The different colors on plates 46a and 46b indicate the dependence on the initial conditions. The integration path is closed, and we see that the values after integration on this circle are the same as the initial conditions: the solution is not many-valued for this integration path and these initial conditions.

3 Applications to CODEs

3.1 Global Error Visualization

Only results of reliable numerical integration methods are of interest. There are several ways to ensure the quality of results, one of which is the estimation of numerical errors. The local error [6], i.e. the error committed during one integration step, is frequently used to control the quality of an iteration of the numerical integration method. But its numerical values and variations are generally useless. The *global error* [2,6], i.e. the error resulting from several integration steps, is more representative than the local one.

Several methods may be used to estimate the global error during an integration. One interesting point about this estimation is its *localization*, because the estimation is performed at each integration step, and thus its values can be correlated to the solution's value.

We estimated the global error during the integration of the Eq. (10) on a 1-radius circle, centered at point 2, with initial conditions taken along a circle γ of radius ρ and centered at the origin.

Color plate 46c, p. 391, visualizes the global error. The estimated global error is displayed with a "basic" quantification function, which processes the min's and the max's of the global error's modulus, *for each solution.* To each value a color is assigned, from blue (low global error's modulus) to green (high global error's modulus). The corresponding palette is drawn on the left of the figure. The figure highlights two points:

- This quantification method is certainly not suited for judging the quality of the results, because it does not provide comparisons between solutions that belong to different initial conditions. However it is useful for the analysis of one solution, and allows us to visually trace the global error's behavior during *one integration.*
- The spotted zone of this surface confirms the previous remark : inside these zones, the global error is completely *meaningless* (the computation gave results of the order of the ε-machine), and should therefore not be studied.

Color plate 46d, p. 391, shows the global error by depicting a global quantification function. Global quantification function means that the min's and the max's used for coloring the points of the surface are performed on the *whole* set of estimated global errors. This provides a *localization* of the global error. The localization may have two possible interpretations:

- Time localization: The green spots on the surface represent the time-zones of high global error, which means that we can visually estimate the time intervals where the integration may produce the poorest results (because the time is displayed on the vertical axis of the figure).
- Spatial localization: Because the integration time is also the parameter of the integration path, we can localize the complex plane zones where x belongs and $y(x)$ has the highest estimated global error.

The last point is very important. We can deform the integration path (leaving the endpoints invariant), and thus avoid the critical zones of high global error. If the global error on the new path does not increase, the computed values are more reliable.

3.2 Compact Visualizations

Another interesting possibility in this visualization model is to embed the solutions curves in other spaces than $\mathbb{C}$. The Riemann sphere [4] is a compact Riemann surface of genus 0. In general, this sphere has two identified poles, say 0 and ∞, and we can construct two projective mappings (also known as *stereo-graphic projections*) as shown on Figure 3. Given a point M on the Riemann sphere, M_1 is the image of M by the first projection from the pole N of the sphere, and M_2 is the image of M by the second projection from the pole S.

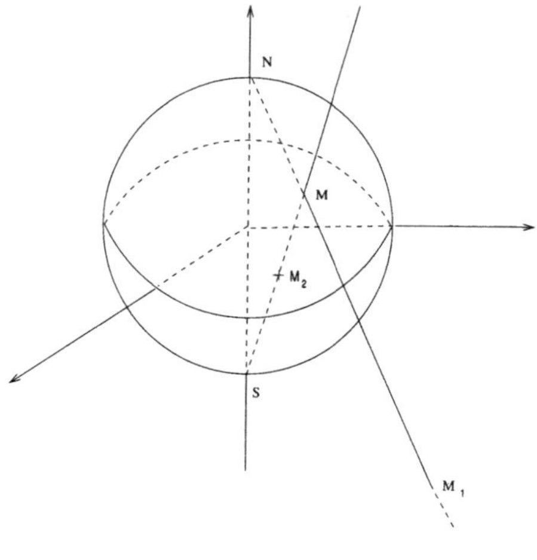

Fig. 3. The Riemann sphere

A given point in the complex plane can thus be lifted up to the Riemann sphere by choosing one of the two inverse mappings, which allows to represent the complex plane compactly. Concerning CODE visualization it is of interest to depict on one figure the behavior of a solution both for finite values and in the vicinity of the infinite point. By doing so, a complete solution curve may be traced as a trajectory on the sphere.

Two such examples are provided on color plates 46e and 46f, p. 391. The first one results from the very simple equation

$$y' = xy\,, \tag{11}$$

and the second one from Eq. (10).

For numerical integration in the vicinity of the infinite point, we applied the variable substitution $u = \frac{1}{t}$. As initial conditions we took $(1, y_0)$, with y_0 on a circle of radius 1 and centered at 0. The lifting of this circle on the Riemann sphere is the equatorial circle.

Color plate 46e, p. 391, shows a compact visualization of the flow of Eq. (11), whose solution is the function

$$y(x) = y_0 e^{-t_0^2/2} e^{t^2/2}\,.$$

Two such solutions for different initial conditions are exchangeable by a simple rotation. We can verify this assertion visually. On this figure we also see the behavior of $y(x)$ for x in the vicinity of 0 (near the sphere point labeled S) and the behavior for x close to ∞ (near the point labeled N). Due to the compactification of the complex plane, it is possible to provide a global visualization of the solution.

In color plate 46f, p. 391, a compact visualization of the flow of Eq. (10) is shown. Here the aim is to provide an idea of the solution's behavior for small and large x on the same figure. Since the sphere point equivalent to the complex zero is the south pole, the solution curves converge to zero when the values of x increase. This complex phase portrait can thus be interpreted with dynamical arguments, such as a motion from an initial position of the solution to zero, for many solutions arising from different initial conditions.

4 Conclusion

The methods in this paper can be extended to the study of N-order complex differential systems, which leads to the visualization of phase portraits of real $2N \times 2N$ dynamical systems. Sophisticated methods exist for such problems [10], and the study of these systems can benefit from such methods.

As a conclusion, CODEs are very useful for "real" problems (real in the mathematical sense, but also in the sense of real-world ones), but the difficulties involved in their visualization can be off-putting. An attempt has been made in this paper, showing that with efficient visualization methods these difficulties can be reduced and helpful interpretations can be obtained.

References

1. R. Aïd, *Intégration numérique d'équations différentielles ordinaires complexes sans spécification a priori du chemin.* Technical report, LMC-IMAG, Grenoble, submitted 1997.
2. R. Aïd and L. Levacher, *Numerical investigations on global error estimation for ordinary differential equations.* Journal on Computational and Applied Mathematics, 1997.
3. C. M. Bender and S. A. Orszag, *Advanced Mathematical Methods for Scientists and Engineers.* McGraw-Hill, 1978.
4. H. Cartan, *Théorie élémentaire des fonctions analytiques d'une ou plusieurs variables complexes.* Hermann, 1961.
5. M. Fedoriouk, *Méthodes asymptotiques pour les équations différentielles ordinaires linéaires.* Mir, 1987.
6. E. Hairer, S.P. Norsett, and G. Wanner, *Solving Ordinary Differential Equations I Nonstiff Problems.* Springer-Verlag, 1980.
7. J. H. Hubbard and B.H. West, *Differential Equations : A Dynamical Systems Approach.* Springer-Verlag, 1995.
8. F. Richard-Jung, *Le phénomène de Stokes en image - RT 65.* Technical report, LMC-Imag, 1991.
9. G. Springer, *Introduction to Riemann Surfaces.* Addison-Wesley, 1957.
10. D. Stalling, M. Zöckler, and H.C. Hege, *Fast display of illuminated field lines.* IEEE Transactions on Visualization and Computer Graphics, **3**:2, April 1997, pp. 118-128.
11. L. Testard, *Visualisation et calculs en nombres complexes.* PhD thesis, INPG, 1997 (submitted).

Appendix: Color Plates

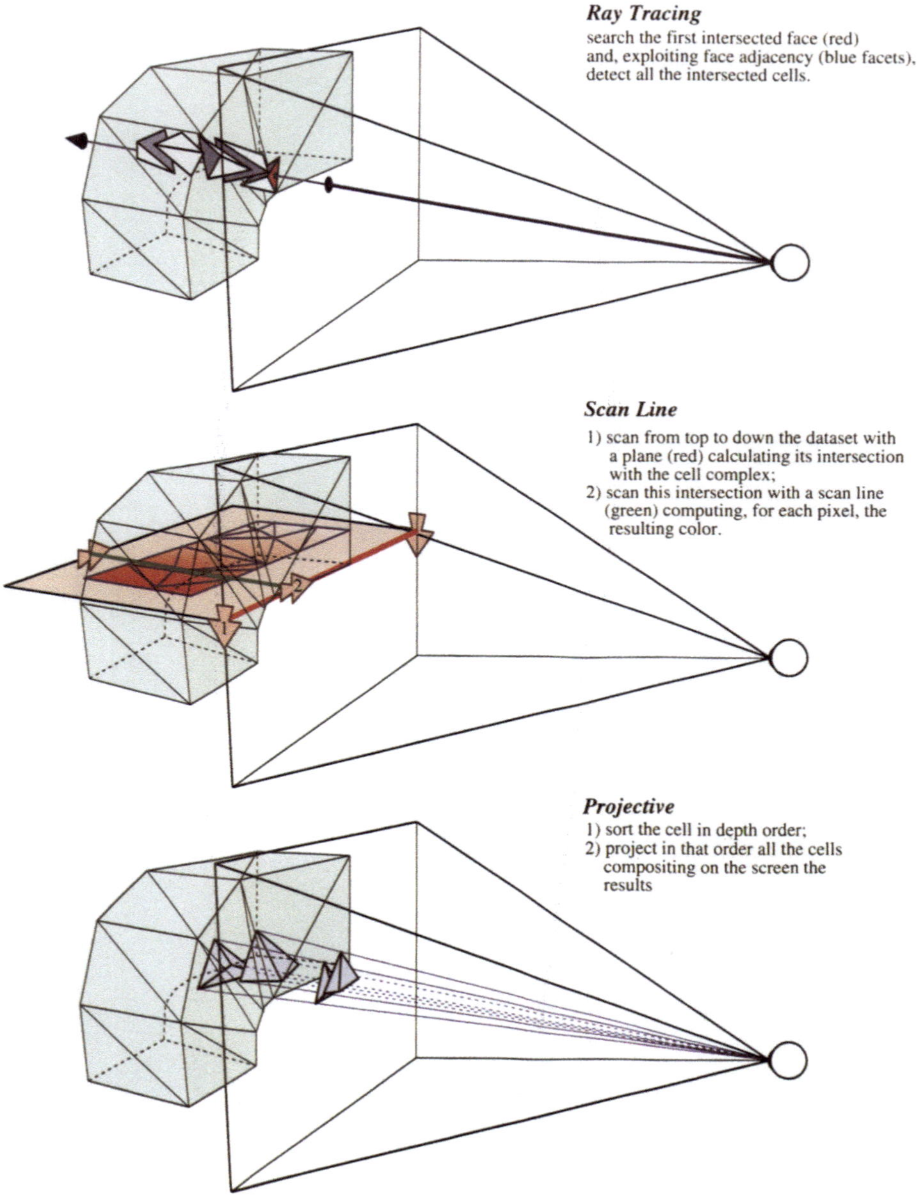

Fig. 1. Tetrahedral visualization approaches. *(Cignoni, Montani, Scopigno, p. 3)*

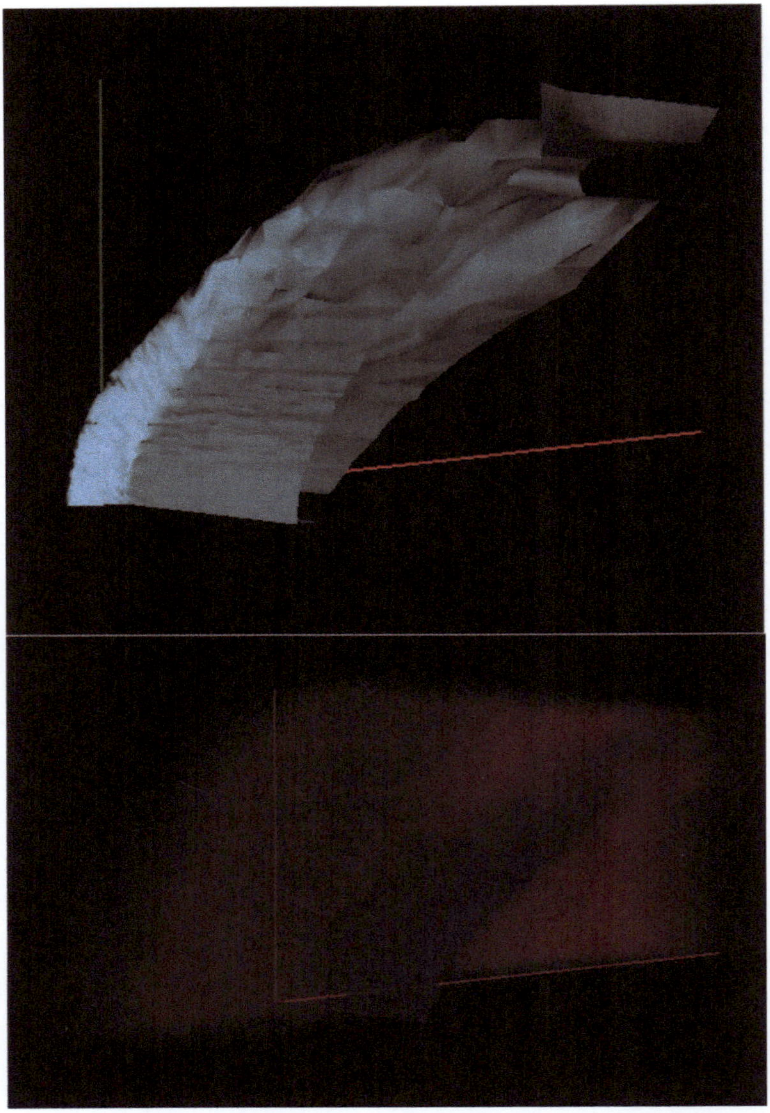

Fig. 2. Isosurface fitting and direct volume rendering. *(Cignoni, Montani, Scopigno, p. 3)*

Fig. 3. MAM/VRS Object viewer during fade-in animation. *(Döllner, Hinrichs, p. 163)*

Fig. 4. Six stages in the $p = 3$ minimax eversion, starting in the top left with a double cover of a Boy's surface critical for the Willmore energy W. (Because the two sheets have opposite orientations, we see their red and blue sides; at this stage they are in the same place, so give mottled colors when rendered.) Next, the two sheets have pulled apart; here the white double locus between red surfaces is a four-fold cover of the Boy's surface double locus, while the part between red and blue is the extra curve seen in Fig. 3, p. 245. In the middle row, the surface starts to be simplified by minimization of W. At right, a pair of triple points has just disappeared in front; because of the eversion's three-fold symmetry, this happens in three places. At the bottom, the three fingers have straightened and almost lost contact. In the last picture, the fingers are retracting; the double locus where the north pole pushes through the south pole will soon disappear. *(Francis, Sullivan, Hartman, p. 237)*

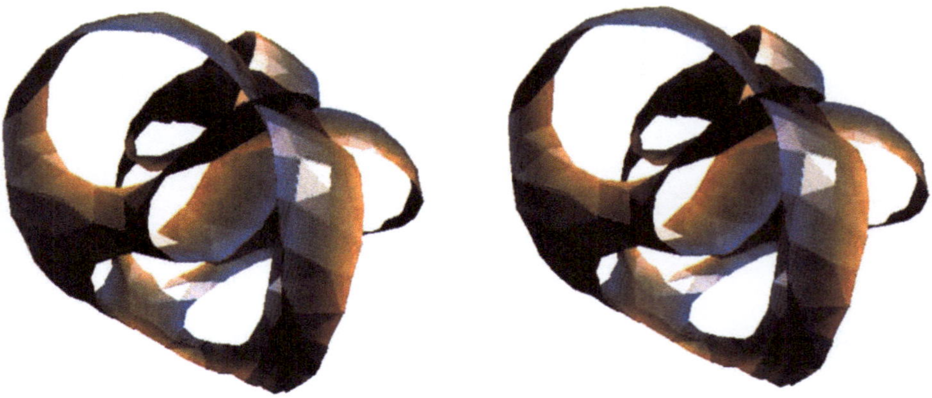

Fig. 5. A slice of the double locus surface for the $p = 2$ minimax eversion, near $t = 0$, with the t direction projected out, drawn in cross-eyed stereo. Six sheets of this surface cross where the halfway model has a quadruple point; the five isthmus events are saddles here. Compare Fig. 4, p. 246, which shows three individual levels from which this surface was pieced together. *(Francis, Sullivan, Hartman, p. 246)*

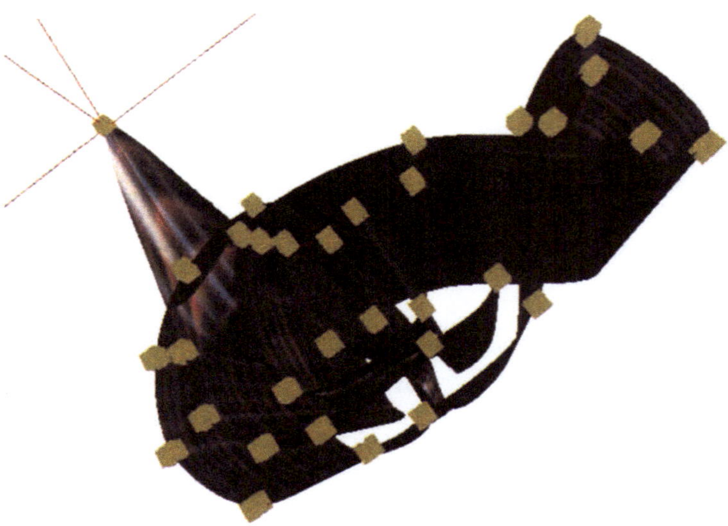

Fig. 6. The double cover of a Boy surface with sheets pulled apart, M_1, can be given by splines. Here we see the slab $M_1[-1, +1]$. The graphical user interface for `illilevel` lets us interactively move control points, as at the upper left. *(Francis, Sullivan, Hartman, p. 252)*

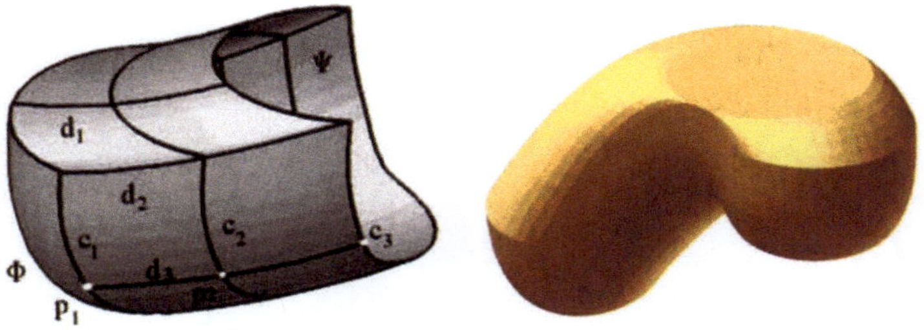

(a) (b)

Fig. 7. Interpolation of a set of lines $\{c\}$ with a set of lines $\{d\}$ (a), resulting sweep object (b). *(Gläser, Gröller, p. 89)*

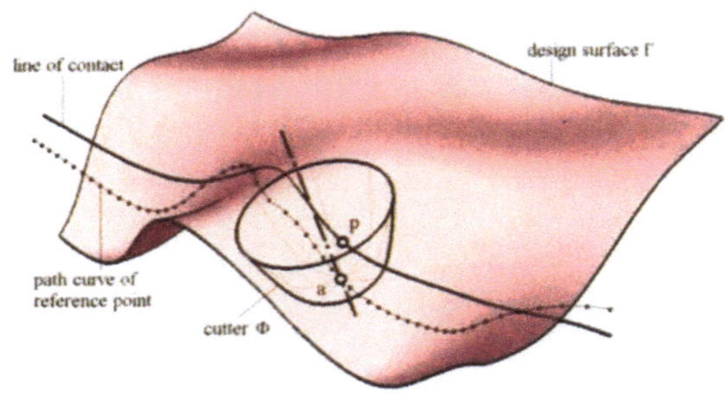

Fig. 8. Path of a fixed point on the cutter's axis when the cutter is moved along the design surface. *(Gläser, Gröller, p. 89)*

Fig. 9. Object with several color coded cutter paths Ψ_k. *(Gläser, Gröller, p. 89)*

Fig. 10. A constant mean curvature surface with finite absolute total curvature, this 30-unduloid of genus one has only cylindrical ends. At the top left, we see a fundamental domain having a 96° angle opposite the cylindrical end. At the top right, four copies leave a 24° gap. Finally, at the bottom, thirty copies close up to form the upper half of the complete surface. *(Große-Brauckmann, Kusner, Sullivan, p. 111)*

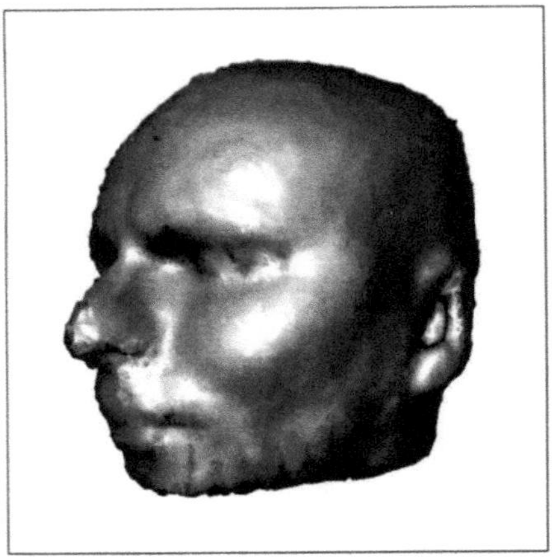

(a) Head 1, H^1 norm

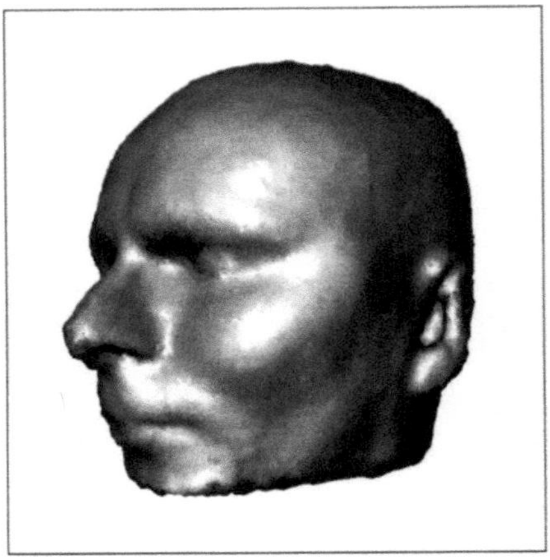

(b) Head 2, H^1 norm

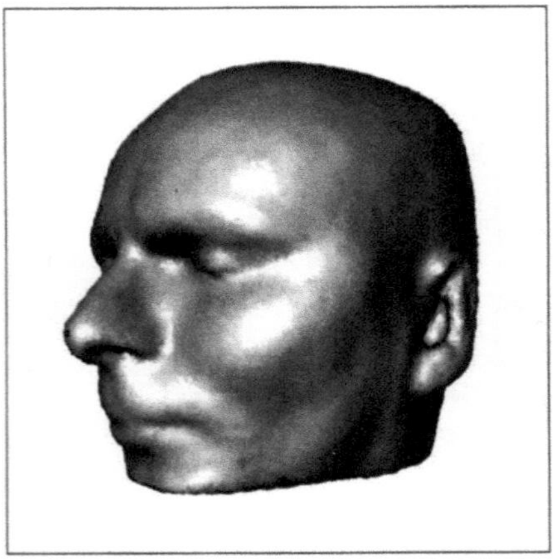

(c) Head 3, H^1 norm

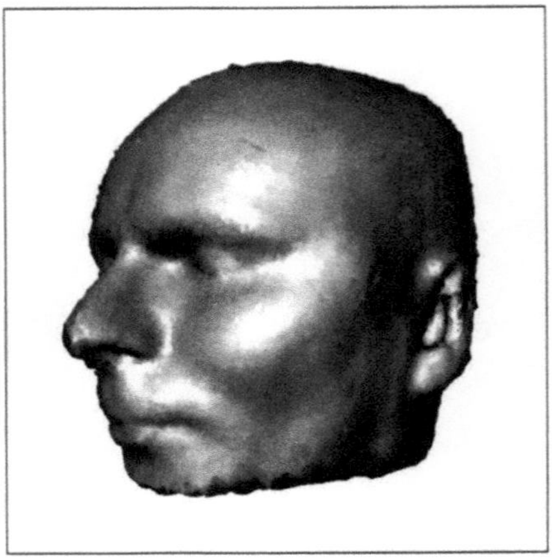

(d) Head 4, L_2 norm

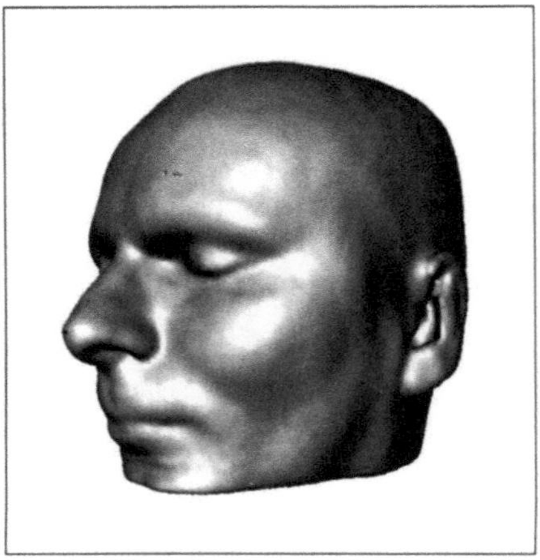

(e) Marching cubes

Fig. 11. Iso-surfaces from a MRI scan of a human head. *(Grosso, Ertl, p. 19)*

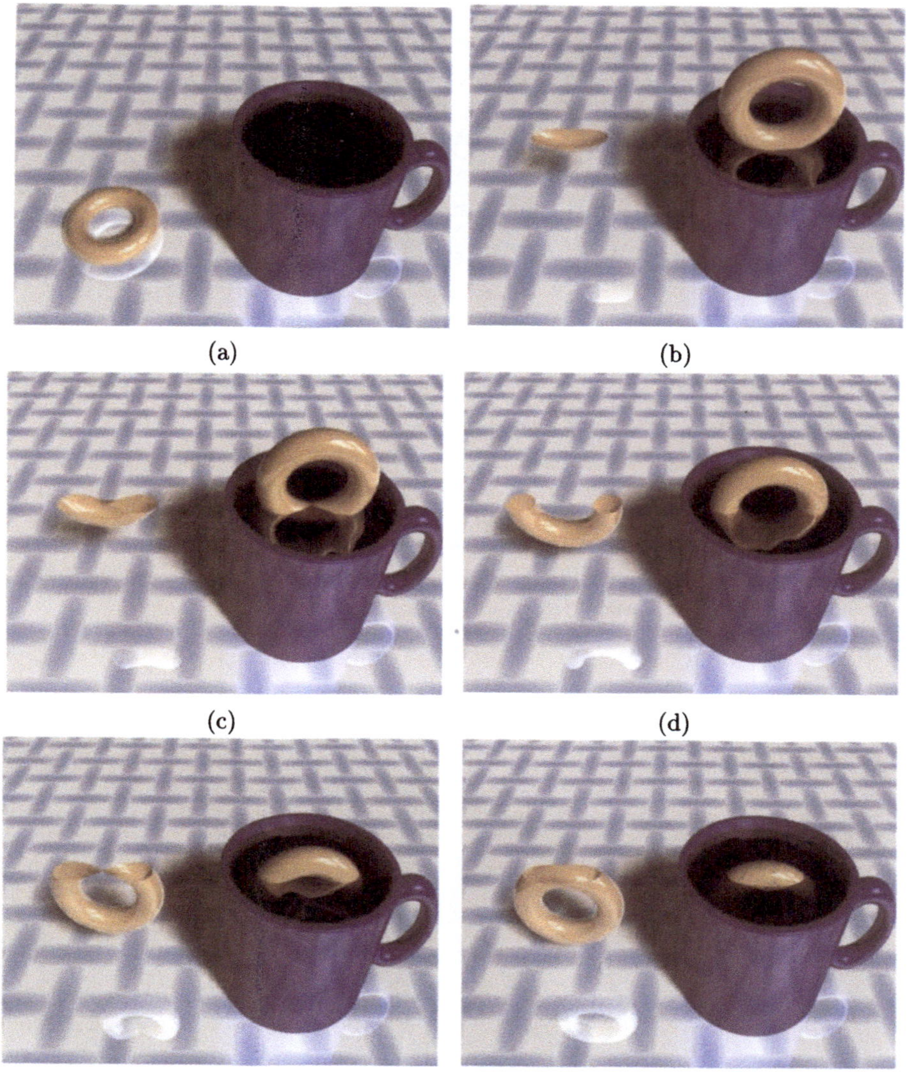

Fig. 12. Dunking a doughnut. A shiny doughnut and a cup of coffee (a). The dunked portion of the doughnut's surface changes from the empty set to a shape homeomorphic to a disk (b). The dunked portion changes (c) from a disk to a truncated cylinder (d). The dunked portion changes (e) from a cylinder to a truncated torus (f). *(Hart, p. 261)*

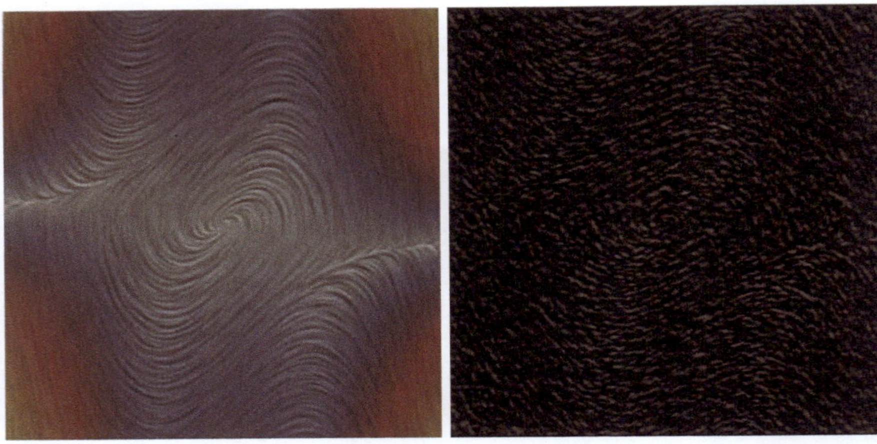

Fig. 13. The left image shows the solution of the ODE $x' = y,\ y' = (1 - x^2)y - x$ using line integral convolution with a triangular filter kernel. This filter produces smoother results than a box filter. Right: contrast enhanced difference between the left image and one generated with a box filter (width of the box filter is 3/4 of the triangle filter). Note, that the error structures are much smaller than the characteristic feature length of the LIC texture. Therefore, LIC images generated by box filters appear noisier. *(Hege, Stalling, p. 312)*

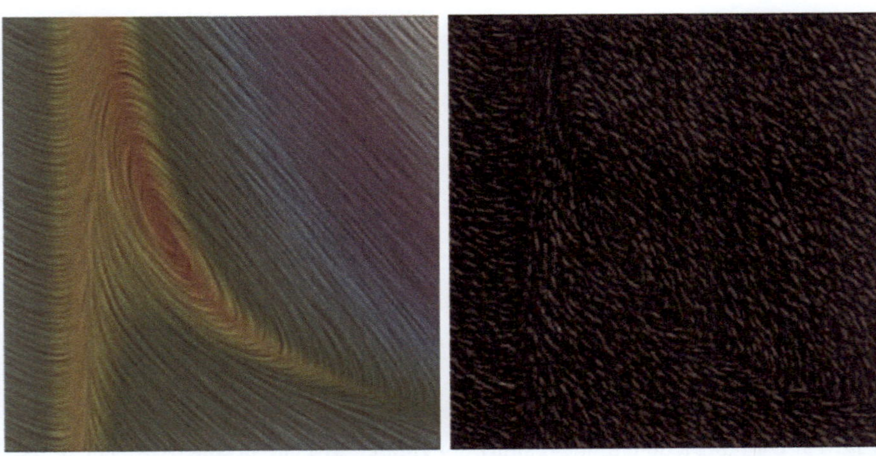

Fig. 14. Like above, a LIC image generated with a triangular filter kernel is shown (left) and compared with a LIC image generated with a box filter by depicting a contrast enhanced difference image (right). The image on the left depicts the solution curves of the ODE $x' = 1 + x^2y - 4x, y' = 3x - x^2y$. Color encodes magnitude of the vector (x', y'). *(Hege, Stalling, p. 312)*

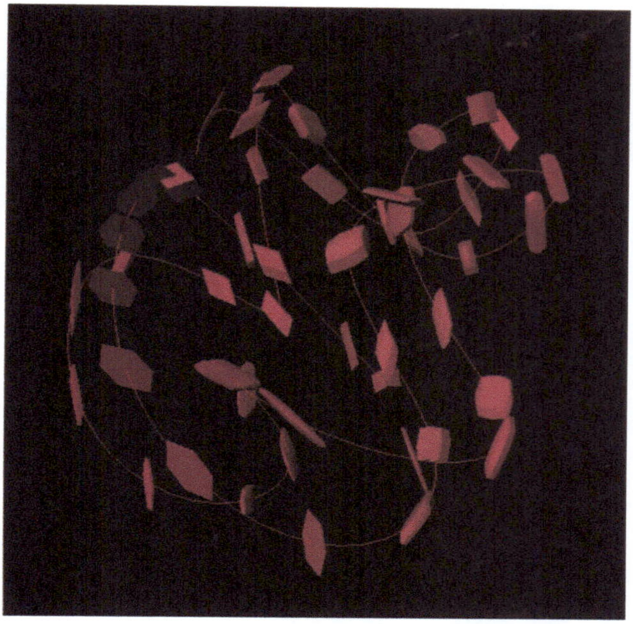

Fig. 15. Zonotope enclosures of order 5 for a solution of Langford's vector field at a total of 50 discrete times. *(Kühn, p. 125)*

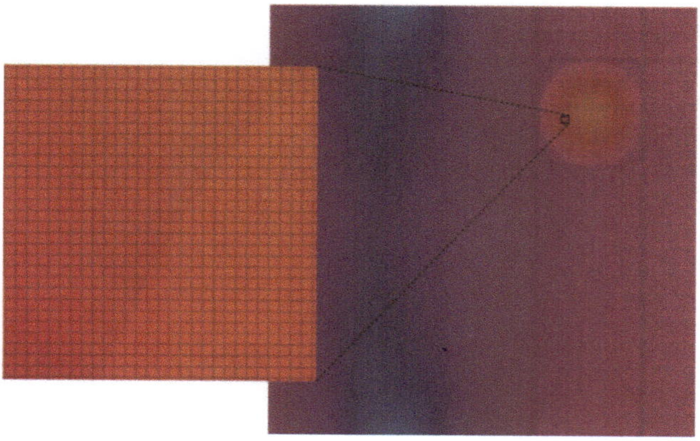

Fig. 16. A slice through a 10 level sparse grid is shown with color shading of the sparse grid function. On the right the grid is traversed up to the 7'th level, which corresponds to a 129^3 full grid. The black box marks a region magnified on the left. It is extracted from the resulting slice of a 10 level traversal, corresponding to a 1025^3 point full grid. *(Heußer, Rumpf, p. 38)*

Fig. 17. An isosurface is extracted from a sparse grid on different grid levels, from left to right the 4, 5, 7 and 9 level case. *(Heußer, Rumpf, p. 39)*

Fig. 18. An isosurface and a magnified local portion, extracted from a numerical 10 level sparse grid data base, which corresponds to a standard 1025^3 point full grid. *(Heußer, Rumpf, p. 41)*

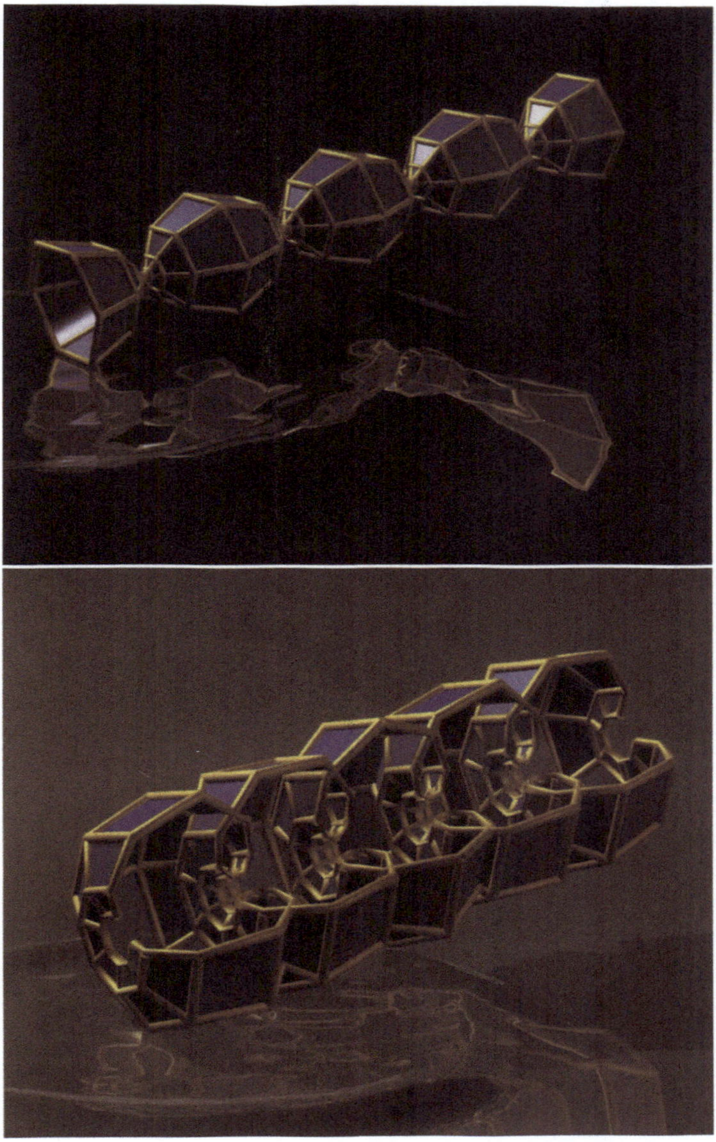

Fig. 19. Two rotational symmetric discrete cmc surfaces. They are related to the billiards in an ellipse and a hyperbola. *(Hoffmann, p. 117)*

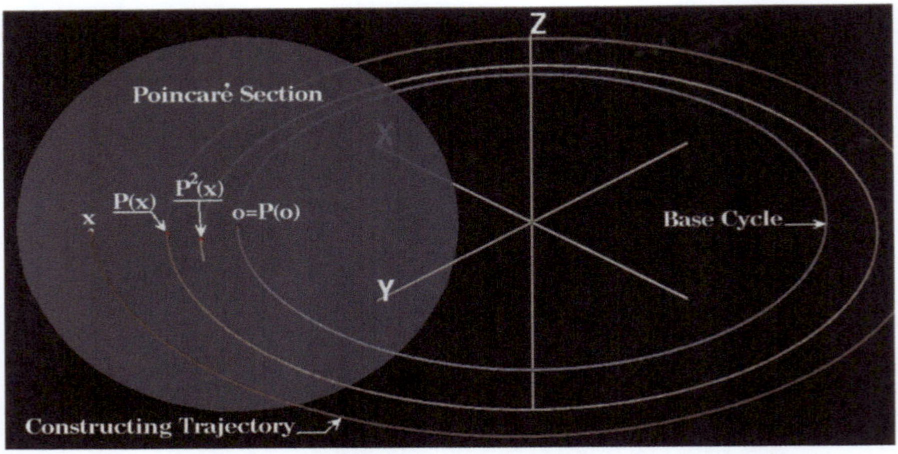

Fig. 20. An illustration of the Poincaré map definition. *(Löffelmann, Kučera, Gröller, p. 317)*

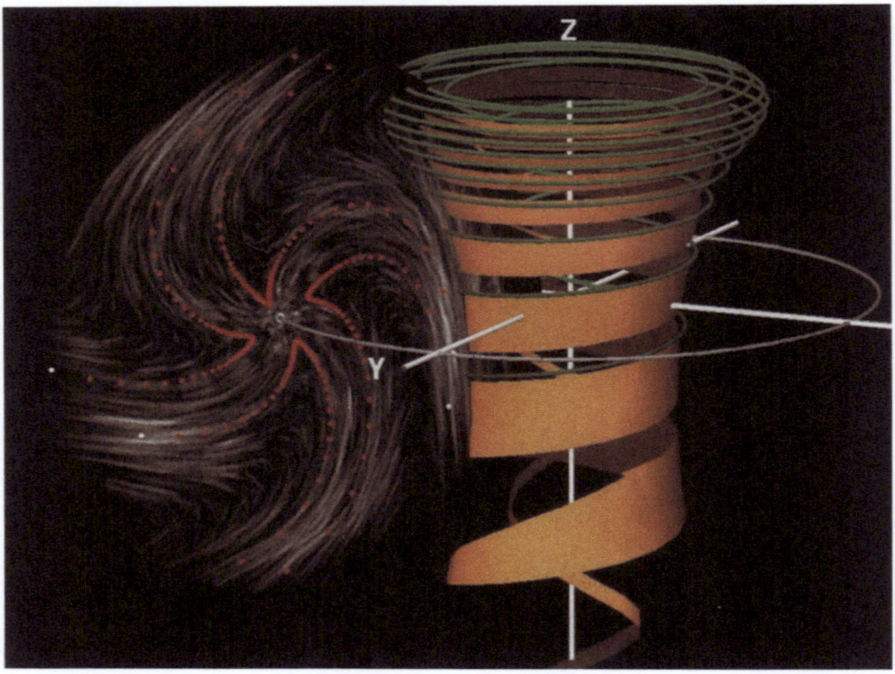

Fig. 21. Enhancing the 2D Poincaré map visualization by the use of 3D flow visualization icons. *(Löffelmann, Kučera, Gröller, p. 323)*

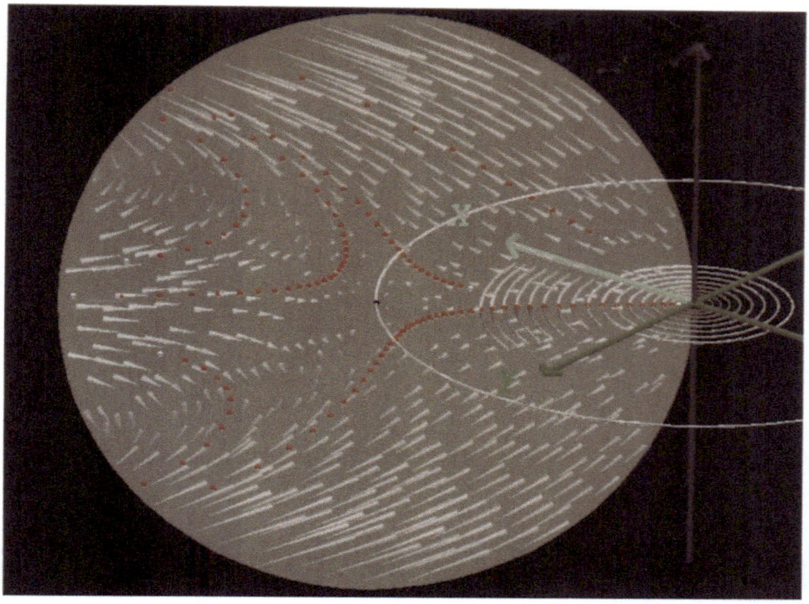

Fig. 22. Visualizing a non-linear saddle cycle. *(Löffelmann, Kučera, Gröller, p. 319)*

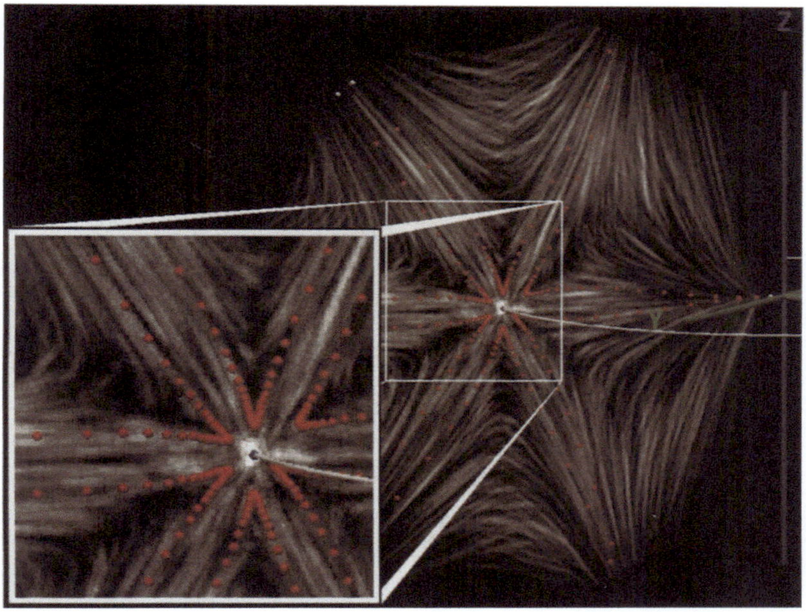

Fig. 23. Visualizing a non-hyperbolic saddle cycle. *(Löffelmann, Kučera, Gröller, p. 318)*

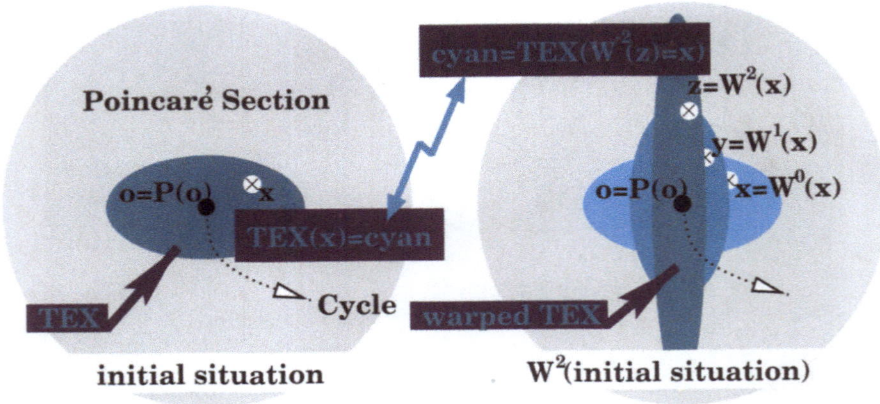

Fig. 24. Evaluating the initial texture after two applications of W. *(Löffelmann, Kučera, Gröller, p. 322)*

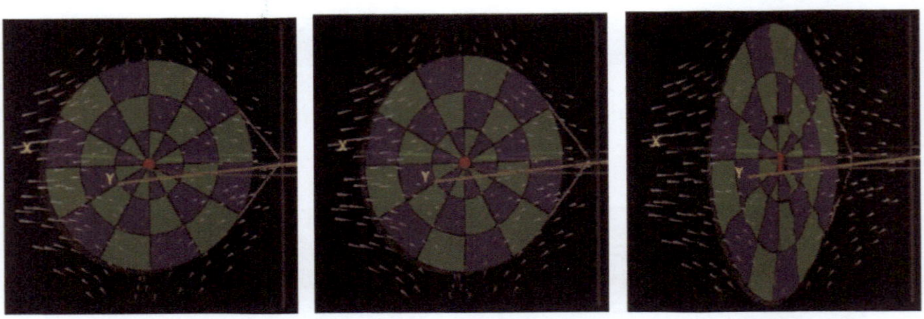

Fig. 25. Images resulting from one, two, and eleven applications of warp function W, i.e. $W(\mathcal{S})$, $W^2(\mathcal{S})$, and $W^{11}(\mathcal{S})$. *(Löffelmann, Kučera, Gröller, p. 325)*

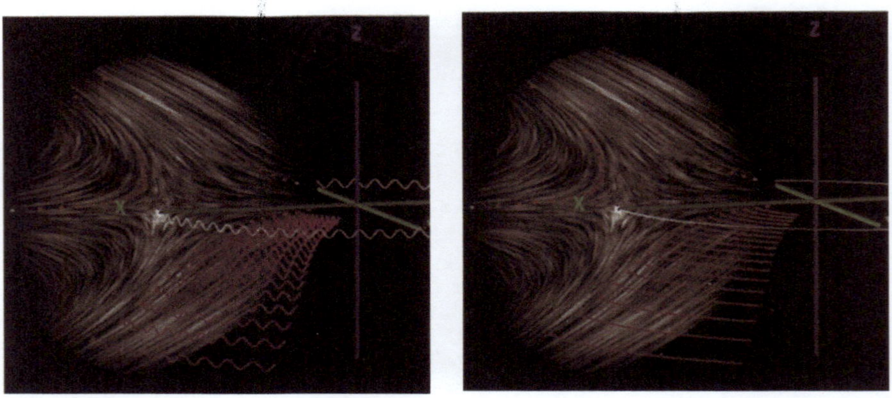

Fig. 26. Visualizing supplementary information in 3D. *(Löffelmann, Kučera, Gröller, p. 323)*

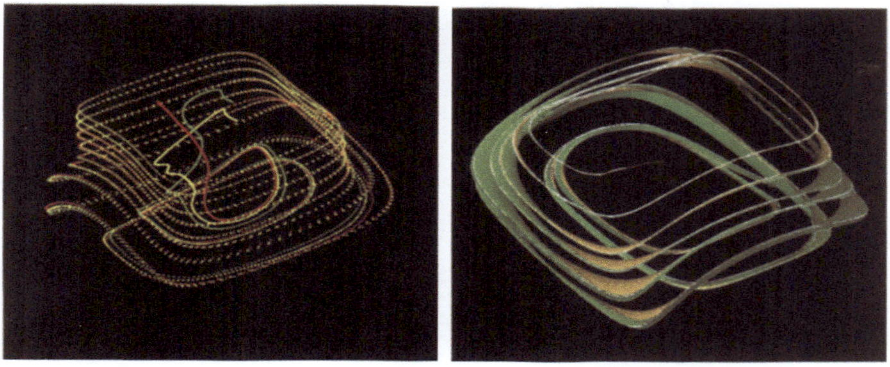

a) b)

Fig. 27. Re-integrations of two problems (left and right), with tolerances reduced by 1/10 and 1/100. Colour is mapped regarding the cost of integration. *(Lopes, Brodlie, p. 329)*

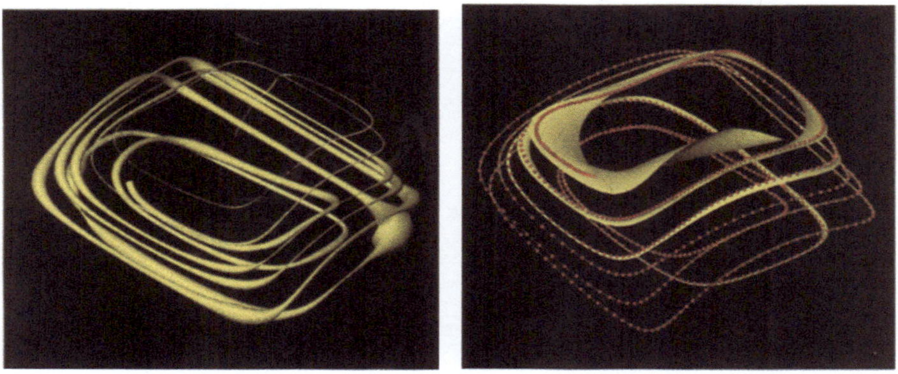

c) d)

Fig. 28. Re-integrations with smaller tolerance (left) and in a backward direction (right). *(Lopes, Brodlie, p. 329)*

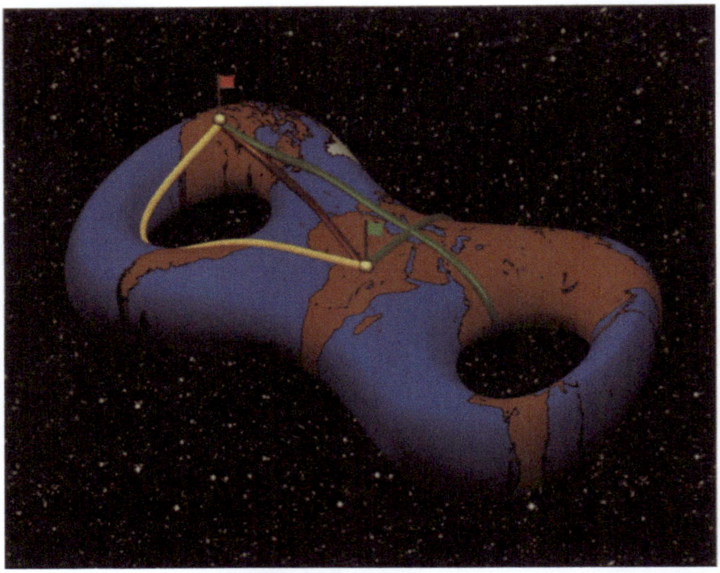

Fig. 29. Geodesics on a smooth surface and different local minimizers. *(Polthier, Schmies, p. 137)*

Fig. 30. Straightest geodesic through the vertex of a polyhedral surface and an unfolding of the adjacent faces to a planar domain. *(Polthier, Schmies, p. 142)*

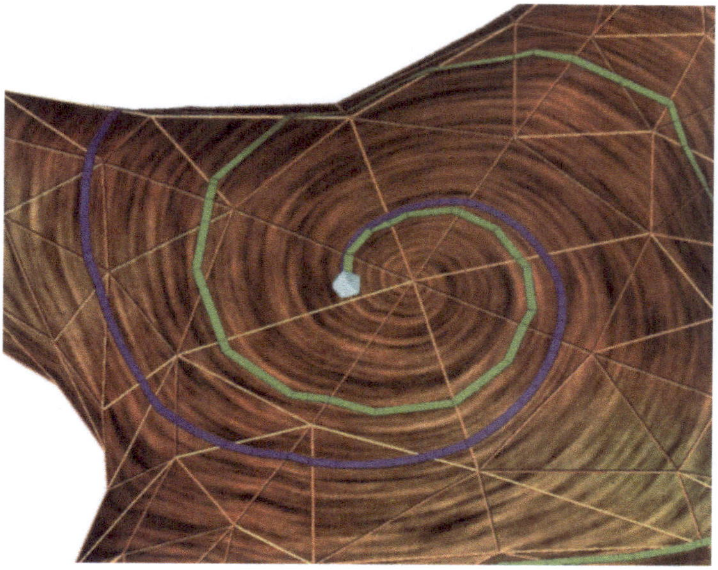

Fig. 31. Comparison of geodesic Euler (blue, stepsize h) and geodesic Runge-Kutta method (green, stepsize 4h) on a polyhedral surface. The particle traces are piecewise straightest geodesics. *(Polthier, Schmies, p. 149)*

Fig. 32. Straightest geodesics uniquely solve the initial value problem for geodesics on polyhedral surfaces and are suitable tool for the numerical study of discrete flows. The front of a point wave on a polyhedral surface follows straightest geodesics. *(Polthier, Schmies, p. 149)*

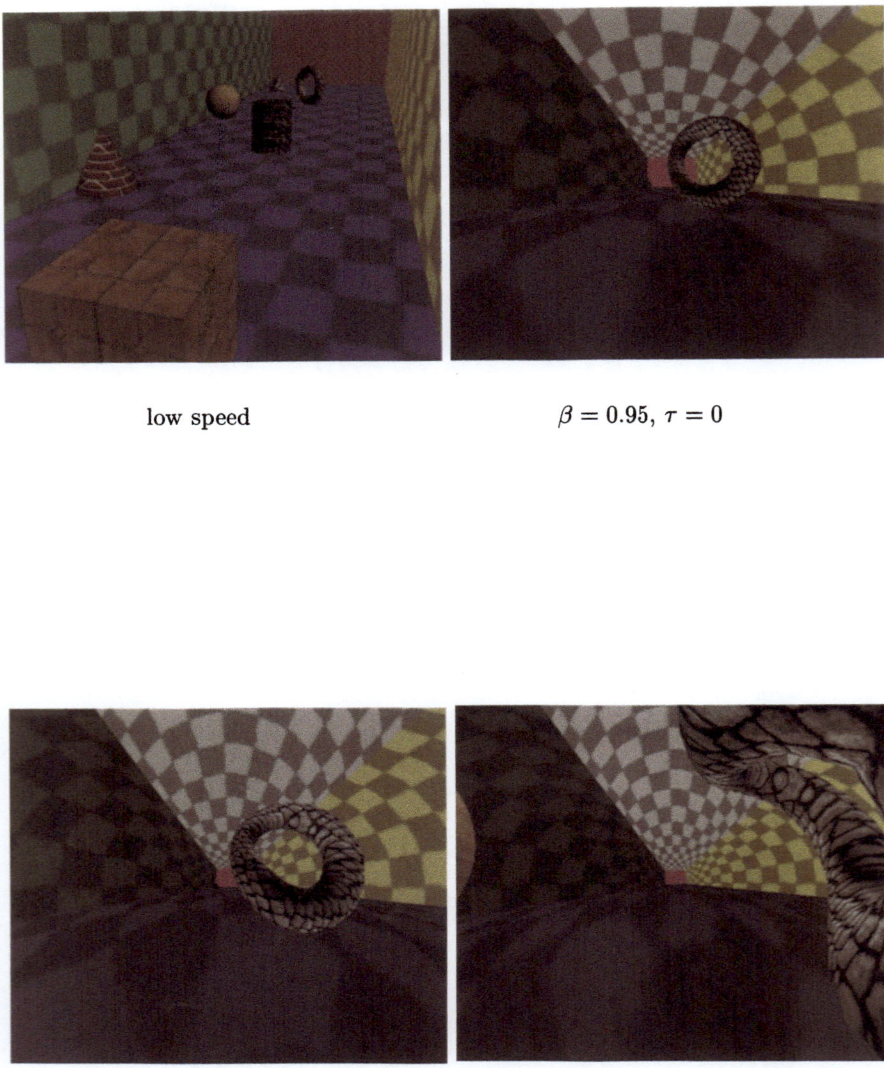

low speed

$\beta = 0.95,\ \tau = 0$

$\beta = 0.983,\ \tau = 2.1$

$\beta = 0.994,\ \tau = 3.7$

Fig. 33. A test scene for the virtual relativistic reality. *(Rau, Weiskopf, Ruder, p. 278)*

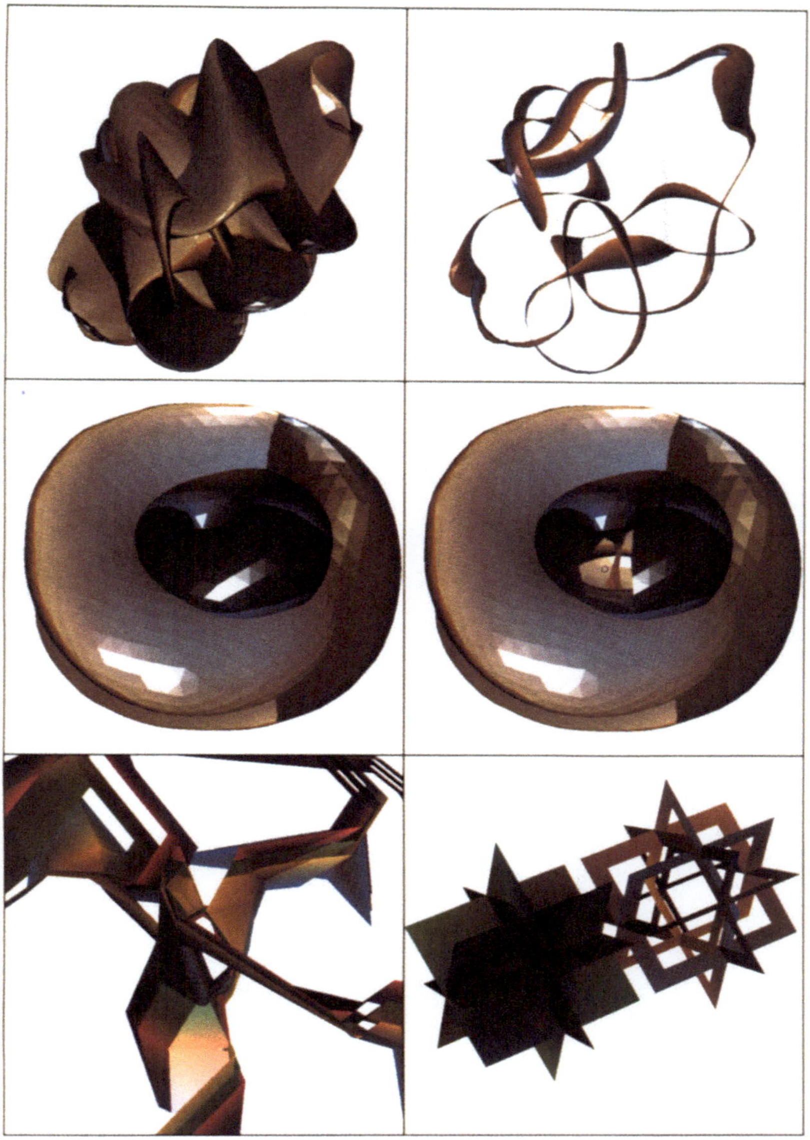

Fig. 34. Some features of Hew. Top row: showing zones. Middle row: illustrating glazing. Bottom row: Use of color to distinguish objects. In lower right are close-ups of two buoys. *(Roseman, p. 284)*

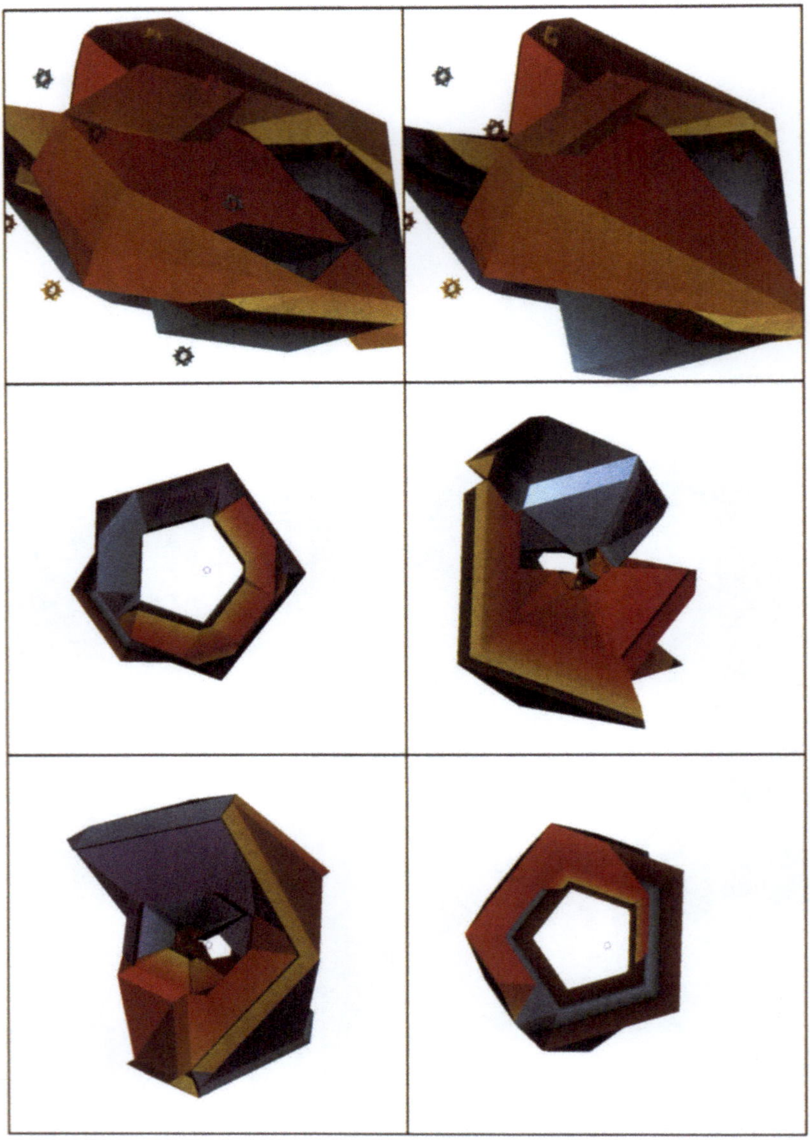

Fig. 35. Six slices of an embedded three manifold (projective 3-space) in R^5 . Top row: two slices shown with array of buoys. Bottom two rows: four slices. Sequence is: middle left, middle right, lower left, lower right. *(Roseman, p. 285)*

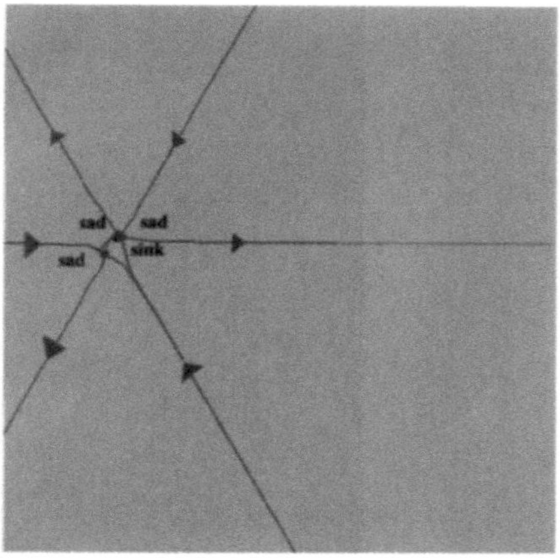

Fig. 36. Linear approximation around monkey saddle. *(Scheuermann, Hagen, p. 350)*

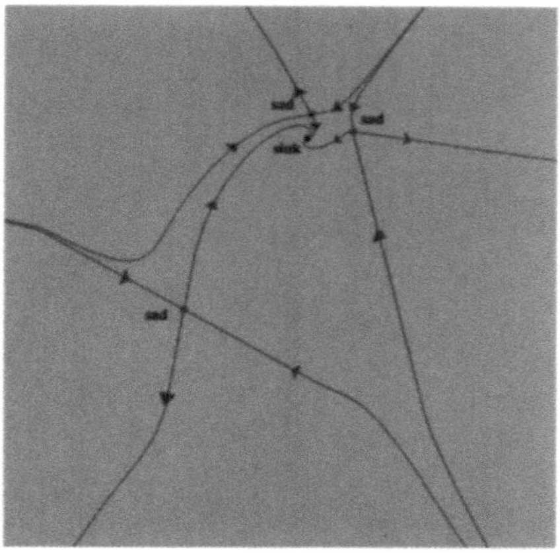

Fig. 37. Zoom of linear approximation around monkey saddle. *(Scheuermann, Hagen, p. 350)*

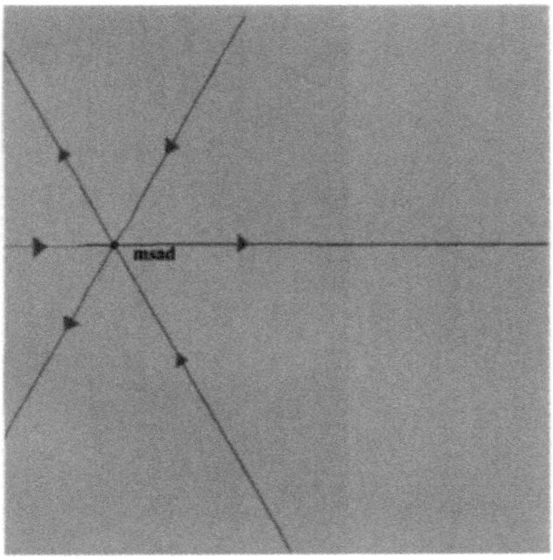

Fig. 38. Clifford approximation around monkey saddle. *(Scheuermann, Hagen, p. 350)*

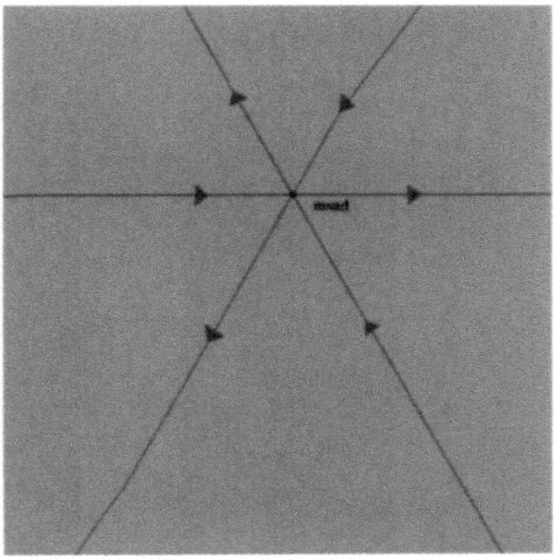

Fig. 39. Zoom of Clifford approximation around monkey saddle. *(Scheuermann, Hagen, p. 350)*

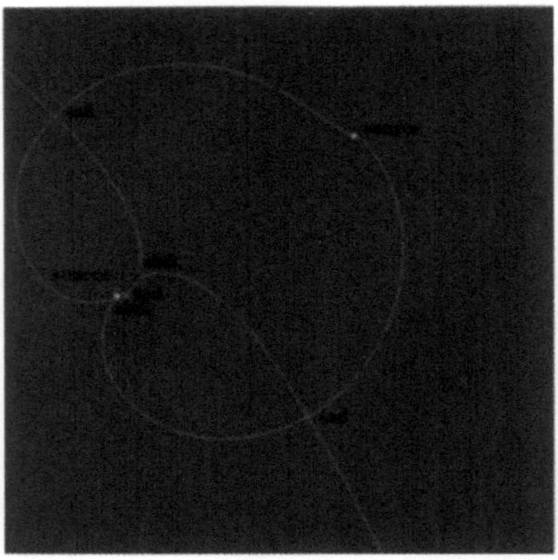

Fig. 40. Linear approximation around close source and sink. *(Scheuermann, Hagen, p. 350)*

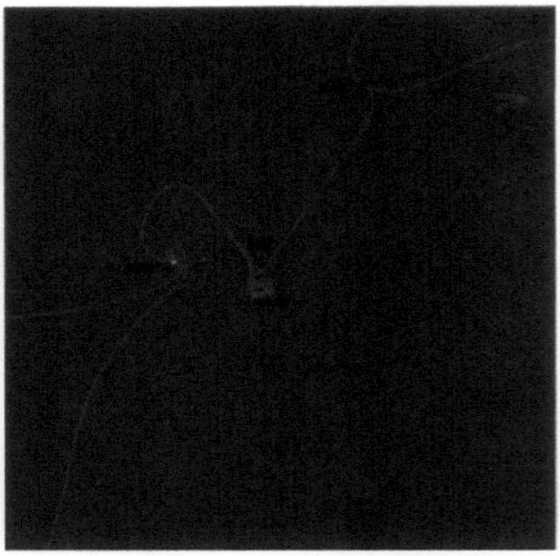

Fig. 41. Zoom of linear approximation around close source and sink. *(Scheuermann, Hagen, p. 350)*

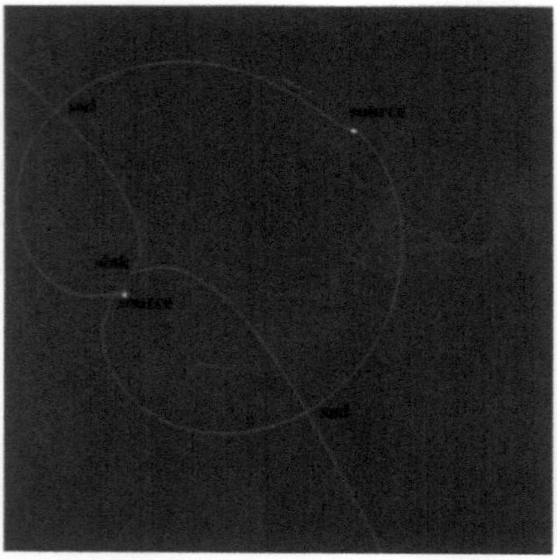

Fig. 42. Clifford approximation around close source and sink. *(Scheuermann, Hagen, p. 350)*

Fig. 43. Zoom of clifford approximation around close source and sink. *(Scheuermann, Hagen, p. 350)*

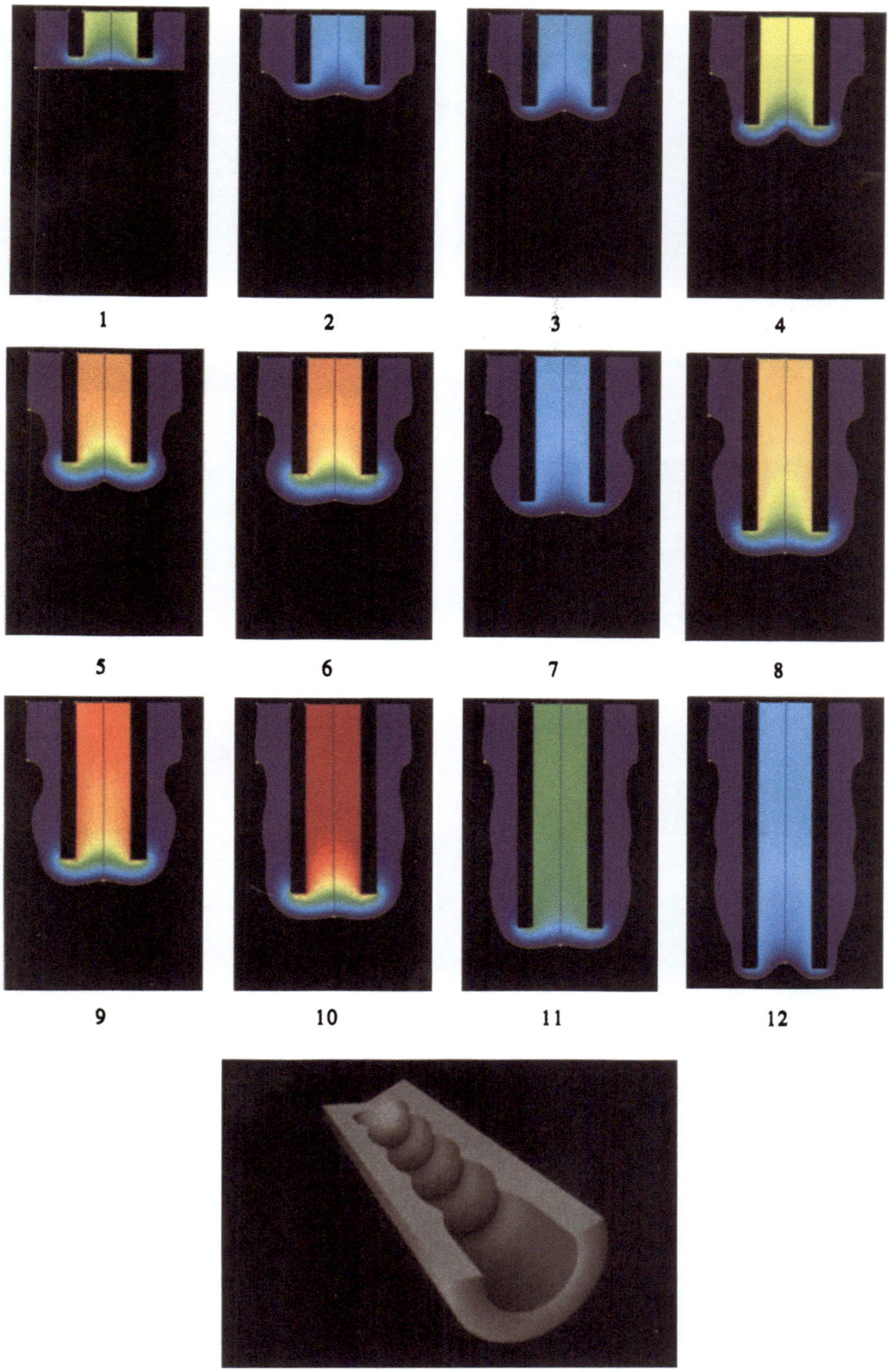

Fig. 44. FEM simulation of an electrochemical drilling process (ECD). Electric potential at different time instants (1-12), and a section in the 3D drilled cavity. *(Telea, van Overveld, p. 218)*

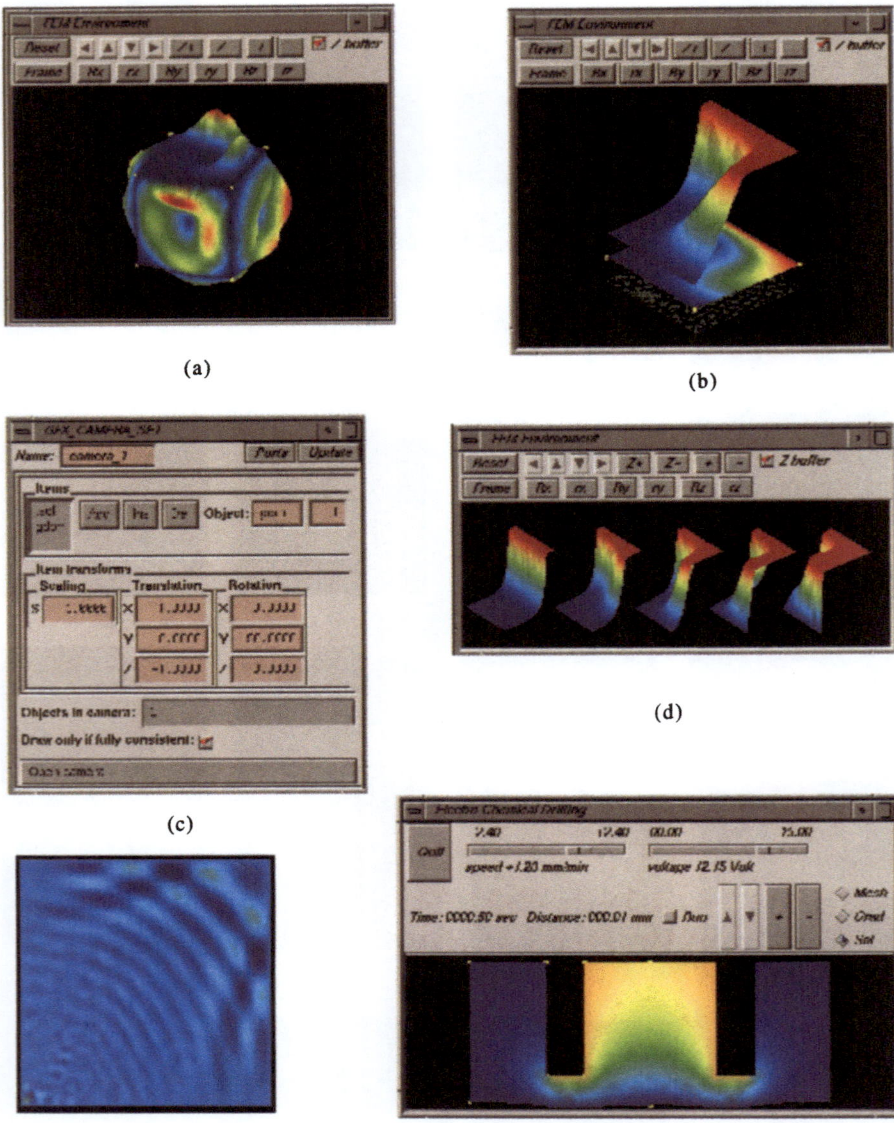

(a) (b) (c) (d) (e) (f)

Fig. 45. Interactive FEM simulations. Diffusion process computed over a 3D cubic domain (a). Temperature solution of a free convection simulation (mesh, solution and solution 3D elevation) (b). Object-oriented interactor for the CAMERA class (c). Temperature during a time dependent free convection simulation (d). Simulation of waves (e). Interface of the ECD FEM simulation (f). *(Telea, van Overveld, p. 218)*

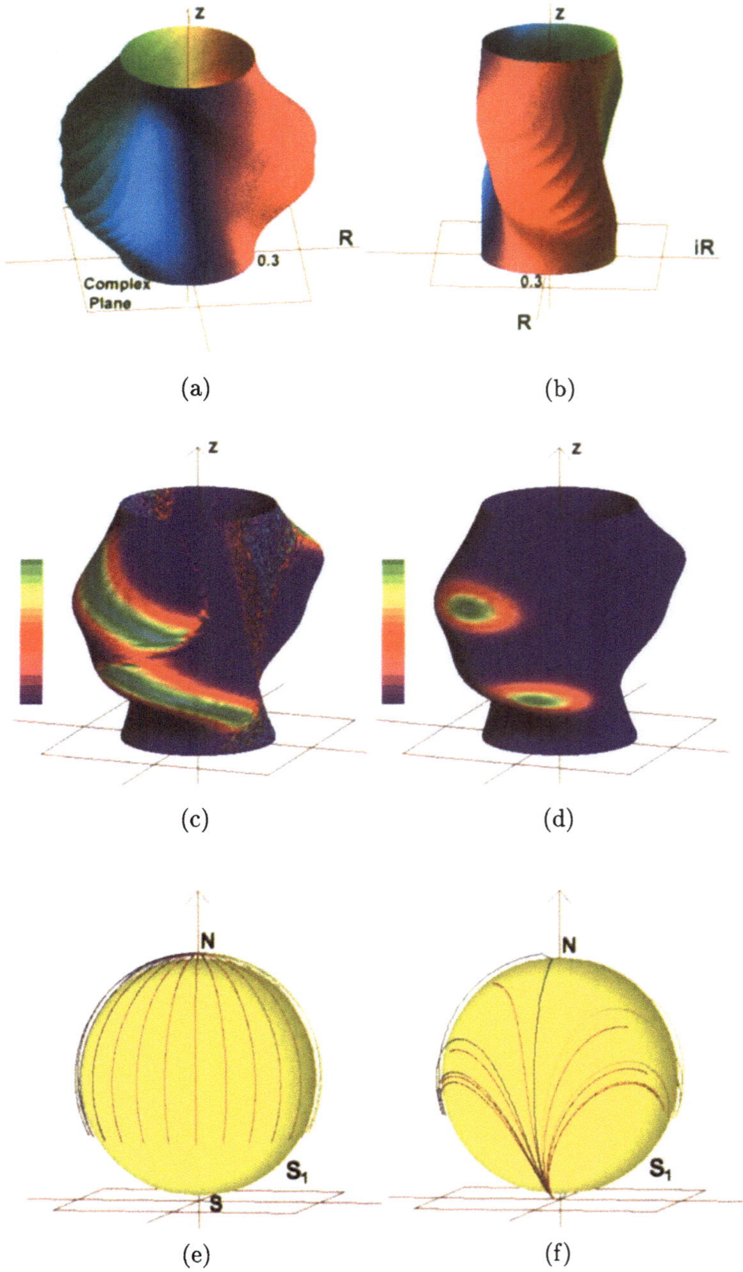

Fig. 46. Visualization of solutions of complex ODEs: (a) flow function of $y' = xy^3$, (b) same, but rotated by 90°, (c) estimated global error of numerical integration, (d) same, but with global quantification function, (e) flow of $y' = xy$ and (e) of $y' = xy^3$, both on a Riemann sphere. *(Testard, p. 359)*

Springer and the environment

At Springer we firmly believe that an international science publisher has a special obligation to the environment, and our corporate policies consistently reflect this conviction.

We also expect our business partners – paper mills, printers, packaging manufacturers, etc. – to commit themselves to using materials and production processes that do not harm the environment. The paper in this book is made from low- or no-chlorine pulp and is acid free, in conformance with international standards for paper permanency.

Springer

MIX
Papier aus verantwortungsvollen Quellen
Paper from responsible sources
FSC® C105338

If you have any concerns about our products,
you can contact us on
ProductSafety@springernature.com

In case Publisher is established outside the EU,
the EU authorized representative is:
Springer Nature Customer Service Center GmbH
Europaplatz 3, 69115 Heidelberg, Germany

Printed by Libri Plureos GmbH
in Hamburg, Germany